ASIC/SoC Functional Design Verification

Ashok B. Mehta

ASIC/SoC Functional Design Verification

A Comprehensive Guide to Technologies and Methodologies

Ashok B. Mehta
Los Gatos, California
USA

ISBN 978-3-319-86620-8 ISBN 978-3-319-59418-7 (eBook)
DOI 10.1007/978-3-319-59418-7

Softcover reprint of the hardcover 1st edition 2017

Printed on acid-free paper

This Springer imprint is published by Springer Nature
The registered company is Springer International Publishing AG
The registered company address is: Gewerbestrasse 11, 6330 Cham, Switzerland

To
My dear wife Ashraf Zahedi
and
My dear parents Rukshmani
and Babubhai Mehta

Preface

Having been a design and verification engineer of CPUs and SoCs for over 20 years, I've come to realize that the design verification field is very exhaustive in its breadth and depth. Knowing only SystemVerilog and UVM may not suffice. Sure, you need to know UVM (Universal Verification Methodology) but also SVA (SystemVerilog Assertions), SFC (SystemVerilog Functional Coverage), CRV (constrained random verification), CDC (clock domain crossing) verification, interconnect NoC (Network on Chip) verification, AMS (analog/mixed signal) verification, low-power verification (UPF), hardware acceleration and emulation, hardware/software co-verification, and static formal (aka static functional aka formal property check) verification technologies and methodologies.

I noticed that there isn't a book that gives a good overview (high level but with sufficient detail) of the technologies and methodologies at hand. Engineers rely on white papers, blogs, and EDA vendor literature to get some understanding of many of these topics. That was the impetus for this book. I have covered all the aforementioned topics in this book.

The book is written such that the reader gets a good comprehensive overview of a given topic but also enough detail to get a good grasp on the topic. The book has a comprehensive bibliography that points to further reading material that the reader can pursue.

The book is meant for managers, decision makers, as well as engineers: managers who want a quick introduction to the technologies and methodologies and engineers who want sufficient detail with applications, which they can then further pursue.

Chapter 1. This chapter introduces the current state of design verification in the industry. Where do the bugs come from? Are they mostly functional bugs?

Chapter 2. This chapter discusses the overall design verification (DV) challenges and solutions. Why is DV still such a long pole in the design cycle? We will discuss a comprehensive verification plan and assess the type of expertise required and how Develop => Simulate => Debug => Cover loop can be improved.

Chapter 3. This chapter discusses what SystemVerilog is. Is it a monolithic language? Or is it an umbrella under which many languages with different syntax and semantics work though a single simulation kernel? How did SystemVerilog came about?

Chapter 4. This chapter describes in detail the architecture of UVM and UVM hierarchy and discusses each hierarchical component (testbench, test, environment, agent, scoreboard, driver, monitor, sequencer, etc.) in detail. Two complete examples of UVM are provided to solidify the concepts.

Chapter 5. This chapter describes constrained random verification (CRV). CRV allows you to constrain your stimulus to better target a design function, thereby allowing you to reach your coverage goal faster with accuracy. From that sense, functional coverage and CRV go hand in hand. You check your coverage and see where the coverage holes are. You then constrain your stimulus to target those holes and improve coverage.

Chapter 6. This chapter discusses SVA (SystemVerilog Assertions) methodology, SVA and functional coverage-driven methodology, and plenty of applications to solidify the concepts. SystemVerilog Assertions (SVA) are one of the most important components of SystemVerilog when it comes design verification. SVA is instrumental in finding corner cases, ease of debug, and coverage of design's sequential logic.

Chapter 7. This chapter discusses differences between code and functional coverage and SFC (SystemVerilog Functional Coverage) fundamentals such as "covergroup," "coverpoint," "cross," "transition," etc. along with complete examples. SystemVerilog Functional Coverage (SFC) is an important component that falls within SystemVerilog.

Chapter 8. This chapter starts with the understanding of metastability and then dives into different synchronizing techniques. It also discusses the role of SystemVerilog Assertions in verification of CDC (clock domain crossing). We will then discuss a complete methodology. CDC has become an ever-increasing problem in multi-clock domain designs. One must solve issues not only at RTL level but also consider the physical timing.

Chapter 9. This chapter discusses challenges and solutions of low-power verification. It goes into sufficient depth of UPF (Unified Power Format) and how it applies to design from Spec to GDSII level. Finally, it discusses low-power estimation and verification at ESL level.

Chapter 10. This chapter discusses static verification, aka formal-based verification. Static verification is an umbrella term, and there are many different technologies that fall under it: for example, logic equivalency check (LEC), clock domain crossing check (CDC), X-state verification, low-power structural checks, ESL ⇔ RTL equivalency, etc. This chapter discusses all these topics and a lot more including state-space explosion problem and the role of SystemVerilog Assertions.

Chapter 11. This chapter discusses ESL (electronic system-level) technology, methodology and OSCI TLM2.0 standard definition, virtual platform examples, and how to use a virtual platform for design verification, among other topics.

Chapter 12. This chapter discusses the methodologies to develop software such that it is ready when hardware is ready to ship. What kind of platform do you need? How does ESL virtual platform play a key role? How do emulators and accelerators fit in the methodology equation?

Chapter 13. This chapter discusses major challenges and solutions of AMS (analog/mixed signal), the current state of affair, analog model abstraction levels, real number modeling, SystemVerilog Assertions-based methodology, etc.

Chapter 14. This chapter discusses challenges, solutions, and methodologies of SoC interconnect verification and a couple of EDA vendor solutions, among other things. Today's SoC contains hundreds of pre-verified IPs, memory controller, DMA engines, etc. All these components need to communicate with each other using many different interconnect technologies (cross-bus based, NoC (Network on Chip) based, etc.).

Chapter 15. This chapter describes a complete product life cycle. Simply designing the ASIC and shipping it is not enough. One needs to also consider the board design into which this ASIC will be used. This chapter describes board-level design issues and correlation between the board-level system design and the ASIC design and verification.

Chapter 16. This chapter discusses, in detail, verification of a complex SoC, namely, a voice over IP network SoC. We will go through a comprehensive verification plan and describe each verification step with VoIP SoC-based real-life design.

Chapter 17. This chapter discusses, in detail, verification of a cache subsystem of a large SoC. We will go through a comprehensive verification plan and describe each verification step with a real-life Cache subsystem SoC verification strategy. This chapter discusses the verification methodology using UVM agents.

Chapter 18. This chapter discusses, in detail, verification of a cache subsystem of a larger SoC. We will go through a comprehensive verification plan and describe each verification step with a real-life Cache subsystem SoC. This chapter discusses the verification methodology using an instruction set simulator (ISS) (as opposed to an UVM agent which is described in Chap. 17).

Acknowledgments

I am very grateful to the many people who helped with the review and editing of the book. The following individuals made significant contributions to the viability of the book:

Cuong Nguyen for a comprehensive and excellent chapter on the Complete Product Design Life Cycle

Vijay Akkati for significant contribution to the chapter on UVM (Universal Verification Methodology)

Frank Lee for his support and encouragement on all things verification

Bob Slee for facilitating close cooperation with EDA vendors

Norbert Eng for educating me on nuances of verification from the beginning of my career

Sandeep Goel for encouragement and impetus for the book

In addition, and of great importance to me, is the encouragement, enthusiasm, and support of my high school friends. Yes, we are still friends after all these years. Affectionately called Class 11B (eleventh grade, B division), they supported my endeavor through and through from the beginning until the end of writing this book. Thank you 11B from the bottom of my heart.

I would also like to express my heartfelt thanks to my brother Shailesh and sisters Raksha and Amita for their never-ending faith and confidence in my endeavors.

And last but certainly not the least, I would like to thank my wife Ashraf Zahedi for her enthusiasm and encouragement throughout the writing of this book and putting up with long nights and weekends required to finish the book. She is the cornerstone of my life, always with a smile and positive attitude to carry the day through the ups and downs of life.

Contents

List of Figures

About the Author

Ashok Mehta has been working in the ASIC/SoC design and verification field for over 30 years. He started his career at Digital Equipment Corporation (DEC) working as a CPU design engineer. He then worked at Data General and Intel (first Pentium architecture verification team) and, after a route of a couple of startups, worked at Applied Micro and currently at TSMC.

He was a very early adopter of Verilog and participated in Verilog, VHDL, iHDL (Intel HDL), and SDF (standard delay format) technical subcommittees. He has also been a proponent of ESL (electronic system-level) designs. At TSMC, he architected and went into production with two industry standard TSMC ESL reference flows that take designs from ESL to RTL while preserving the verification environment for reuse from ESL to RTL.

He holds 17 US patents in the field of SoC and 3DIC design verification.

He is also the author of the second edition of the book *SystemVerilog Assertions and Functional Coverage: A Comprehensive Guide to Languages, Methodologies and Applications* (Springer, June 2016).

Ashok earned an MSEE from the University of Missouri.

In his spare time, he is an amateur photographer and likes to play drums on 1970s rock music driving his neighbors up the wall.☺

Chapter 1
Introduction

ASIC functional design verification has been and continues to be a long pole in the entire design cycle from architecture to GDS-II tape-out. Many excellent methodologies have emerged to tackle this never-ending dilemma. UVM (Universal Verification Methodology) and UPF (Unified Power Format for Low Power) have now become cornerstones of pretty much all functional design verification methodologies, at least for complex SoC designs. It is indeed a robust, configurable, transaction level reusable methodology. Gone are the days when ad hoc Verilog testbenches ruled the verification domain. Reusability was sparse.

Design verification (DV) is a large and complex domain that contains many technologies, languages, and methodologies. A DV engineer cannot get away with SystemVerilog, UVM, and hardware microarchitecture knowledge alone. She/he needs to grasp as much of the DV domain as possible and help with each aspect of the DV project as necessary. A DV engineer should not get pigeonholed in only one of many technologies that fall under DV umbrella.

At the least, the following technologies fall under DV domain:

- UVM (Universal Verification Methodology).
- UPF (Unified Power Format) low-power verification using UPF.
- AMS (analog/mixed signal) verification. Real number modeling, etc.
- SystemVerilog Assertions (SVA) and functional coverage (SFC) languages and methodology.
- Coverage-driven verification (CDV) and constrained random verification (CRV).
- Static verification technologies. Static formal verification (model checking), static + simulation hybrid methodology, X-state verification, CDC (clock domain crossing), etc.
- Logic equivalency check (LEC). Design teams mostly take on this task. But the DV (design verification) team also needs to have this expertise.
- ESL—Electronic System Level (TLM 2.0) virtual platform development (for both software development and verification tests/reference model development).

A.B. Mehta, *ASIC/SoC Functional Design Verification*,
DOI 10.1007/978-3-319-59418-7_1

- Hardware/software co-verification (hint: use virtual platform methodology).
- SoC interconnect (bus-based and NoC—network-on-chip) verification.
- Simulation speedup using hardware acceleration, emulation, and prototyping.

The book focuses on these technologies and gives the reader a good comprehensive overview of the entire design verification paradigm. Sufficient detail is given for each technology to let the reader fully immerse in the technology/methodology under discussion after which they can look up further detail from a reference book or any other resource. A list of recommended books is given at the end of the book.

1.1 Functional Design Verification: Current State of Affair

As is well known in the industry, the design complexity at 16 nm and below node is exploding. Small form factor requirements and conflicting demands of high performance and low power and small area result in ever so complex design architecture. Multi-core, multi-threading, and power, performance, and area (PPA) demands exacerbate the design complexity and functional verification thereof.

The burden lies on functional and sequential functional domain verification to make sure that the design adheres to the specification. Not only is RTL (and virtual platform level) functional verification important but so is silicon validation. Days when engineering teams would take months to validate the silicon in the lab are over. What can you do during pre-silicon verification to guarantee post-silicon validation a first pass success?

The biggest challenge that the companies face is short time-to-market to deliver first pass working silicon of increasing complexity. Functional design verification is the long poll to design tape-out. Here are two key problem statements:

1. Design verification productivity: 40–50% of project resources go to functional design verification. The chart in Fig. 1.1 shows design cost for different parts of a design cycle. As is evident, the design verification cost component is about 40+% of the total design cost. In other words, this problem states that we must increase the productivity of functional design verification and shorten the design ⇔ simulate ⇔ debug ⇔ cover loop. This is a productivity issue, which needs to be addressed.

Continuing with the productivity issue, the chart in Fig. 1.2 shows that the compounded complexity growth rate per year is 58% while the compounded productivity growth rate is only 21%. There is a huge gap between what *needs* to get done and what *is* getting done. This is another example of why the productivity of design cycle components such as functional design verification must be improved.

2. Design coverage: The second problem statement states that more than 50% of designs require respin due to functional bugs. One of the factors that contribute to this is the fact that we did not objectively determine *before* tape-out that we had really *covered* the entire design space with our testbench. The motto "If it's not verified, it will not work" seems to have taken hold in design cycle. Not knowing if you have indeed covered the entire design space is the real culprit toward escaped bugs and functional silicon failures.

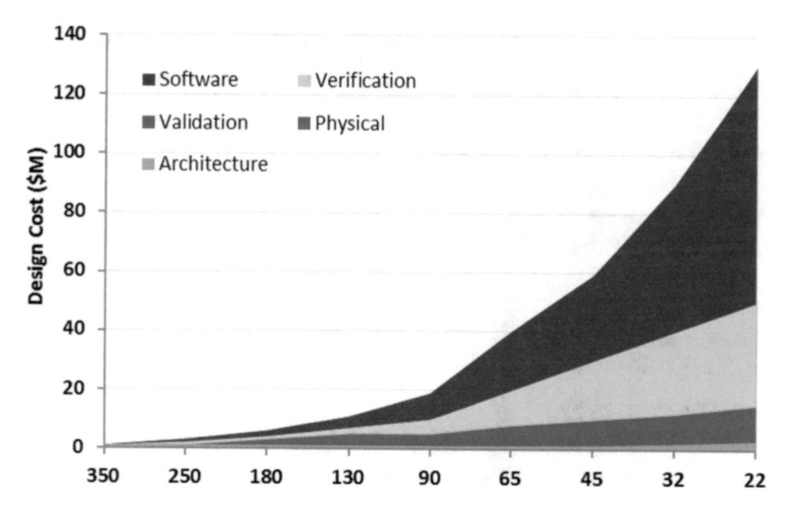

Fig. 1.1 Verification cost increases as the technology node shrinks

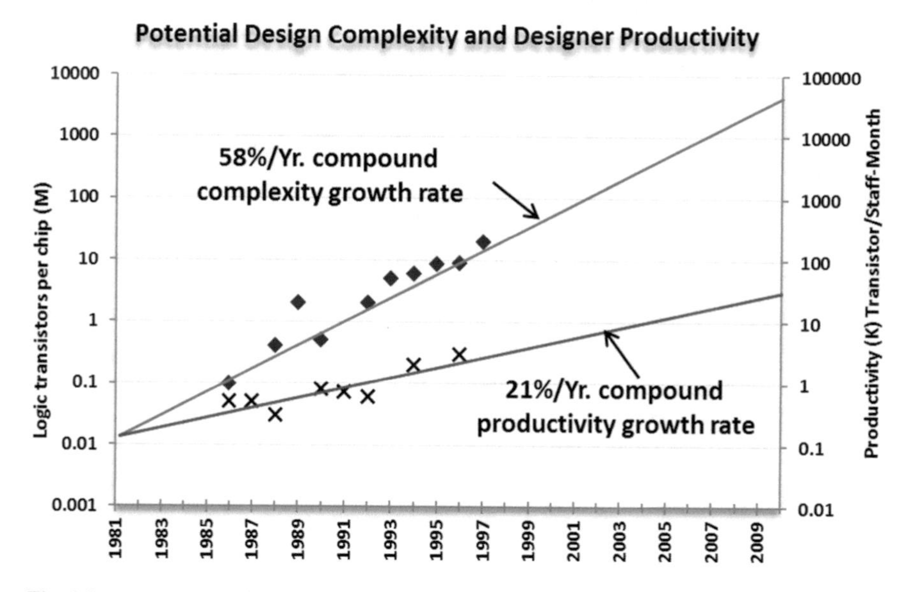

Fig. 1.2 Design productivity and design complexity

1.2 Where Are the Bugs?

As Fig. 1.3 shows, up to 80% of designs require a respin due to a functional bug. That's a huge problem. You better deploy an all compassing functional verification methodology that goes beyond UVM.

In the next chapter, we'll see how to solve these challenges.

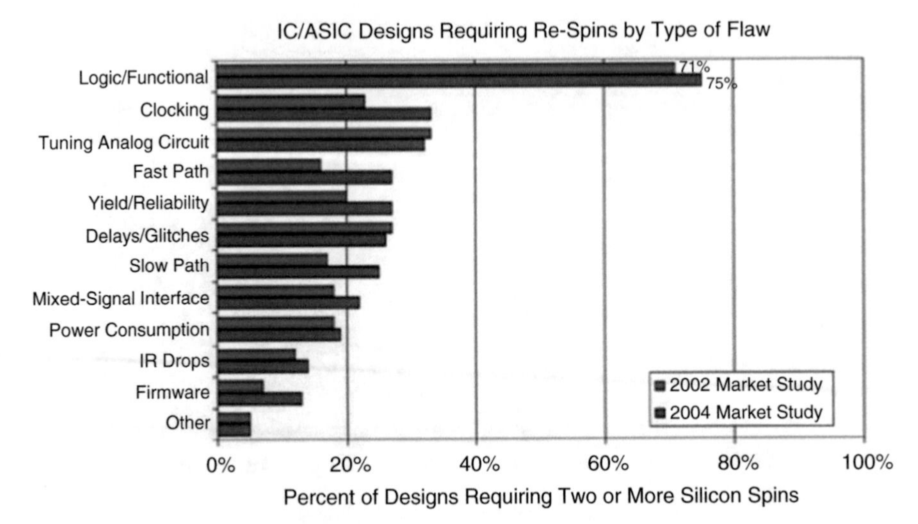

Fig. 1.3 ASIC respin causes (Collett)

Chapter 2
Functional Verification: Challenges and Solutions

Chapter Introduction
This chapter will discuss the overall design verification (DV) challenges and solutions. Why is DV still such a long pole in the design cycle? We will discuss a comprehensive verification plan and see the type of expertise required at each step of verification and how to improve the develop => simulate => debug => cover loop.

2.1 Verification Challenges and Solutions

So, what are the specific challenges that SoC design verification faces? Why is it such a long pole? Let us look at the higher-level challenges and their solutions. After that, we'll go through a detailed verification plan that the author has successfully deployed in many successful projects.

Here's a simplified, but to the point, functional verification cycle (Fig. 2.1). The cycle consists of four phases:

1. Development: verification plan, DV architecture, testbench, and tests development.
2. Simulation: software simulation, acceleration, emulation, etc.
3. Debug: transaction level, signal level, etc. This will be a big component if you did not deploy assertions, for example.
4. Cover: functional, code, and SVA coverage that feeds back to the development stage.

Each of these four phases poses significant challenges. Obviously, we need to reduce the time to complete and improve efficiency and robustness of the tasks at each stage:

1. Reduce time to develop and improve robustness.
2. Reduce time to simulate and improve simulation accuracy and throughput.

A.B. Mehta, *ASIC/SoC Functional Design Verification*,
DOI 10.1007/978-3-319-59418-7_2

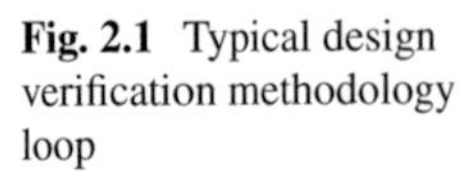

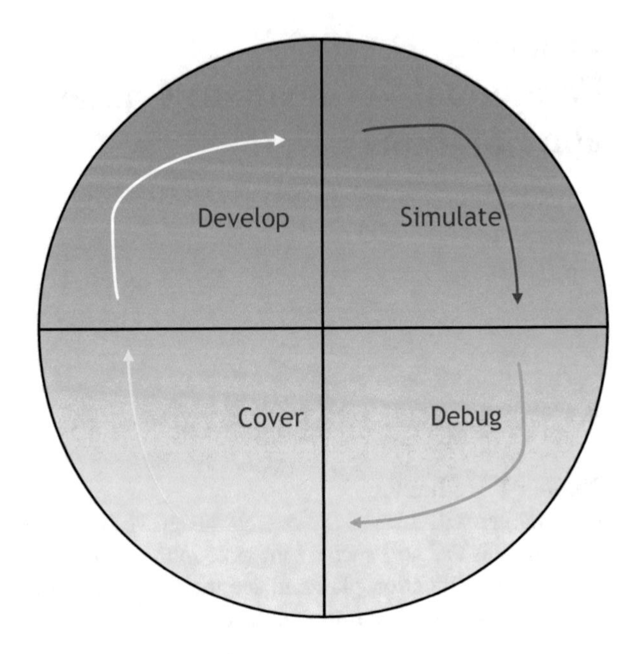

Fig. 2.1 Typical design verification methodology loop

3. Reduce time to debug and improve efficiency.
4. Reduce time to "comprehensive" cover.

Let us solve the challenge of reducing time and improving robustness at each stage of the functional verification cycle.

2.1.1 Reduce Time to Develop

Development time includes functional verification plan development, verification environment creation, DV architecture development, testbench development, and tests development. Of these, the tests and testbench development are the most time consuming. Here is a strategy to reduce time to develop.

1. Raise abstraction level of tests. Use TLM (Transaction Level Modeling) methodologies such as UVM, SystemVerilog/C++/DPI, etc. The higher the abstraction level, the easier it is to model and maintain verification logic. Modification and debug of transaction level logic is much easier, further reducing time to develop testbench, reference models (scoreboard), peripheral models, and other such verification logic.
2. Use constrained random verification (CRV) methodologies to reach exhaustive coverage with fewer tests. CRV is discussed in detail in Chap. 5. Fewer tests mean less time to develop and debug.

3. Develop verification components (e.g., UVM agents) that are reusable. Make them parameterized for adoptability in future projects. UVM is discussed in detail in Chap. 4.
4. Use SystemVerilog Assertions (SVA) to reduce time to develop complex sequential and combinatorial checks. As we will see, assertions are intuitive and much simpler to model, especially for complex temporal domain checks. Verilog code for a given assertion will be much lengthier, hard to model, and hard to debug. SVA indeed reduces time to develop and debug. SVA is discussed in detail in Chap. 6.

Verification Plan Development

Instead of using the age old .docx- or .xls-based plans, use the new technology/tools available from EDA vendors that allow an automated and coverage-driven verification plan development in an organized way. The plan needs to correlate to the design specification in an intuitive and comprehensive manner. Changes in design specs should be automated to reflect in the verification plan.

Nothing is fully automated in life (unfortunately). Even with automation, you need to carefully layout what is it that you need to include in a verification plan. Verification plan is only as good as your knowledge of the design architecture and to some extent microarchitecture. Detailed attention needs to be given to issues such as asynchronous FIFO, state transitions, live locks, dead locks, etc. Relaying only on end-to-end verification can miss corner cases. All this needs to be specified in a verification plan. A comprehensive verification plan will go a long way in reducing time and effort consumed by the later stages of verification.

A comprehensive verification plan is presented in Sect. 2.2.

2.1.2 Reduce Time to Simulate

- Higher-level abstractions simulate much faster than pure RTL testbench which is modeled at signal level. Use transaction level testbench (e.g., UVM, TLM 2.0 transaction level models). TLM reduces time to develop, debug, and simulate. Electronic system level (i.e., TLM 2.0 level modeling) has come a long way for practical deployment in verification environments. Refer to Chap. 11 for complete detail on ESL and TLM 2.0 methodology.
- Deploy well-thought-out hardware acceleration, emulation, or FPGA prototype methodologies. Develop transaction level testbenches that interact directly with the accelerated or emulated design. Chapter 12 is devoted to deploying hardware acceleration, emulation, and virtual prototyping to speed up simulation as well as develop software.
- Use coverage-driven verification (CDV) methodologies to reduce the number of tests to simulate to reach the defined coverage goals. Refer to Chap. 7 on functional coverage to understand CDV.

2.1.3 Reduce Time to Debug

- Use SystemVerilog Assertion-based verification (ABV) methodology to quickly reach to the source of the bug. As we will see (Chap. 6), assertions are placed at various places in design to catch bugs where they occur. Traditional way of debug is at IO level. You detect the effect of a bug at primary output. You then trace back from primary output until you find the cause of the bug resulting in lengthy debug time. In contrast, an SVA points directly at the source of the failure (e.g., a FIFO overflow assertion will point directly to the FIFO overflow logic in RTL that failed) drastically reducing the debug effort.
- Use transaction level methodologies to reduce debugging effort (and not get bogged down into signal level granularity).
- Constrained random verification allows for fewer tests. They also narrow down the cone of logic to debug. CRV indeed reduces time to debug.

2.1.4 Reduce Time to Cover: Check How Good Is Your Testbench

- Use SystemVerilog *functional coverage* language to measure the *intent* of the design. How well have your testbench verified the "intent" of the design. For example, have you verified all transition of write/read/snoop on the bus? Have you verified that a CPU1-snoop occurs to the same line while a CP2-write invalid occurs to the same line? Code coverage will not help with this. We will cover functional coverage in plenty detail in Chap. 7.
- Use *cover* feature of SystemVerilog Assertions to cover complex *sequential* domain specification of your design. As we will see in Chap. 7, "cover" helps with making sure that you have exercised low-level sequential domain conditions with your testbench. *If an assertion does not fire, that does not necessarily mean that there is no bug*. One of the reasons for an assertion to *not* fire is that you probably never stimulated the required condition (antecedent) in the first place. If you do not stimulate a condition, how would you know if there is indeed a bug in the design? "Cover" helps you determine if you have indeed exercised the required temporal domain condition.
- Use code coverage to cover *structural* coverage (yes, code coverage is still important as the first line of defense even though it simply provides structural coverage). As we will see in detail in the section on SV functional coverage, structural coverage does not verify the *intent* of the design, it simply sees that the code that you have written has been exercised (e.g., have you verified all "case" items of a "case" statement, or toggled all possible assigns, conditional expressions, states, etc.). Nonetheless, code coverage is still important as a starting point to measure coverage of the design.

2.2 A Comprehensive Verification Plan

Following verification plan is typical of large SoC projects. Let us also establish during each step the type of expertise required. That will tell us the need for a well-diversified DV team. Each of the following verification plan point is discussed in detail throughout the rest of the book. This is mainly a higher-level snapshot. Chapters 15 (Voice over IP SoC) and 16 (Cache Memory Subsystem) will showcase real-life verification plans based on the outline described below.

Here is the outline of a comprehensive verification plan.

1. ***Identify Subsystems Within Your SoC***

   - For example, audio subsystem, memory subsystem, graphics subsystem, etc.
   - You start verification of subsystems and then move onto concurrent subsystems leading to full system verification.
   - Subsystem also allows you to determine a methodology for subsystem-block level stimulus/response methodology. Block level verification can/will be imported to subsystem level verification.

     – Expertise: Hardware design architecture and microarchitecture.

2. ***Determine Subsystem Stimulus and Response Methodology***

   - For example, graphics subsystem will require a "way" to feed the external bus with a single/multi-frame input. What should be the format for this input data? How would you measure response? How many different UVM agents would you need? What type of scoreboard? Reference Model?

     – Expertise: Hardware design architecture and microarchitecture. UVM, SVA, C/C++, SystemC/TLM2.0.

3. ***Stimulus Traffic Generation Requirements***

   - For example, what type of traffic would you generate for a video subsystem? Will that be single frame? Multi-frame? Do you require a live video stream? How will that simulate on RTL? How would you verify without a live stream?
   - How will you generate Ethernet traffic? An external software stack (won't find all the bugs) or a constrained random packet generator (much better choice)?

     – Expertise: SoC design verification, UVM, SVA, C/C++, SystemC/TLM2.

4. ***Subsystem Response Checking Methodology***

   - How will you check the output of a video engine/subsystem?
   - How will you detect a corrupt Ethernet packet transmission?
   - How will you check for SoC interrupt generation logic (i.e., did the logic generate an interrupt when a corrupt Ethernet packet was received)?
   - For a CPU, how will you check for the CPU architectural state integrity at the end of an instruction?

- What about reference model generation? Do you need it? What will be the sync point between transaction level reference model output and signal level RTL output?
- What is the methodology for deploying assertions both at the microarchitectural level and block/SoC IO level interface?
  - Expertise: Reference model generation, UVM, SVA, TLM2.0, etc.

5. ***SoC Interconnect: Determine Verification of Either an NoC (Network on Chip), Cache Coherent NoC- or Bus-Based Interconnect***
   - Again, how would you generate real traffic to verify the interconnect?
   - Would you create "stub" models (i.e., BFM/UVM agents) to act as initiators and targets of the interconnect? What kind of stimulus would you provide to these UVM agents to stress the interconnect?
   - What would be your methodology for performance measurement of the interconnect?
     - Expertise: Knowledge of NoC and bus interconnect. UVM, SVA, SFC, NoC generation tools.
6. ***Low-Power Verification***
   - Identify power subdomains. Power switches. Isolation cells. Retention cells.
   - Create UPF generation methodology. Can it be automated?
   - Determine stimulus strategy to verify each individual power domain turned ON/OFF. Verify the same with multiple power domains.
     - Expertise: UPF, low-power technology, power domains, power switches, isolation cells, retention cells, single- and multi-power domain concurrency.
7. ***Static Formal or Static + Simulation Hybrid Methodology***
   - Static formal (as of the writing of this book) does not work at full SoC level for a large SoC. So, start with identifying control logic paths where assertions can be written to limit the logic cone(s). Identify critical clock domain crossing blocks, critical state machines, asynchronous logic interfaces, etc. Static formal has the problem of state/space explosion for larger logic blocks. Static formal exercises all possible permutations of combinational and sequential domain tests to see that the assertion holds. So, what's the strategy to partition the blocks?
   - If static formal does not work, identify tools that allow static + simulation hybrid simulation. This methodology will simulate inputs to the logic cone, determine valid input values, and then apply static formal to see if assertions hold.
     - Expertise: Static formal and static + hybrid formal tools.
8. ***SystemVerilog Assertions (SVA) Methodology***

Deploying SystemVerilog Assertions is one of the most important strategies you can deploy to reduce time to cover, develop, and debug. Carefully plan on writing assertions for microarchitecture, subsystem, system, IO interface, inter-block interface, and critical state machines.

- Assertions need to be added to RTL by designers while block and SoC level interface assertions need to be added by the DV engineers.
- How will you know that you have added enough assertions? Rule of thumb is if a test fails and none of the assertions fire, you haven't placed assertions in the failing path.

  – Expertise: SystemVerilog Assertions language and methodology. Note that UVM does not cover SVA language.

9. ***Functional Coverage***

- Determine logic that needs to be *functionally* covered.
- How will you leverage code coverage with SystemVerilog functional coverage?
- What's the strategy to constraint stimulus to achieve desired functional coverage?
- How will you determine that you have specified all required coverpoints and covergroups? This is the hardest (and sometimes subjective) question to answer. Continue to add functional coverpoints as the project progresses. Do not consider the very first coverage plan to be an end in all.

  – Expertise: UVM, SystemVerilog functional coverage language. UVM does not cover SFC language.

10. ***Software/Hardware Co-verification?***

- Think about deploying advanced methodologies such as TLM2.0 (ESL) ⇔ RTL. This allows you to speed up software running on a virtual platform of the SoC. TLM2.0 is transaction level and so is UVM. So, integration of UVM testbench with software virtual platform will not have significant challenges. You will be able to run software code with such methodology. If TLM2.0 virtual platform is not your cup of tea, have a plan to deploy hardware acceleration or emulation to do hardware-software co-verification.

  – Expertise: TLM2.0 SystemC. C++ expertise. SystemC TLM2.0 is an entirely different language orthogonal to SystemVerilog. Simulation acceleration, emulation, and FPGA prototyping.

11. ***Simulation Regressions: Hardware Acceleration or Emulation or FPGA Prototyping***

- What are the pros/cons of acceleration vs. emulation vs. prototyping?
- Acceleration will have better debug capabilities than emulation.

  – But the speed maybe in a few MHz at best.

  - Does acceleration provide enough speed for software development?
  - If the testbench is still in SystemVerilog (i.e., outside the acceleration box), will the SystemVerilog ⇔ acceleration maintain required speed?
  - What about memories? What about multiple clocks? What is the debug strategy?
  - How about assertions? Will they compile into acceleration hardware?
  - How will functional coverage be measured?

  Expertise: Knowledge of hardware acceleration. RTL to acceleration netlist mapping, multi-clock domain, etc.

- Emulation will be orders of magnitude faster than acceleration.
  - BUT emulation cannot start until RTL is ready. Since that is the case, will it be too late for software development?
  - How easy/hard will it be to debug since internal node visibility maybe poor.
  - What about assertions and functional coverpoints? Will they be emulated?

  Expertise: Knowledge FPGA (or not)-based emulation technology. ASIC RTL mapping to FPGA logic or acceleration hardware, clocks, SCHEMI interface, compile times, etc. UVM does not cover this.

12. ***Virtual Platform Methodology***

- This is the ESL/TLM2.0 methodology. Do you need it? How will you use a virtual platform as a reference model to check SoC response?
- There are significant advantages to this methodology.
  - You will be able to develop software before the RTL is ready.
  - You will be able to create and verify tests before the RTL is ready.
  - The virtual platform can act as a reference model to match the architectural state of the SoC at transaction boundaries.

  Expertise: TLM2.0 SystemC standard language. C++ expertise.

13. ***AMS (Analog/Mixed Simulation)***

- What will be the verification strategy to verify analog ⇔ digital boundary crossing?
- How will you generate analog behavioral models? Simulation with Spice will be extremely slow and won't be practical.
- How will you guarantee analog behavioral model is 100% accurate to Spice model?
- Do you need to deploy real number modeling?
- How will you do low-power verification on analog⇔digital boundary?
  - Expertise: Real number modeling (RNM), Verilog-A language, AMS technology, analog⇔digital interface, analog behavioral modeling (non-RNM), etc.

Chapter 3
SystemVerilog Paradigm

Chapter Introduction
This chapter will discuss what is SystemVerilog. Is it a monolithic language? Or is it an umbrella under which many languages with different syntax and semantics work with a single simulation kernel? How did SystemVerilog came about?

3.1 SystemVerilog Language Umbrella

Let us look at the SystemVerilog language and its subcomponents (sublanguages). Many users consider SystemVerilog as one monolithic language without realizing that it is a culmination of different languages all working in concert around a common simulation engine (single unified simulation thread).

As shown in Fig. 3.1, SystemVerilog is not a monolithic language. It has the following components (sublanguages):

1. SystemVerilog Object-Oriented language for functional verification. This is the subset that transaction-level methodology such as UVM uses.
2. SystemVerilog for Design (knowledge of OOP language subset is not required here)
3. SystemVerilog Assertions (SVA) language
4. SystemVerilog Functional Coverage (FC) language

Note that SystemVerilog, SystemVerilog Assertions (SVA), and SystemVerilog Functional Coverage (SFC) are three totally *orthogonal* languages. In other words, SVA syntax/semantic is totally different from that of SystemVerilog and SFC. Similarly, SFC is totally orthogonal to SystemVerilog and SVA, and finally SystemVerilog for OOP and design is totally different from SVA and SFC.

In any design, there are three main components of verification: (1) stimulus generators to stimulate the design, (2) response checkers to see that the device adheres to the device specifications, and (3) coverage components to see that we have indeed

A.B. Mehta, *ASIC/SoC Functional Design Verification*,
DOI 10.1007/978-3-319-59418-7_3

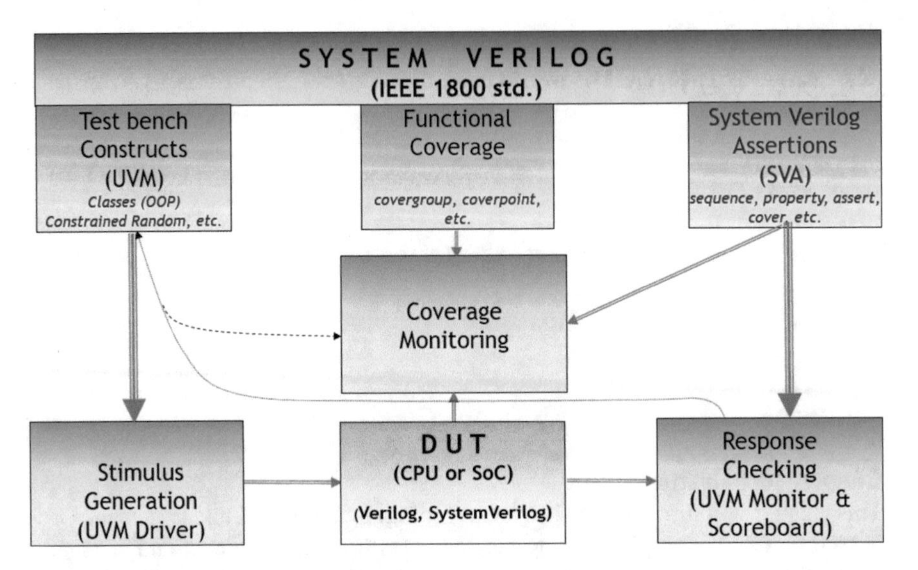

Fig. 3.1 SystemVerilog language paradigm

structurally and functionally covered everything in the DUT per the device specifications.

1. *Stimulus Generation*. This entails creating different ways in which a DUT needs to be exercised. For example, a peripheral (e.g., USB) may be modeled as a Bus Functional Model (or a UVM (Universal Verification Methodology) agent) to drive traffic via SystemVerilog transactions to the DUT. This is where the SystemVerilog TLM subset comes into play.
2. *Response Checking*. Now that you have stimulated the DUT, you need to make sure that the device has responded to that stimulus per the device specs. Here is where SVA, UVM monitors, scoreboards, reference models, and other such techniques come into picture. SVA will check to see that the design not only meets high-level specifications but also low-level combinatorial and sequential design rules.
3. *Functional Coverage*. How do we know that we have exercised everything that the device specification dictates? Code coverage is one measure. But code coverage is only *structural*. For example, it will point out if a conditional has been exercised. But code coverage has no idea if the conditional itself is correct, which is where functional coverage comes into picture. Functional coverage gives an objective measure of the design coverage, e.g., have we verified all different cache access transitions (e.g., write followed by read from the same address) to L2 from CPU? Code coverage will not give such measure.

3.2 SystemVerilog Language Evolution

Industry recognized the need for a standard language that allowed the design *and* verification of a device be built avoiding multi-language cumbersome environments. Enter Superlog, which was a language with high-level constructs required for functional verification. Superlog was donated (along with other language donations) to create SystemVerilog 3.0 from which evolved SystemVerilog 3.1, which added new features for design. But over 70% of the new language subset was dedicated to functional verification. We can only thank the Superlog inventor (the same inventor as that for Verilog, namely, Phil Moorby) and the Accelera technical subcommittees for having a long-term vision to design such a robust all-encompassing language. No multi-language solutions were required any more. No more reinventing of the wheel with each project was required anymore. Refer to Fig. 3.2.

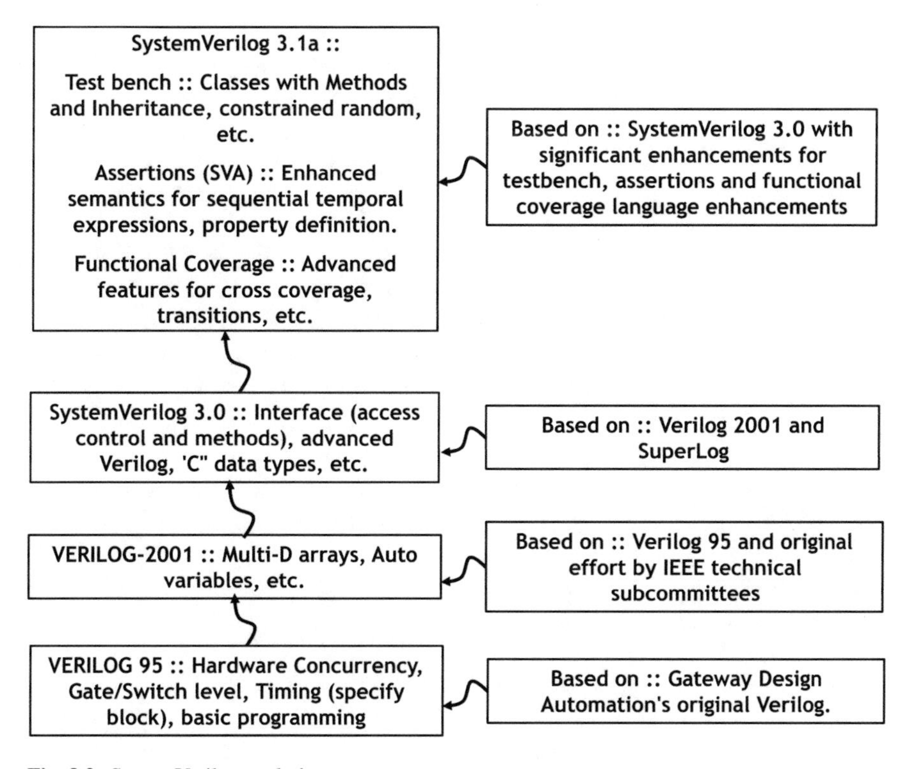

Fig. 3.2 SystemVerilog evolution

Chapter 4
UVM (Universal Verification Methodology)

Chapter Introduction
This chapter will describe in detail the architecture of UVM, UVM hierarchy, and discuss each hierarchical component (testbench, test, environment, agent, scoreboard, driver, monitor, sequencer, etc.) in detail. We will also go through two complete examples to solidify the concepts.

4.1 What Is UVM?

Ah, isn't this now the mother of all verification methodologies? Let us start with what UVM is and then delve into detail and examples.

UVM is a transaction-level methodology (TLM) designed for testbench development. It is a class library that makes it easy to write configurable and reusable code. You do need to understand the basic concepts of OOP (object-oriented programming), but the designers of UVM did all the hard work. They created the so-called class library whose components can be used to develop a testbench. You don't need to be an OOP expert. Reusable in the sense that once you put together the required code/infrastructure in place, using UVM class library, you will be able to carry forward that to the next project. Only the driver (in UVM agent), the scoreboard, and the basic transaction (sequence) and sequence library that contains tests need to change. With pure SystemVerilog, testbenches are written ad hoc since there is no coding standard there. These testbenches are then not quite reusable. The code is also hard to understand and maintain.

UVM is class based and communicates between these classes via transactions. This helps keep communication interface among class components separate from the implementation detail in the UVM agent driver. If you don't understand some of this, hold on. We'll go into the UVM hierarchy and examples to solidify the concepts. The UVM class library provides generic utilities, such as component

A.B. Mehta, *ASIC/SoC Functional Design Verification*,
DOI 10.1007/978-3-319-59418-7_4

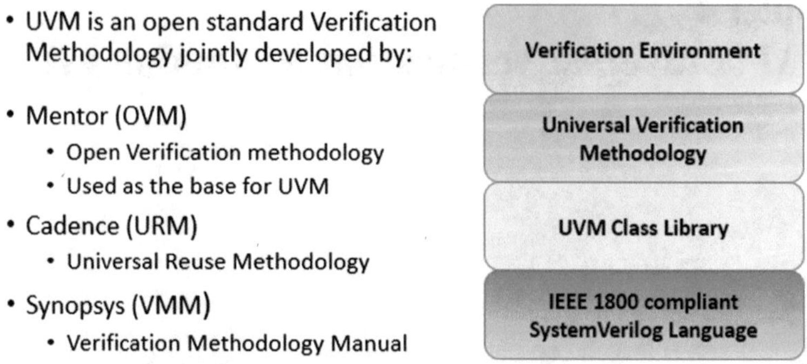

- UVM is an open standard Verification Methodology jointly developed by:
- Mentor (OVM)
  - Open Verification methodology
  - Used as the base for UVM
- Cadence (URM)
  - Universal Reuse Methodology
- Synopsys (VMM)
  - Verification Methodology Manual

Verification Environment

Universal Verification Methodology

UVM Class Library

IEEE 1800 compliant SystemVerilog Language

Fig. 4.1 Evolution of UVM

hierarchy, transaction library model (TLM), configuration database, etc., which enable the users to create virtually any structure they want for the testbench.

UVM provides a set of transaction-level communication interfaces and channels that you can use to connect components at the transaction level. The use of TLM interfaces isolates each component from changes in other components throughout the environment. When coupled with the phased, flexible-build infrastructure in UVM, TLM promotes reuse by allowing any component to be swapped for another, as long as they have the same interfaces. In addition, TLM provides the basis for easily encapsulating components into reusable components, called verification components, to maximize reuse and minimize the time and effort required to build a verification environment.

But I am getting carried away here. Let's see how UVM came to being.

As noted in Fig. 4.1, there were three competing methodologies in the industry. Open Verification Methodology (OVM) from Mentor, Universal Reuse Methodology (URM) from Cadence, and Verification Methodology Manual (VMM) from Synopsys. Each one provided a base class library and TLM level modeling capabilities. Each one claimed to allow you to write reusable code.

The end result was that the customer base was confused on which one to adopt. That resulted in most people not even adopting any of the methodologies. That in turn, left almost 70% of SystemVerilog language simply unused, which wasted the effort of the originators of the language.

After intense debate on unifying all three base class libraries and methodologies, Accelera came forward and decided to pick OVM and build an industry standard methodology around it that was "universal." Most of the features of the language came from OVM, and thus UVM was born. The users felt confident in adopting this universal methodology and not get tied to a specific vendor. All vendors now support this methodology and the standard class library that comes with the tool. And oh boy, did the user community adopt it. It's everywhere in all major companies and with all those who develop large SoCs. It is still to proliferate (as of this writing) to companies that develop "smaller" less complex chips.

Now let us go through a high-level description of the UVM hierarchy and all the components that lie underneath it. But before that we must understand polymorphism since this technology is the bedrock of UVM.

4.2 Polymorphism

Polymorphism in its simplest form means using one object as an instance of a base class. For example, if you have the class "Creature" and from that you derive the class "elephant," you can treat "elephant" as if it is the instance of class "Creature".

Polymorphism is derived from Greek word, where "poly" means many and "morph" means forms, i.e., it's an ability to appear in many forms. [((Modh))]

The OOP (object-oriented programming) term for multiple routines sharing a common name is "polymorphism." Polymorphism allows the use of super-class handle to hold sub-class object and to access the methods of those sub-classes from the super-class handle itself.

To achieve this, functions and/or tasks in SV are declared as virtual functions/ tasks to allow sub-classes to override the behavior of the function/task. So, we can say, Polymorphism = Inheritance + virtual method.

Example:

```
class vehicle; // Parent class
  virtual function vehicle_type();    // Virtual function
     $display("vehicle");
  endfunction

  virtual task color();  // Virtual task
     $display("It has color");
  endtask
endclass

class four_wheeler extends vehicle;    //child class
   function vehicle_type();
     $display("It's a four wheeler");
   endfunction

   task color();
     $display("It has different colors");
   endtask
endclass

class BENZ extends four_wheeler;    // "grand" child class
   function vehicle_type();
     $display("It's a BENZ");
   endfunction
```

```
   task color();
     $display("It is Black");
   endtask
endclass

program polymorphism;// program block
  initial begin
    vehicle vehcl;
    four_wheeler four_whlr;
    BENZ benz;

    four_whlr=new ();
    benz=new ();

     vehcl=four_whlr; // No need to create an object by calling a
new method
    vehcl.vehicle_type(); // accessing child class method by calling
base class method

vehcl=benz;
     vehcl.vehicle_type(); // accessing "grand" child by calling
base class method

     four_whlr=benz;
    four_whlr.color(); // accessing "grand" child method by calling
child class method
  end
endprogram
```

Output:
It's a four wheeler.
It's a BENZ.
It is black.

4.3 UVM Hierarchy

Figure 4.2 [(Accelera, Universal Verification Methodology (UVM) 1.2 User's guide)] shows a simple UVM hierarchy (aka testbench architecture). Following components make up the hierarchy.

- UVM Testbench
- UVM Test
- UVM Environment
  - The top-level environment contains one or more environments. Each environment contains an agent for a specific DUT interface. The environment may contain a scoreboard. A scoreboard compares the outputs with expected outputs.

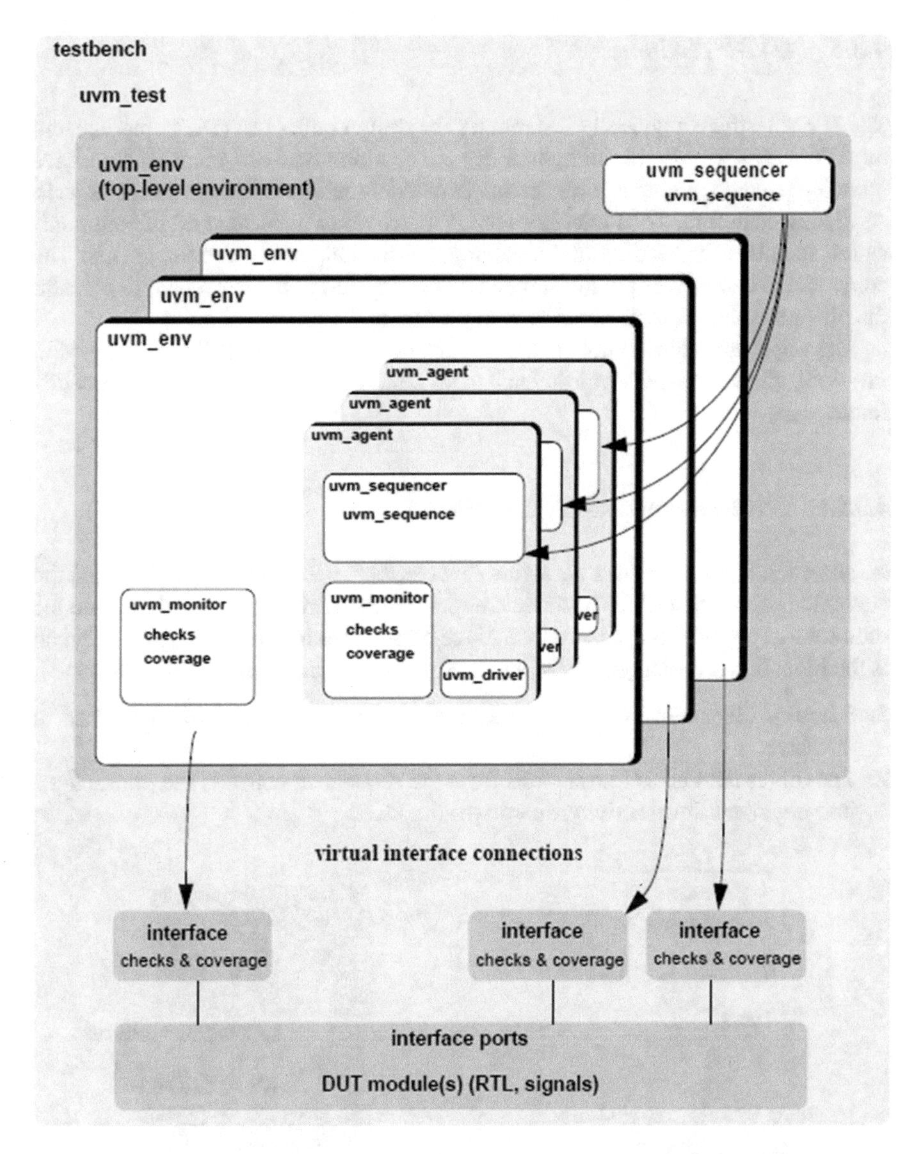

Fig. 4.2 UVM hierarchy (Accelera, Universal Verification Methodology (UVM) 1.2 User's guide)

Either the expected outputs are computed within a scoreboard or they can use (via DPI) an external reference model or any other way of comparing simulated output with expected output.

- UVM Agent
  - UVM agent comprises of a sequencer, a driver, and/or a monitor.

Let us look at each component at high level.

4.3.1 *UVM Testbench*

The UVM testbench typically instantiates the design under test (DUT) module and the UVM Test class and configures the connections between them. If you have module-based components, they are instantiated under the UVM testbench as well. As discussed before, TLM interfaces in UVM provide a consistent set of communication methods for sending and receiving transactions between components. The components themselves are instantiated and connected in the testbench, to perform the different operations required to verify a design.

The important point to note is that the UVM Test is dynamically instantiated at run-time, allowing the UVM testbench to be compiled once and run with many different tests.

4.3.1.1 UVM Transaction-Level Testbench

As shown in Fig. 4.3, UVM transaction-level testbench instantiates the DUT and the agent that drives the DUT. The agent comprises of the driver, the sequencer, and the monitor. An optional scoreboard to analyze data is also instantiated. This testbench is the most basic, utilizing a UVM agent. The components of this testbench are:

1. Stimulus Generator (sequencer) that creates transaction-level traffic to feed to the driver.
2. The driver then takes transactions from the sequencer, converts the transactions into pin signal-level activity, and drives the DUT.

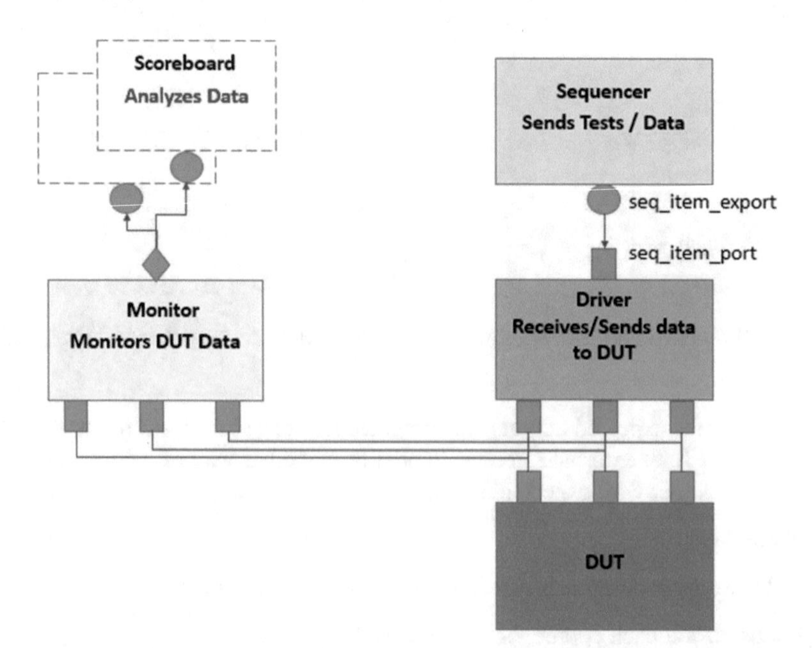

Fig. 4.3 UVM transaction-level testbench

3. A monitor that snoops the signal-level activity and converts them back into transactions that can then be fed to a scoreboard.
4. The scoreboard that gets the monitored transactions from the monitor compares them with expected transactions (response transactions).

4.3.2 UVM Test

The UVM Test is the top-level UVM component in the UVM testbench. Note that in Fig. 4.2, testbench seems to be the top-level component. But testbench merely instantiates the design under test (DUT) and the UVM Test class and configures the connection between them. A test is a class that encapsulates test-specific instructions written by the test writer.

The UVM Test typically:

1. Instantiate the top-level environment.
2. Configure the environment (via factory overrides or the configuration database).
3. Apply stimulus by invoking UVM sequences through the environment (typically one per DUT interface) to the DUT.

Tests in UVM are classes that are derived from an *uvm_test* class. Using classes allows inheritance and reuse of tests. Typically, a base test class is defined that instantiates and configures the top-level environment and is then extended to define scenario-specific configurations such as which sequences to run, coverage parameters, etc. The test instantiates the top-level environment just like any other verification component.

Typically, there is one base UVM Test with the UVM environment instantiation and other common items. Individual tests will extend this base test and configure the environment differently or select different sequences to run.

4.3.3 UVM Environment

The UVM environment is a component that groups together other verification components that are interrelated. Typical components that are usually instantiated inside the UVM environment are UVM agents, UVM scoreboards, or even other UVM environments. The top-level UVM environment encapsulates all the verification components targeting the DUT.

The top-level environment is a container that defines the reusable component topology within the UVM Tests. The top-level environment instantiates and configures the reusable verification IP and defines the default configuration of that IP as required by the application. Multiple tests can instantiate the top-level environment class and determine the nature of traffic to generate and send for the selected configuration. Additionally, the tests can override the default configuration of the top-level environment to better achieve their goals.

As noted above, the top-level UVM environment can instantiate other UVM environments. Typically for each interface to the DUT, you will find a separate environment per interface. For example, PCIe environment, USB environment, etc. Some of these IP environments can be grouped together into cluster environments (e.g., an IP interface environment, CPU environment, etc.).

4.3.4 UVM Agent

The UVM agent is a hierarchical component that groups together other verification components that are dealing with a specific DUT interface. Agent includes a UVM sequencer to manage stimulus flow, a UVM driver to apply stimulus to the DUT interface, and a UVM monitor to monitor the DUT interface. UVM agents might include other components, like coverage collectors, protocol checkers, and a TLM model.

To reiterate, the UVM agent is the component that drives the signal-level interface of the DUT. At a minimum, it will have a sequencer and a driver and an optional monitor (even though, I am not sure what you would do without a monitor to convert the DUT signal-level activity into transactions that can be sent to an analysis port to a scoreboard for pass/fail determination). The sequences are fed to the sequencer which then sends these sequences (aka tests or transactions) to the driver. The driver converts a transaction into signal-level DUT interface detail (e.g., PCIe interface). This is depicted in Fig. 4.4.

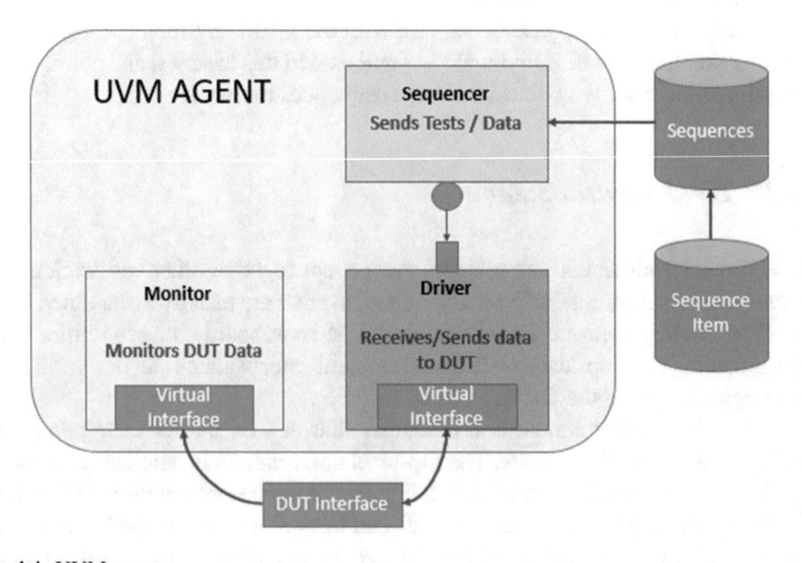

Fig. 4.4 UVM agent

Note that the agent can operate in an active mode or a passive mode. In the active mode, it can generate the stimulus (i.e., the driver drives DUT input and senses DUT outputs). In the passive mode, the driver and the sequencer remain silent (disabled) and only the monitor remains active. Monitor simply monitors the outputs of DUT; it cannot control the IO of the DUT. You can dynamically configure an agent in either an active mode or a passive mode. Monitor is an unidirectional interface, while driver is a bidirectional interface. This is depicted in Fig. 4.4.

4.3.5 UVM Sequence Item

UVM sequence item (i.e., a transaction) is the fundamental lowest denominator object in the UVM hierarchy. It is the definition of the basic transaction that will then be used to develop UVM sequences. The sequence item defines the basic transaction (e.g., an AXI transaction) data items and/or constrains imposed on them. While the driver deals with signal activities at the bit level, it doesn't make sense to keep this level of abstraction as we move away from the DUT, so the concept of transaction was created.

UVM sequence items, i.e., transactions are the smallest data transfers that can be executed in a verification model. They can include variables, constraints, and even methods for operating on themselves.

Here's an example of a sequence item:

```
class lpi_seq_item extends uvm_sequence_item;
        `uvm_object_utils(lpi_seq_item)

        //Data members
        rand bit        slp_req0;
        rand bit        slp_req1;
        rand bit        wakeup_req0;
        rand bit        wakeup_req1;
        rand bit        ss_wakeup;
        rand bit        ss_sleep;

        //UVM Constructor
        function new (string name="lpi_seq_item");
        super.new (name);
        endfunction

        //constraints on data members
        constraint slp_wakeup_reqs {
        (((slp_req0 || slp_req1) && (wakeup_req0 || wakeup_req1))
!= 1); };
endclass: lpi_seq_item
```

4.3.6 UVM Sequence

After a basic *uvm_sequence_item* has been created, the verification environment will need to generate sequences using the sequence item to be sent to the sequencer. Sequences are an ordered collection of transactions (sequence items); they shape transactions to our needs and generate as many as we want. Since the variables in the transaction (sequence item) are of type "rand," if we want to test just a specific set of addresses in a master-slave communication topology, we could restrict the randomization to that set of values instead of wasting simulation time in invalid (or redundant) values.

Sequences are extended from uvm_sequence, and their main job is generating multiple transactions. After generating those transactions, there is another class that takes them to the sequencer (discussed next).

For example, using the example of sequence item shown in Sect. 4.3.5, you can write a sequence as follows.

```
class lpi_basic_seq extends uvm_sequence #(lpi_seq_item);
         `uvm_object_utils(lpi_basic_seq)

         rand int num_of_trans;

         function new (string name="lpi_basic_seq");
           super.new (name);
          endfunction

          extern task body();
endclass: lpi_basic_seq

task lpi_basic_seq::body();
         lpi_seq_itemseq_item;
         seq_item = lpi_seq_item::type_id::create("seq_item");

         for (int i = 0; i < num_of_trans; i++)
         begin
          `uvm_info(get_type_name(),$psprintf("in seq for count =
%d", i, ,UVM_LOW)

          start_item(seq_item);

          if(!seq_item.randomize())
           begin
             `uvm_error("body","Randomization failed for seq_item")
end

               `uvm_info(get_type_name(),$psprintf("obj is req0 = %d,
req1 = %d, sleep0 = %d, sleep1 = %d", seq_item.wakeup_req0, seq_item.
wakeup_req1, seq_item.slp_req0, seq_item.slp_req1) ,UVM_LOW)
```

```
        finish_item(seq_item);
end
endtask: body
```

The explanation of the code is as follows:

lpi_basic_seq is extended from the *uvm_sequence* which is parameterized with the *lpi_seq_item*. We also declare a rand variable *num_of_trans*.

Then we define the task body which is the gist of the sequence. In the task *body*(), we first define a *seq_item* of type *lpi_seq_item* which is what the task body will work on.

The "for" loop will iterate for a random number of times (since *num_of_trans* is declared "rand"). It will first *start_item* (start the sequence) which is a call that *blocks* until the driver accesses the transaction being created.

Then it checks to see if *seq_item* can indeed be randomized. If not, it will generate an error and abort and will print the transaction information and do a *finish_item* which is also a blocking call which blocks until the driver has completed the operation of the current transaction.

Note that, UVM sequences can operate hierarchically with one sequence, called a parent sequence, invoking another sequence, called a child sequence. To operate, each UVM sequence is eventually bound to a UVM sequencer. Multiple UVM sequence instances can be bound to the same UVM sequencer.

4.3.7 UVM Sequencer

The sequencer controls the flow of request and response sequence items between sequences and the driver. UVM sequencer is a simple component that serves as an arbiter for controlling transaction flow from multiple stimulus sequences.

The sequencer and driver use TLM interface to communicate.

uvm_sequencer and uvm_driver base classes have seq_item_export and seq_item_port defined respectively. The user needs to connect them using TLM connect method.

Example:

```
driver.seq_item_port.connect(sequencer.seq_item_export);
```

Examples shown in Sects. 4.7.4 and 4.8.4 further shed light on a sequencer.

4.3.8 UVM Driver

Driver is where the TLM transaction-level world meets the DUT signal/clock/ pin-level world. Driver *receives* sequences from the sequencer, converts the received sequences into signal-level activities, and drives them on the DUT interface as per the interface protocol. Or the driver *pulls* sequences from the sequencer and sends them to the signal-level interface. This interaction will be observed and evaluated by another block, the monitor, and as a result, the driver's functionality should only be limited to send the necessary data to the DUT. Note that nothing prevents the Driver from monitoring the transmitted/received data from DUT—but that violates the rules of modularity. Also, if you embed the monitor in the driver, you can't turn the monitor ON/OFF.

The driver has a TLM port to receive transactions from the sequencer and access to the DUT interface to drive the DUT signals.

Driver is written by extending uvm_driver.

uvm_driver is inherited from uvm_component; Methods and TLM port (seq_ item_port) are defined for communication between sequencer and driver.

The uvm_driver is a parameterized class; and it is parameterized with the type of the request sequence_item and the type of the response sequence_item.

The UVM driver is discussed in detail with the UVM example discussed in Sect. 4.7.2.

4.3.8.1 uvm_driver Methods

get_next_item

This method blocks until a REQ sequence_item is available in the sequencer.

try_next_item

This is a non-blocking variant of the get_next_item () method. It will return a null pointer if there is no REQ sequence_item available in the sequencer.

item_done

The non-blocking item_done () method completes the driver–sequencer handshake, and it should be called after a get_next_item () or a successful try_next_item () call.

put

The put () method is non-blocking and is used to place a RSP sequence_item in the sequencer.

Examples shown in Sects. 4.7.2 and 4.8.5 further shed light on a driver.

4.3.9 UVM Monitor

Monitor, in a sense, is reverse of the driver. It takes the DUT signal/pin-level activities and converts them back into transactions to be sent out to the rest of the UVM testbench (e.g., to the scoreboard) for analysis. Monitor broadcasts the created transactions through its *analysis port*. Note that comparing of the received output from the DUT to that with expected output is normally done in the scoreboard and not directly in the monitor (even though there is nothing that prevents that from happening).

The reason is to preserve modularity of the testbench. Monitor, as the name suggests, monitors the DUT signals and coverts them to transactions. That's it. It's the job of the scoreboard (or any other component for that matter) to receive the broadcasted transaction from the Driver and do the comparison with the expected outputs.

The UVM monitor can perform internally some processing on the transactions produced (such as coverage collection, checking, logging, recording, etc.) or can delegate that to dedicated components connected to the monitor's analysis port.

Example shown in Sects. 4.7.5 and 4.8.6 provide examples on building a UVM monitor.

4.3.10 UVM Scoreboard

The scoreboard simply means that it is a checker (not to be confused with SystemVerilog SVA "checker"). It checks the response of the DUT against expected response. The UVM scoreboard usually receives transactions from the monitor through UVM agent analysis ports and the transactions through a reference model to produce expected transactions and then compares the expected output versus the received transaction from the monitor.

There are many ways to implement a scoreboard. For example, if you are using a reference model, you may use SystemVerilog–DPI API to communicate with the scoreboard, pass transactions via DPI to the reference model, convert reference model response into transactions, and compare the DUT output transaction with the one provided by the reference model. Reference model can be a C/C++ model or a TLM2.0 SystemC model or simply another SystemVerilog model.

4.4 UVM Class Library

Figure 4.5 [(Accelera, Universal Verification Methodology (UVM) 1.2 User's guide)] shows the building blocks of UVM class library that you can use to quickly build well constructed, reusable, configurable components and testbenches. The library contains base classes, utilities, and macros.

The advantages of using the UVM class library [(Accelera, Universal Verification Methodology (UVM) 1.2 User's guide)] include:

(a) A robust set of built-in features—The UVM class library provides many features that are required for verification, including complete implementation of printing, copying, test phases, factory methods, and more.
(b) Correctly implemented UVM concepts—Each component can be derived from a corresponding UVM class library component. Using these base class elements increase the readability of your code since each component's role is predetermined by its parent class.

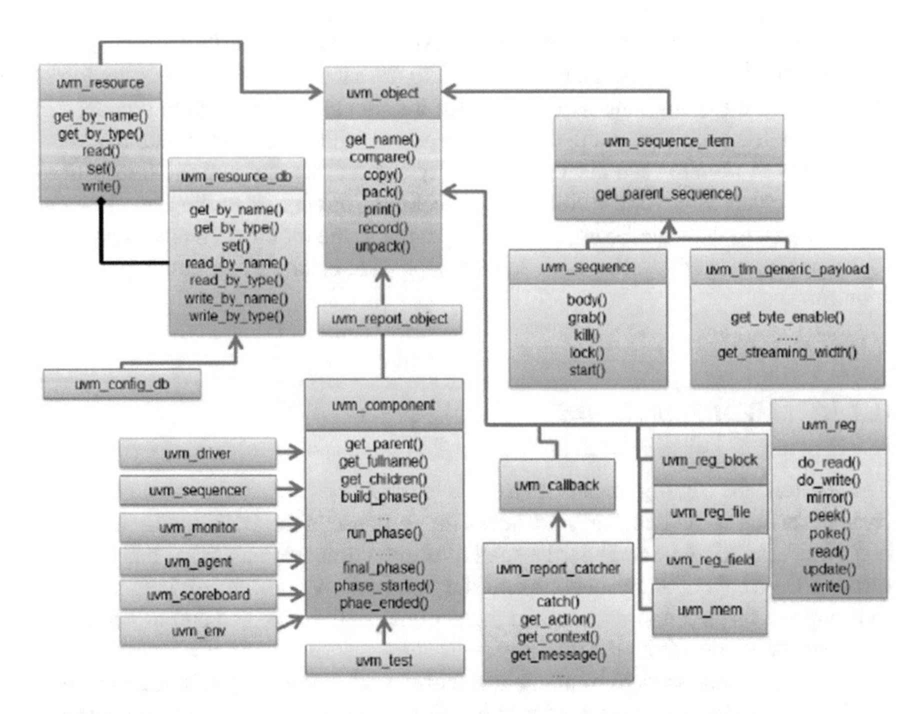

Fig. 4.5 UVM class library hierarchy (Accelera, Universal Verification Methodology (UVM) 1.2 User's guide)

The UVM class library also provides various utilities to simplify the development and use of verification environments. These utilities support configurability by providing a standard resource sharing database.

They support debugging by providing a user–controllable messaging utility for failure reporting and general reporting purposes. They support testbench construction by providing a standard communication infrastructure between verification components (TLM) and flexible verification environment construction (UVM factory). Finally, they also provide macros for allowing more compact coding styles.

4.5 UVM Transaction-Level Communication Protocol: Basics

4.5.1 Basic Transaction-Level Communication

The fundamental transaction-level interfaces have the concept of a "port" and an "export." A TLM "port" defines a set of methods (aka the Application Programming Interface—API) to be used for a connection. A TLM "export" implements these methods. Connecting a "port" to an "export" allows the implementation to be executed when the "port" method is called.

Figure 4.6 shows an example of a single producer connected to a single consumer.

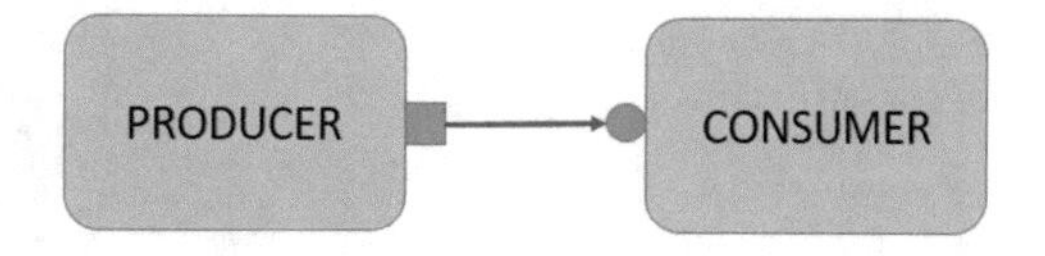

Fig. 4.6 UVM producer port to consumer export

The square on the producer block is called a "port," and the circle on the consumer block is called an "export." The producer sends out transactions and method calls (e.g., the method "put" in the following example) on its "port" (or per the language semantics, "put_port").

Here's sample code for a producer (Accelera, UVM 1.2 User's Guide, 2015):

```
class producer extends uvm_component;
uvm_blocking_put_port #(my_trans) put_port; // 1 parameter
      function new (string name, uvm_component parent);
             put_port = new ("put_port", this);
             ...
             endfunction

           virtual task run();
           my_trans myT;
           for (int i = 0; i < N; i++) begin
                     // Generate myT.
                     put_port.put (myT);
          end
          endtask
endclass
```

The uvm_blocking_put_port simply means that the producer will block until the consumer's "put" implementation is complete. As you can see, the producer sends the transaction through its put_port with "put" method. This "put" method is provided as part of the uvm_blocking_put_port class. The "put" method is implemented by the consumer. The semantics of the put operation are defined by TLM. In this case, the put() call in the producer will block until the consumer's put implementation is complete. Other than that, the operation of producer is completely independent of the put implementation (uvm_blocking_put_imp). The modularity provided by TLM fosters an environment in which components may be easily reused since the interfaces are well defined.

Here's an implementation of the consumer.

```
class consumer extends uvm_component;
uvm_blocking_put_imp  #(my_trans,  consumer)  put_export;  //  2
parameters
...
     task put (my_trans t);
            . case(t.kind)
```

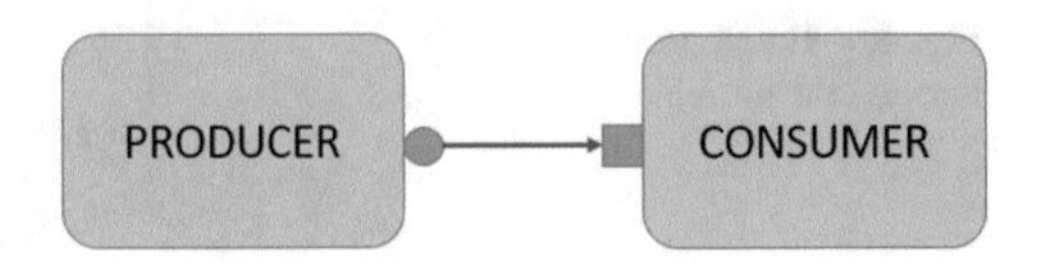

Fig. 4.7 Producer export to consumer port

```
                    BURST_READ: // Do burst read.
                    BURST_WRITE: // Do burst write.
            endcase
      endtask
endclass
```

To reiterate, whenever the producer invokes the "put" method, the consumer's implementation (task *put*) will be executed.

Converse of "put" is "get" (Fig. 4.7). In other words, here the consumer *gets* a transaction from the producer. The consumer requests transactions from the producer via its "get" port. Note that now the "square box" (i.e., the *port)* is on the Consumer and the circle (i.e., the *export*) is on the producer side. This means that consumer calls the "get" method which is implemented by the producer.

Here's a simple example (Accelera, Universal Verification Methodology (UVM) 1.2 User's guide).

```
class get_consumer extends uvm_component;
      uvm_blocking_get_port #(my_trans) get_port;
      function new (string name, uvm_component parent);
            get_port = new ("get_port", this);
            ...
      endfunction

      virtual task run();
      my_trans myT;
                  for (int i = 0; i < N; i++) begin
                  // Generate t.
                  get_port.get(myT);
                  end
      endtask
endclass
```

As with put (), the get_consumer's *get*() call will block until the get_producer's method completes. In TLM terms, put () and get() are blocking methods.

```
class get_producer extends uvm_component;
uvm_blocking_get_imp #(my_trans, get_producer) get_export;
...
      task get(output my_trans t);
```

Fig. 4.8 Producer to consumer via TLM FIFO

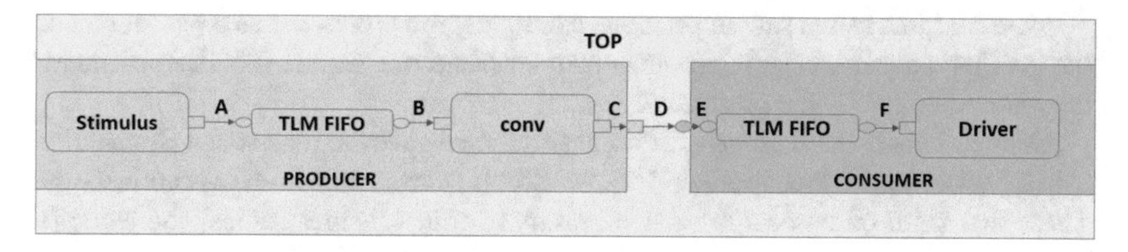

Fig. 4.9 Stimulus to driver hierarchical connection

```
            simple_trans tmp = new();
            // Assign values to tmp.
            t = tmp;
        endtask
endclass
```

Note: In both these examples, there is a single process running, with control passing from the port to the export and back again. The direction of data flow (from producer to consumer) is the same in both examples.

But what if you want the producer and the consumer to operate independently? In the example above, the consumer will be active only when its put () method is called by the producer. UVM provides the uvm_tlm_fifo channel to facilitate such communication. The uvm_tlm_fifo implements all the TLM interface methods, so the producer *puts* the transaction into the uvm_tlm_fifo, while the consumer independently *gets* the transaction from the FIFO, as shown in Fig. 4.8.

When the producer puts a transaction into the FIFO, it will block if the FIFO is full; otherwise, it will put the object into the FIFO and return immediately. The get operation will return immediately if a transaction is available (and will then be removed from the FIFO); otherwise, it will block until a transaction is available. Thus, two consecutive get() calls will yield different transactions to the consumer. The related peek() method returns a copy of the available transaction without removing it. Two consecutive peek() calls will return copies of the same transaction.

4.5.2 *Hierarchical Connections*

Ok, let us take the above examples a bit further. What if you want to make connections across hierarchical boundaries? Let us consider the hierarchical design shown in Fig. 4.9.

Here again you have the producer and the consumer. But there is a hierarchy under each of these components. The idea is to transfer a "stim" transaction (stimulus transaction) to the eventual "drv" (driver) component.

The producer contains "stim," "fifo," and "conv," while the consumer contains "fifo" and "drv".

Note that just as we saw in previous examples, the producer has a put_port (i.e., the port puts a transaction) and the consumer has a put_export (i.e., it implements the method called by the producer).

Connections C and E are of a different sort than what have seen so far. Connection C is a port-to-port connection, and connection E is an export-to-export connection. These two kinds of connections are necessary to complete hierarchical connections. Connection C imports a port from the outer component to the inner component. Connection E exports an export upwards in the hierarchy from the inner component to the outer one. Ultimately, every transaction-level connection must resolve so that a port is connected to an export. However, the port and export terminals do not need to be at the same place in the hierarchy. We use port-to-port and export-to-export connections to bring connectors to a hierarchical boundary to be accessed at the next higher level of hierarchy.

4.5.3 Analysis Ports and Exports

Ok so far, we have seen "port" and "export." Let us look at a third type of interface called "analysis_port." This port is represented as a *diamond* (Fig. 4.10). As we discussed above, the key distinction between the two types of TLM communication is that the put/get ports typically require a corresponding export to supply the implementation. For analysis, however, the emphasis is on a particular component, such as a monitor, being able to produce a stream of transactions, regardless of whether there is a target actually connected to it. Modular analysis components are then connected to the analysis_port, each of which processes the transaction stream in a particular way.

Let us look at the analysis port and export separately.

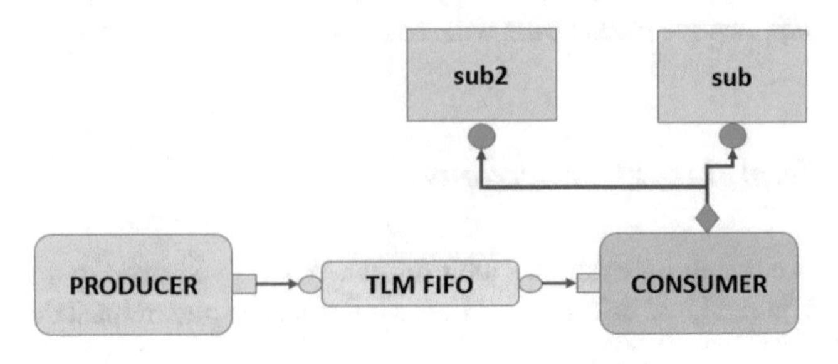

Fig. 4.10 Analysis port

In UVM semantics, the analysis port is called "uvm_analysis_port." It's a specialized TLM port whose interface consists of a single function called "write ()." The analysis port contains a list of analysis_exports that are connected to it. Refer to Fig. 4.10. When the component calls analysis_port.write(), the analysis_port cycles through the list of connected exports and calls the write() method of each connected export. If nothing is connected, the write() call simply returns. Thus, an analysis port may be connected to zero or many analysis exports, but the operation of the component that writes to the analysis port does not depend on the number of exports connected. Because write() is a void function, the call will always complete in the same delta cycle, regardless of how many components (for example, scoreboards, coverage collectors, and so on) are connected.

Components to which *export* is connected implement the write() function. If multiple exports are connected to an analysis port, the port will call the write() of each export, in order. Since all implementations of write() must be functions, the analysis port's write() function completes immediately, regardless of how many exports are connected to it. When multiple subscribers are connected to an analysis_port, each is passed a pointer to the same transaction object, the argument to the write() call. Each write() implementation must make a local copy of the transaction and then operate on the copy to avoid corrupting the transaction contents for any other subscriber that may have received the same pointer. UVM also includes an analysis_fifo, which is a uvm_tlm_fifo that also includes an analysis export, to allow blocking components access to the analysis transaction stream. The analysis_fifo is unbounded, so the monitor's write() call is guaranteed to succeed immediately. The analysis component may then get the transactions from the analysis_fifo whenever it pleases.

4.6 UVM Phases

As shown in Fig. 4.11, following phases make up the UVM Phases. Each phase is described below. Basically, there are three phases overall. The build phase builds top-level testbench topology and the connect phase connects environment topology. The run phase does exactly what the name suggests, namely, run the test! All phases "under" the run umbrella run in zero time, except of course the run() phase. And finally, the cleanup phase gathers details on the final DUT state, processes the simulation results, and does simulation results analysis and reporting.

The following provides at finer granularity the phases at play in UVM.

4.6.1 Build Phases

The build phases are executed at the start of the UVM testbench simulation, and their overall purpose is to construct, configure, and connect the testbench component hierarchy [(MentorGraphics)].

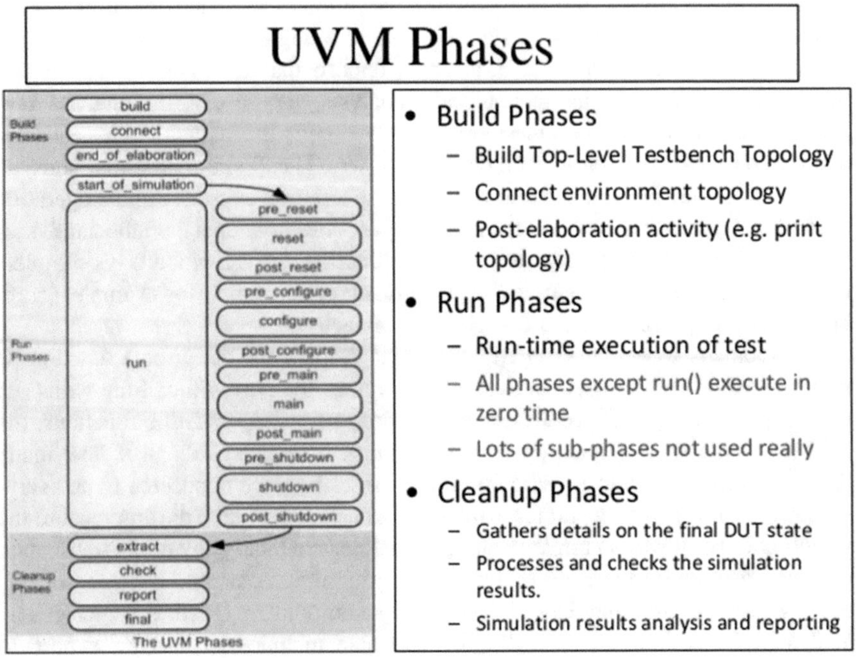
UVM Phases
Build Phases
build
connect
end_of_elaboration
start_of_simulation
pre_reset
reset
post_reset
pre_configure
configure
post_configure
pre_main
main
post_main
pre_shutdown
shutdown
post_shutdown
Run Phases
run
extract
check
report
final
Cleanup Phases
The UVM Phases
• Build Phases
– Build Top-Level Testbench Topology
– Connect environment topology
– Post-elaboration activity (e.g. print topology)
• Run Phases
– Run-time execution of test
– All phases except run() execute in zero time
– Lots of sub-phases not used really
• Cleanup Phases
– Gathers details on the final DUT state
– Processes and checks the simulation results.
– Simulation results analysis and reporting

Fig. 4.11 UVM phases (MentorGraphics)

All the build phase methods are functions and therefore execute in zero simulation time.

4.6.1.1 Build

Once the UVM testbench root node component is constructed, the build phase starts to execute. It constructs the testbench component hierarchy from the top of the hierarchy downwards. The construction of each component is deferred so that each layer in the component hierarchy can be configured by the level above. During the build phase, uvm_components are indirectly constructed using the UVM factory.

4.6.1.2 Connect

The connect phase is used to make TLM connections between components or to assign handles to testbench resources. It must occur after the build method has put the testbench component hierarchy in place and works from the bottom of the hierarchy upwards.

4.6.1.3 end_of_elaboration

The end_of_elaboration phase is used to make any final adjustments to the structure, configuration, or connectivity of the testbench before simulation starts. Its implementation can assume that the testbench component hierarchy and interconnectivity is in place. This phase executes bottom up.

4.6.2 Run-Time Phases

The testbench stimulus is generated and executed during the run-time phases which follow the build phases. After the start_of_simulation phase, the UVM executes the run phase and the phases pre_reset through to post_shutdown in parallel. The run phase was present in the OVM and is preserved to allow OVM components to be easily migrated to the UVM. The other phases were added to UVM to give finer run-time phase granularity for tests, scoreboards, and other similar components. It is expected that most testbenches will only use reset, configure, main, and shutdown and not their pre- and post variants.

4.6.2.1 start_of_simulation

The start_of_simulation phase is a function which occurs before the time-consuming part of the testbench begins. It is intended to be used for displaying banners, testbench topology, or configuration information. It is called in bottom-up order.

4.6.2.2 Run

The run phase occurs after the start_of_simulation phase and is used for the stimulus generation and checking activities of the testbench. *The run phase is implemented as a task, and all uvm_component run tasks are executed in parallel.* Transactors such as drivers and monitors will nearly always use this phase.

4.6.2.3 pre_reset

The pre_reset phase starts at the same time as the run phase. Its purpose is to take care of any activity that should occur before reset, such as waiting for a power-good signal to go active. I do not anticipate much use for this phase.

4.6.2.4 Reset

The reset phase is reserved for DUT or interface-specific reset behavior. For example, this phase would be used to generate a reset and to put an interface into its default state.

4.6.2.5 post_reset

The post_reset phase is intended for any activity required immediately following reset. This might include training or rate negotiation behavior. I do not anticipate much use for this phase.

4.6.2.6 pre_configure

The pre_configure phase is intended for anything that is required to prepare for the DUT's configuration process after reset is completed, such as waiting for components (e.g., drivers) required for configuration to complete training and/or rate negotiation. It may also be used as a last chance to modify the information described by the test/environment to be uploaded to the DUT. I do not anticipate much use for this phase.

4.6.2.7 Configure

The configure phase is used to program the DUT and any memories in the testbench so that it is ready for the start of the test case. It can also be used to set signals to a state ready for the test case start.

4.6.2.8 post_configure

The post_configure phase is used to wait for the effects of configuration to propagate through the DUT or for it to reach a state where it is ready to start the main test stimulus. I do not anticipate much use for this phase.

4.6.2.9 pre_main

The pre_main phase is used to ensure that all required components are ready to start generating stimulus. I do not anticipate much use for this phase.

4.6.2.10 Main

This is where the stimulus specified by the test case is generated and applied to the DUT. It completes when either all stimulus is exhausted or a time-out occurs. Most data throughput will be handled by sequences started in this phase.

4.6.2.11 post_main

This phase is used to take care of any finalization of the main phase. I do not anticipate much use for this phase.

4.6.2.12 pre_shutdown

This phase is a buffer for any DUT stimulus that needs to take place before the shutdown phase. I do not anticipate much use for this phase.

4.6.2.13 Shutdown

The shutdown phase is used to ensure that the effects of stimulus generated during the main phase have propagated through the DUT and that any resultant data has drained away.

4.6.2.14 post_shutdown

Perform any final activities before exiting the active simulation phases. At the end of the post_shutdown phase, the UVM testbench execution process starts the cleanup phases. I do not anticipate much use for this phase.

4.6.3 Cleanup Phases

The cleanup phases are used to extract information from scoreboards and functional coverage monitors to determine whether the test case has passed and/or reached its coverage goals [(MentorGraphics)]. The cleanup phases are implemented as functions and therefore take zero time to execute. They work from the bottom upwards in the component hierarchy.

4.6.3.1 Extract

The extract phase is used to retrieve and process information from scoreboards and functional coverage monitors. This may include the calculation of statistical information used by the report phase. This phase is usually used by analysis components.

4.6.3.2 Check

The check phase is used to check that the DUT behaved correctly and to identify any errors that may have occurred during the execution of the testbench. This phase is usually used by analysis components.

4.6.3.3 Report

The report phase is used to display the results of the simulation or to write the results to file. This phase is usually used by analysis components.

4.6.3.4 Final

The final phase is used to complete any other outstanding actions that the testbench has not already completed.

Here's a very simple example of a basic *uvm_component* showing different UVM phases.

```
class generic_component extends uvm_component;
    `uvm_component_utils(generic_component)

    function new(string name, uvm_component parent);
       super.new(name, parent);
    endfunction: new

    function void build_phase(uvm_phase phase);
       super.build_phase(phase);

       //Code for constructors goes here
    endfunction: build_phase

    function void connect_phase(uvm_phase phase);
        super.connect_phase(phase);

       //Code for connecting components goes here
    endfunction: connect_phase
```

```
    task run_phase(uvm_phase phase);
       //Code for simulation goes here
    endtask: run_phase

    function void report_phase(uvm_phase phase);
       //Code for showing simulation results goes here
    endfunction: report_phase

endclass: generic_component
```

4.7 UVM Example: One

We now present a UVM example, which describes building sequence_item, sequence, UVM agent, UVM sequencer, UVM driver, etc.

4.7.1 Modeling a Sequence Item

A sequence item is essentially a transaction upon (using) which sequences can be built. So, they are transaction objects used as stimulus to the DUT. The UVM class library provides the uvm_sequence_item base class. Every user-defined sequence item should be derived directly or indirectly from this base class.

UVM has built-in automation for many service routines that a data item needs. For example, you can use:

- Print() to print a data item
- Copy() to copy the contents of a data item
- Compare() to compare two similar objects

UVM allows you to use a built-in, mature, and consistent implementation of these routines.

Here's an example.

```
class bus_seq_item extends uvm_sequence_item;

// Request data properties are rand
rand logic [31:0] addr;
rand logic [31:0] write_data;
rand bit read_not_write;
rand int delay;

// Response data properties are NOT rand
bit error;
logic [31:0] read_data;
```

```
//Factory registration
`uvm_object_utils(bus_seq_item)

       function new (string name = "bus_seq_item");
               super.new (name);
       endfunction

// Delay between bus cycles is constrained
constraint at_least_1 {delay inside {[1:20]};}
// 32 bit aligned transfers
constraint align_32 {addr[1:0] == 0;}

endclass bus_seq_item
```

General rules of thumb for creating a sequence_item:

Review your DUT interface properties and functionality, and add those as variables in the class where you define your sequence_item. In the example above, bus_seq_item creates a sequence_item for a simple Read/Write interface. It defines, addr, write_data, read_data, read_not_write, delay, and error fields. This is the user defined sequence_item derived from uvm_sequence_item. You also notice that for the stimulus fields, you may "rand" the fields to drive constraint random stimulus. Note that the outputs from the DUT (i.e., the response fields) are not "rand" (that wouldn't make sense, would it?). Note the two constraints that are also part of the class bus_seq_item. Hence, whenever stimulus is driven to the DUT, it will be constrained within the limits specified.

4.7.1.1 Inheritance and Constraint Layering

Continuing with the above example, you may want to adjust the sequence item generation by adding more constraints to the bus_seq_item class definition. In SystemVerilog, this is done using inheritance. The following example shows how you can extend the class bus_seq_item and constrain write_data.

```
class write_data_constraint extends bus_seq_item;

constraint write_constraint {write_data[3:0] == 4'b0000; }
`uvm_object_utils(write_data_constraint)

// Constructor
        function new (string name = "write_data_constraint");
                super.new (name);
        endfunction: new

endclass
```

4.7.2 Building UVM Driver

4.7.2.1 Driver Basics

Driver is the main signal-level interface to the DUT. It drives data items per a specific interface protocol (e.g., PCIe) and drives and senses the interface. It gets its instructions, i.e., the transactions from the sequencer which it then converts to protocol-specific signal-level activity. The UVM class library provides the uvm_driver base class, from which all driver classes should be extended, either directly or indirectly. The driver has a TLM port through which it communicates with the sequencer.

Note that you cannot drive the DUT signals directly from the SystemVerilog class-based environment (i.e., class-based code). You need to declare a virtual interface in the driver to connect to the DUT.

4.7.2.2 Driver Example

```
//By default - response and request type are assumed same if not
provided
class simple_driver extends uvm_driver #(simple_item, simple_rsp);

simple_item s_item;
virtual dut_if vif;

// UVM automation macros for general components
`uvm_component_utils(simple_driver)

// Constructor
function new (string name = "simple_driver", uvm_component parent);
          super.new (name, parent);
endfunction: new


//BUILD Phase
function void build_phase(uvm_phase phase);
         super.build_phase(phase);
        if(!uvm_config_db#(virtual dut_if)::get(this,"","vif",vif))
begin
               `uvm_fatal("NOVIF", {"virtual interface must be set
for: ", get_full_name(),".vif"});
        end
endfunction: build_phase


//RUN Phase
task run_phase(uvm_phase phase);
        forever begin
```

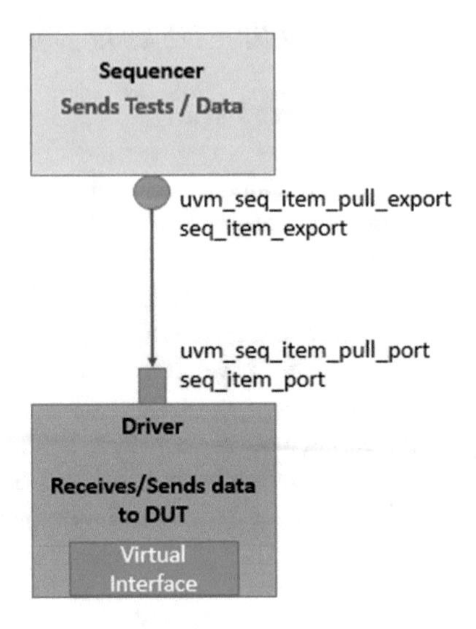

Fig. 4.12 Sequencer driver connection

```
        // Get the next data item from sequencer (may block).
        seq_item_port.get_next_item(s_item);
           fork begin
           // Execute the item.
                   drive_and_respond(s_item);
           end
           join_none
           seq_item_port.item_done(); // Consume the request.
           end
endtask: run

task drive_and_respond (input simple_item item);
         // Add your logic here.
endtask: drive_and_respond

endclass: simple_driver
```

4.7.3 Basic Sequencer and Driver Interaction

Figure 4.12 shows a simple diagram depicting interaction between a driver and a sequencer.

In this example, we are creating a simple_driver from the base class uvm_driver. In addition, we show a basic way for driver to interact with the sequencer. That is done using the tasks get_next_item() and item_done(). As demonstrated in the above example, the driver uses get_next_item() to fetch the next randomized item to be sent to the DUT. After sending it to the DUT, the driver signals the sequencer that the item was processed using item_done(). Note that get_next_item() is blocking until an item is provided by the sequencer.

In addition to the get_next_item() task, the uvm_seq_item_pull_port class provides another task, try_next_item(). This task will return in the same simulation step if no data items are available for execution. You can use this task to have the driver execute some idle transactions, such as when the DUT has to be stimulated when there are no meaningful data to transmit. The following example shows a revised implementation of the run_phase() task. This time using try_next_item() to drive idle transactions as long as there is no real data item to execute:

```
task run_phase(uvm_phase phase);

        forever
        begin
        // Try the next data item from sequencer (does not block).
        seq_item_port.try_next_item(s_item);
               if (s_item == null)
               begin
                 // No data item to execute, send an idle transaction.
                 ...
               end

               else
               begin
                 // Got a valid item from the sequencer, execute it.
                 ...
                 // Signal the sequencer; we are done.
                 seq_item_port.item_done();
              end
       end
endtask: run
```

In some protocols, such as pipelined protocols, the driver may operate on several transactions at the same time. The sequencer–driver connection, however, is a single item handshake which shall be completed before the next item is retrieved from the sequencer. In such a scenario, the driver can complete the handshake by calling item_done() without a response and provide the response by a subsequent call to put_response() with the real response data.

In some sequences, a generated value depends on the response to previously generated data. By default, the data items between the driver and the sequencer are copied by reference, which means that the changes the driver makes to the data item will be visible inside the sequencer. In cases where the data item between the driver and the sequencer is copied by value, the driver needs to return the processed response back to the sequencer. Do this using the optional argument to item_done(),

```
seq_item_port.item_done(rsp);
or using the put_response() method,

seq_item_port.put_response(rsp);
or using the built-in analysis port in uvm_driver.

rsp_port.write(rsp);
```

Note: Before providing the response, the response's sequence and transaction id must be set to correspond to the request transaction using rsp.set_id_info(req).

Note: put_response() is a blocking method, so the sequencer must do a corresponding get_response(rsp).

4.7.4 Building UVM Sequencer

As noted previously, the sequencer controls the flow of request and response sequence items between sequences and the driver. UVM sequencer is a simple component that serves as an arbiter for controlling transaction flow from multiple stimulus sequences.

Sequences are extended from uvm_sequence, and their main job is generating multiple transactions. After generating those transactions, there is another class that takes them to the driver: the sequencer. The code for the sequencer is usually very simple and in simple environments, the default class from UVM is enough to cover most of the cases.

```
class simple_sequencer extends uvm_sequencer #(simple_item, simple_rsp);

`uvm_component_utils(simple_sequencer);

       function new (input string name, uvm_component parent=null)'
              super.new (name, parent);
       endfunction

endclass: simple_sequencer
```

4.7.5 Building UVM Monitor

```
class simple_monitor extends uvm_monitor;
  `uvm_component_utils(simple_monitor)

  //Interface
  virtual monitor_if monitor_vif;

  //Constructor
```

```
        function new (string name="simple_monitor", uvm_component
parent=null);
           super.new (name, parent);
           endfunction

   function void build_phase (uvm_phase phase);
   endfunction: build_phase

function void connect_phase (uvm_phase phase);
endfunction: connect_phase

extern task run_phase (uvm_phase phase);
endclass: simple_monitor

task simple_monitor::run_phase(uvm_phase phase);
//your code here
endtask: run_phase
endclass: simple_monitor
```

4.7.6 UVM Agent: Connecting Driver, Sequencer, and Monitor

```
class simple_agent extends uvm_agent;


uvm_active_passive_enum is_active;


simple_sequencer sequencer;
simple_driver driver;
simple_monitor monitor;

// Use build() phase to create agents' subcomponents.
virtual function void build_phase (uvm_phase phase);
         super.build_phase(phase)
         monitor = simple_monitor::type_id::create("monitor", this);

         if (is_active == UVM_ACTIVE) begin
                  // Build the sequencer and driver.
                  sequencer  =  simple_sequencer::type_id::create
("sequencer", this);
                  driver=simple_driver::type_id::create("driver",
this);
end
endfunction: build_phase
```

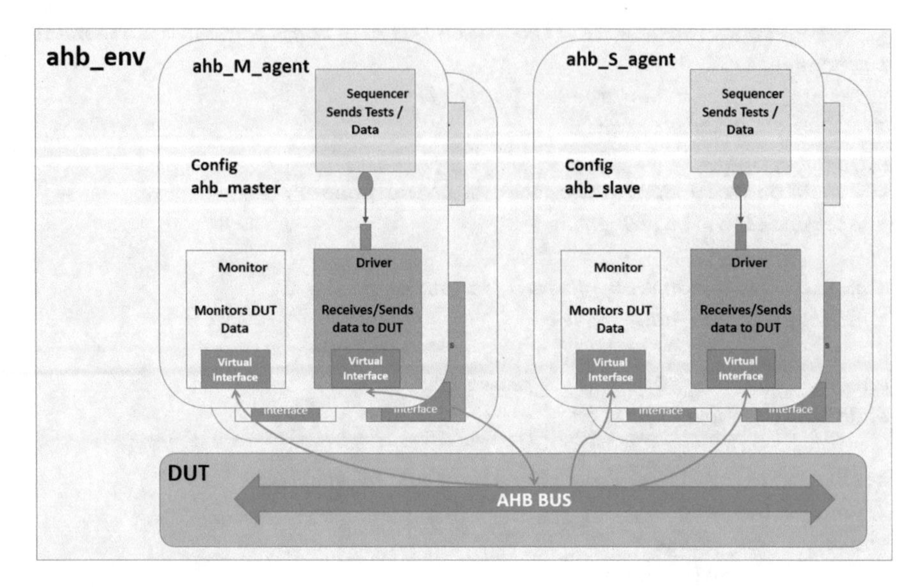

Fig. 4.13 UVM environment

```
//Use connect phase to connect components together

virtual function void connect_phase (uvm_phase phase);
        if (is_active == UVM_ACTIVE) begin
                driver.seq_item_port.connect(sequencer.
seq_item_export);
        end
endfunction: connect_phase
endclass: simple_agent
```

In this example, we declare the driver, monitor, and sequencer. We then "create" the monitor and sequencer using *create*(). The "if" condition tests the is_active property to determine whether the driver and the sequencer are created in this agent. Since we have (is_active == UVM_ACTIVE), we indeed create the driver and the sequencer using the *create*() call. Note that the *create*() call should always be called from the *build_phase()* method to create any multi-hierarchical component. We again check for the is_active flag and if it is indeed active, then only we connect the driver and the sequencer using the *.connect()* method.

4.7.7 Building the Environment

A typical UVM environment is shown in Fig. 4.13.

The environment class is the top container of reusable components. It instantiates and configures all of its subcomponents. Most verification reuse occurs at the environment level where the user instantiates an environment class and configures it and its agents for specific verification tasks. For example, a user might need to change the number of masters and slaves in a new environment as shown below (Accelera, Universal Verification Methodology (UVM) 1.2 User's guide).

```
class ahb_env extends uvm_env;
int num_masters;
ahb_M_agent masters[];

`uvm_component_utils_begin(ahb_env)
`uvm_field_int(num_masters, UVM_ALL_ON)
`uvm_component_utils_end

virtual function void build_phase(phase);
         string inst_name;
         super.build_phase(phase);
                 if (num_masters ==0))
                 `uvm_fatal("NONUM",{"'num_masters' must be set"};
                  masters = new[num_masters];
                  for (int i = 0; i < num_masters; i++) begin
                  $sformat(inst_name, "masters[%0d]", i);
                  masters[i]  = ahb_M_agent::type_id::create(inst_
name,this);
             end
        // Build slaves and other components.
endfunction
function new (string name, uvm_component parent);
         super.new (name, parent);
endfunction : new
endclass
```

A complete example of an environment is presented in Sect. 4.8.12

4.7.8 UVM Top-Level Module (Testbench) Example

The top_tb example below has added an SPI (Serial Peripheral Interface) and intr_if (Interrupt Interface) which is not shown in any of the diagrams above. This top_tb example assumes three interfaces, namely, spi_if, apb_if, and intr_if which are typical peripheral interfaces of an SoC.

```
module top_tb;


import uvm_pkg::*;
import spi_test_lib_pkg::*;
// PCLK and PRESETn
logic PCLK;
logic PRESETn;

// Instantiate the interfaces:


        apb_if APB(PCLK, PRESETn); // APB interface
        spi_if SPI(); // SPI Interface
        intr_if INTR(); // Interrupt

//Instantiate DUT
        spi_top DUT(
                // APB Interface:
                .PCLK(PCLK),
                .PRESETN(PRESETn),
                .PSEL(APB.PSEL[0]),
                .PADDR(APB.PADDR[4:0]),
                .PWDATA(APB.PWDATA),
                .PRDATA(APB.PRDATA),
                .PENABLE(APB.PENABLE),
                .PREADY(APB.PREADY),
                .PSLVERR(),
                .PWRITE(APB.PWRITE),

                // Interrupt output
                .IRQ(INTR.IRQ),

                // SPI signals
                .ss_pad_o(SPI.cs),
                .sclk_pad_o(SPI.clk),
                .mosi_pad_o(SPI.mosi),
                .miso_pad_i(SPI.miso)
        );

// UVM initial block:
// Virtual interface wrapping & run_test()
initial begin
        uvm_config_db #(virtual apb_if)::set( null , "uvm_test_top" ,
"APB_vif" , APB);
```

```
        uvm_config_db #(virtual spi_if)::set( null , "uvm_test_top"
, "SPI_vif" , SPI);
          uvm_config_db #(virtual intr_if)::set( null , "uvm_test_
top" , "INTR_vif", INTR);
          run_test();
end
// Clock and reset initial block:
//
initial begin
          PCLK = 0;
          PRESETn = 0;
          repeat(8) begin
          #10ns PCLK = ~PCLK;
end
PRESETn = 1;
forever begin
          #10ns PCLK = ~PCLK;
end
endmodule: top_tb
```

Another example of a Testbench is provided in Sect. 4.8.13

4.8 UVM Example: Two

This is a complete example of a low power interface block. This example will guide you to create your own simple UVM testbench, test, environment, agent, sequencer, driver, and monitor. The code is self-explanatory with embedded comments.

4.8.1 DUT: lpi.sv

```
module lpi(
     input wire clk,
     input wire rst_n,
     input wire slp_req0,
     input wire slp_req1,
     input wire wakeup_req0,
     input wire wakeup_req1,
     output wire ss_wakeup,
     output wire ss_sleep
     );
localparam
```

```
        S_IDLE     = 4'h1,
        S_ON      = 4'h2,
        S_SLEEP     = 4'h4,
        S_UP      = 4'h8;

reg [3:0]sm, nxt_sm;
wire wakeup = wakeup_req0 | wakeup_req1;
wire sleep =slp_req0 &slp_req1;
always @(posedge clk or negedge rst_n) begin
   if (!rst_n) begin
       sm <= #1 S_IDLE;
end
else
  sm              <= #1 nxt_sm;
end
always @* begin
 case(sm)
     S_IDLE : nxt_sm= wakeup ? S_ON: S_IDLE;
     S_ON : nxt_sm= sleep ? S_SLEEP: S_ON;
     S_SLEEP: nxt_sm= wakeup ? S_ON : S_SLEEP;
 endcase // case (sm)
end

assign ss_wake = (nxt_sm == S_ON) ? 1'b1 : 1'b0;
assign ss_slp = (nxt_sm == S_SLEEP) ? 1'b1 : 1'b0;

endmodule // lpi
```

4.8.2 lpi_if.sv

```
interface lpi_if(input bit lpi_clk, lpi_rstn);


        //bit is_active;
        logic          slp_req0;
        logic          slp_req1;
        logic          wakeup_req0;
        logic          wakeup_req1;
        logic          ss_wake;
        logic          ss_sleep;

endinterface: lpi_if
```

4.8.3 lpi_seq_item.sv

```
class lpi_seq_item extends uvm_sequence_item;
          `uvm_object_utils(lpi_seq_item)

          //Data members
          rand bit     slp_req0;
          rand bit     slp_req1;
          rand bit     wakeup_req0;
          rand bit     wakeup_req1;
          rand bit     ss_wakeup;
          rand bit     ss_sleep;

          //UVM methods
          function new (string name="lpi_seq_item");
           super.new (name);
          endfunction

          constraint slp_wakeup_reqs {
                    (((slp_req0 || slp_req1) && (wakeup_req0 ||
wakeup_req1)) != 1); };
endclass: lpi_seq_item
```

4.8.4 lpi_sequencer.sv

```
class lpi_sequencer extends uvm_sequencer#(lpi_seq_item);
 `uvm_sequencer_utils(lpi_sequencer)

    function new (string name="lpi_sequencer", uvm_component par-
ent = null);
         super.new (name, parent);
           `uvm_update_sequence_lib
    endfunction

endclass: lpi_sequencer
```

4.8.5 lpi_driver.sv

```
class lpi_driver extends uvm_driver#(lpi_seq_item);
  `uvm_component_utils(lpi_driver)
```

```
  //Virtual interface
  virtual lpi_iflpi_vif;
  lpi_seq_item seq_item;

function new (string name="lpi_driver", uvm_component parent=null);
    super.new (name, parent);
  endfunction

  task run_phase(uvm_phase phase);


   // Initialize all interface values asynchronously at the start
of run_phase
   lpi_vif.slp_req0 <= 0;
   lpi_vif.slp_req1 <= 0;
   lpi_vif.wakeup_req0<= 0;
   lpi_vif.wakeup_req1<= 0;
forever
  begin
   seq_item_port.get_next_item(seq_item);
   `uvm_info(get_name(), "Sendingtransaction\n", UVM_LOW)
   seq_item.print();
   @(posedge lpi_vif.lpi_clk);
   lpi_vif.slp_req0 <= seq_item.slp_req0;
   lpi_vif.slp_req1 <= seq_item.slp_req1;
   lpi_vif.wakeup_req1 <= seq_item.wakeup_req0;
   lpi_vif.wakeup_req0 <= seq_item.wakeup_req1;

   seq_item_port.item_done();
   end

endtask: run_phase


endclass:lpi_driver
```

4.8.6 lpi_monitor.sv

```
class lpi_monitor extends uvm_monitor;
  `uvm_component_utils(lpi_monitor)

  //Interface
  virtual lpi_if lpi_vif;
```

```
  //UVM specific methods
 function new (string name="lpi_monitor", uvm_component parent=null);
  super.new (name, parent);
  endfunction

  function void build_phase (uvm_phase phase);
  endfunction : build_phase

  function void connect_phase (uvm_phase phase);
  endfunction : connect_phase

  extern task run_phase (uvm_phase phase);
  endclass:lpi_monitor

  task lpi_monitor::run_phase(uvm_phase phase);
  endtask : run_phase
```

4.8.7 lpi_agent.sv

```
class lpi_agent extends uvm_agent;


        uvm_active_passive_enum is_active;


        //Component
        lpi_driver lpi_driver_h;
        lpi_sequencerlpi_sequencer_h;
        lpi_monitorlpi_monitor_h;

        //Interface
        virtual lpi_if lpi_vif;

        `uvm_component_utils_begin(lpi_agent)
        `uvm_field_enum(uvm_active_passive_enum, is_active, UVM_ALL_ON)
        `uvm_component_utils_end

        //UVM methods
        function new (string name, uvm_component parent=null);
          super.new (name, parent);
  endfunction
```

```
  function void build_phase(uvm_phase phase);
            super.build();

            //Retrieve interface from config db
           if(!(uvm_config_db #(virtual lpi_if)::get(null, "", "lpi_
vif", lpi_vif)))
            begin
         `uvm_fatal(get_name(),"Can't retrieve lpi_vif from config db\n")
         end
         //Build driver and sequencer
         lpi_driver_h = lpi_driver::type_id::create("lpi_driver_h",
this);
         lpi_sequencer_h=   lpi_sequencer::type_id::create("lpi_
sequencer_h",this);
         `uvm_info(get_name(), "lpi agent is active now\n", UVM_LOW)
         `uvm_info(get_name(),  "lpi  agent  is_active  setting
finish\n", UVM_LOW)

         //Build monitor
         lpi_monitor_h = lpi_monitor::type_id::create("lpi_monitor_h",
this);

        endfunction:build_phase


        function void connect_phase(uvm_phase phase);

          lpi_driver_h.seq_item_port.connect(lpi_sequencer_h.
seq_item_export);
         lpi_driver_h.lpi_vif= this.lpi_vif;
         lpi_monitor_h.lpi_vif = this.lpi_vif;
         endfunction:connect_phase

endclass:lpi_agent


package lpi_agent_pkg;


  import uvm_pkg::*;

       `include "uvm_macros.svh"
       `include"lpi_seq_item.sv"
       `include"lpi_sequencer.sv"
       `include"lpi_driver.sv"
```

```
        `include"lpi_basic_seq.sv"
        `include"lpi_monitor.sv"
        `include"lpi_agent.sv"
        `include"lpi_env.sv"
        `include"lpi_top_v_sequencer.sv"
        `include"lpi_top_env.sv"

endpackage : lpi_agent_pkg
```

4.8.8 lpi_basic_sequence.sv

```
class lpi_basic_seq extends uvm_sequence #(lpi_seq_item);
          `uvm_object_utils(lpi_basic_seq)

          rand int num_of_trans;


          //UVM specific methods
          function new (string name="lpi_basic_seq");
            super.new (name);
          endfunction

          extern task body();
endclass:lpi_basic_seq

task lpi_basic_seq::body();
        lpi_seq_itemseq_item;
        seq_item = lpi_seq_item::type_id::create("seq_item");

        for(int i = 0; i < num_of_trans; i++)
        begin
          `uvm_info(get_type_name(),$psprintf("in seq for count =
%d", i) ,UVM_LOW)
         start_item(seq_item);
         if(!seq_item.randomize())begin
        `uvm_error("body","Randomization failed for seq_item")
        end
       `uvm_info(get_type_name(),$psprintf("obj is req0 = %d, req1
= %d, sleep0 = %d,sleep1 = %d", seq_item.wakeup_req0, seq_item.
wakeup_req1, seq_item.slp_req0, seq_item.slp_req1) ,UVM_LOW)
        finish_item(seq_item);
  end
endtask: body
```

4.8.9 lpi_basic_test.sv

```
class lpi_basic_test extends uvm_test;
        `uvm_component_utils(lpi_basic_test)
        lpi_basic_seqlpi_basic_seq_h;
        lpi_top_env m_lpi_top_env;
        virtual lpi_if lpi_vif;

        //Constructor
        function new (string name = "lpi_basic_test", uvm_compo-
nent parent);
        super.new (name, parent);
        endfunction

// Build Phase
       function void build_phase(uvm_phase phase);
       super.build_phase(phase);
      m_lpi_top_env= lpi_top_env::type_id::create("m_lpi_top_env",
this);
       lpi_basic_seq_h= lpi_basic_seq::type_id::create("lpi_basic_
seq_h", this);
        set_config_int("m_lpi_top_env.lpi_env_h.lpi_agent_h", "is_
active", UVM_ACTIVE);
       endfunction:build_phase

//RUN Phase
       task run_phase(uvm_phase phase);
       super.run_phase(phase);
       phase.raise_objection(this, "starting test_seq");

     if(!(uvm_config_db #(virtual lpi_if)::get(null, "", "lpi_vif",
lpi_vif)))
       begin
       `uvm_fatal(get_name(),"Can't retrieve lpi_vif from config db\n")
       end

      if (!lpi_basic_seq_h.randomize() with {num_of_trans == 20;})
        `uvm_fatal(get_name(), "Randomization of lpi_basic_seq_h
Sequence Failed \n")

        `uvm_info(get_name(),"starting seq here \n",UVM_LOW)


        wait (lpi_vif.lpi_rstn == 1);
        `uvm_info(get_name(),"reset done\n",UVM_LOW)
```

```
        lpi_basic_seq_h.start(m_lpi_top_env.lpi_env_h.lpi_
agent_h.lpi_sequencer_h);

        `uvm_info(get_name(),"###################################
########### \n",UVM_LOW)
        `uvm_info(get_name(),"###!!!!!!  Hello  World  !!!!!!###
\n",UVM_LOW)
        `uvm_info(get_name(),"###################################
########### \n",UVM_LOW)

        phase.drop_objection(this,"Finished lpi_basic_test\n");
       endtask

endclass
```

4.8.10 lpi_env.sv

```
class lpi_env extends uvm_env;
  `uvm_component_utils(lpi_env)

//Agent
  lpi_agent      lpi_agent_h;

  //UVM methods
  function new (string name="lpi_env", uvm_component parent);
  super.new (name,parent);
    endfunction

  extern function void build_phase(uvm_phase phase);
  extern function void connect_phase(uvm_phase phase);

endclass:lpi_env

function void lpi_env::build_phase(uvm_phase phase);
       super.build_phase(phase);
       //Build agent
       lpi_agent_h=     lpi_agent::type_id::create("lpi_agent_h",
this);
endfunction:build_phase

function void lpi_env::connect_phase(uvm_phase phase);
endfunction:connect_phase
```

4.8.11 lpi_top_v_sequencer.sv

```
class lpi_top_v_sequencer extends uvm_sequencer;
         `uvm_component_utils(lpi_top_v_sequencer)
         uvm_sequencer_baselpi_sqr;

         function new (string name = "lpi_top_v_sequencer", uvm_
component parent=null);
         super.new (name, parent);
         endfunction

endclass:lpi_top_v_sequencer
```

4.8.12 lpi top environment.sv

```
class lpi_top_env extends uvm_env;
   `uvm_component_utils(lpi_top_env)

    //Global event
    uvm_event_poole_pool = uvm_event_pool::get_global_pool();

    // Instantiate environments
    lpi_top_v_sequencerlpi_top_v_sqr_h;
    lpi_env lpi_env_h;

    // Methods
    extern function new (string name="lpi_top_env", uvm_component
parent);
    extern function void build_phase(uvm_phase phase);
    extern function void connect_phase(uvm_phase phase);
    extern task run_phase(uvm_phase phase);
  endclass: lpi_top_env

function lpi_top_env::new (string name="lpi_top_env", uvm_compo-
nent parent);
      super.new (name, parent);
endfunction

function void lpi_top_env::build_phase(uvm_phase phase);
  string msg = "\n";
```

```
    msg = {msg, "========================================\n"};
    msg = {msg, "*LPI TOP ENV BUILD PHASE SUMMARY*\n"};
    msg = {msg, "========================================\n"};
    // Setting default verbosity to LOW
    if (!$test$plusargs("UVM_VERBOSITY")) begin
    `uvm_info (get_name()," \nUVM_VERBOSITY not defined, using UVM_
LOW \n ", UVM_LOW)
    uvm_top.set_report_verbosity_level_hier(UVM_LOW);
end

super.build_phase(phase);


    lpi_env_h = lpi_env::type_id::create("lpi_env_h",this);
    //*** Build Virtual Sequencer ***
    msg = {msg, "LPI TOP V_SEQUENCER\n"};
    lpi_top_v_sqr_h  =  lpi_top_v_sequencer::type_id::create("lpi_
top_v_sqr_h", this);

  endfunction:build_phase


//Connect Phase
function void lpi_top_env::connect_phase(uvm_phase phase);
    string msg = "\n";
    int i;
    super.connect();
    msg = {msg, "========================================\n"};
    msg = {msg, "*LPI TOP ENV CONNECT PHASE SUMMARY*\n"};
    msg = {msg, "========================================\n"};

   lpi_top_v_sqr_h.lpi_sqr = lpi_env_h.lpi_agent_h.lpi_sequencer_h;
   //Connect other agents sequencers
     uvm_config_db  #(lpi_top_v_sequencer)::set(uvm_top,  "",  "lpi_
top_v_sqr_h", lpi_top_v_sqr_h);
   `uvm_info(get_name(), msg, UVM_LOW)
//-----------------------------------------------------------------
  endfunction:connect_phase

task lpi_top_env::run_phase(uvm_phase phase);
int i;
super.run_phase(phase);
endtask
```

4.8.13 lpi_testbench.sv

```
`include "uvm_macros.svh"
 `default_nettype wire

module lpi_testbench();
import uvm_pkg::*;
reg lpi_dut_clk;
reg dut_rst_n;
//Interface
lpi_iflpi_dut_if (.lpi_clk(lpi_dut_clk));
 //set lpi_if to config DB
  initial
  begin
         uvm_config_db#(virtual lpi_if)::set(null, "", "lpi_vif",
lpi_dut_if);
end

lpi dut_lpi(
          .clk(lpi_dut_clk),
.rst_n(dut_rst_n),
.slp_req0(lpi_dut_if.slp_req0),
.slp_req1(lpi_dut_if.slp_req1),
.wakeup_req0(lpi_dut_if.wakeup_req0),
.wakeup_req1(lpi_dut_if.wakeup_req0),
.ss_wakeup(lpi_dut_if.ss_wakeup),
.ss_sleep(lpi_dut_if.ss_sleep)
);

initial
begin
          lpi_dut_clk = 0;
dut_rst_n = 0;
#1000ns;
dut_rst_n = 1;
end

always
begin
        #10ns  lpi_dut_clk = ~lpi_dut_clk;
end
```

```
initial
begin
        #0; run_test();
end

endmodule
```

4.9 UVM Is Reusable

As discussed before, one of UVM's main advantage is that it is reusable. Reusable in the sense that once you put together the required code/infrastructure in place, using UVM class libraries, you will be able to carry forward that to the next project. Only the driver (in UVM agent) and the basic transaction (sequence) and sequence library that contains tests need to change.

Let us examine Fig. 4.14. This figure shows the UVM agent level implementation. There is a Stimulus Generator, a response "collector," and a scoreboard. A scoreboard is nothing but a monitor/checker that checks to see that the response is in accordance with the specification of the design.

So, what is reusable in this verification environment? Here are the components that are (almost) generic and can be reused.

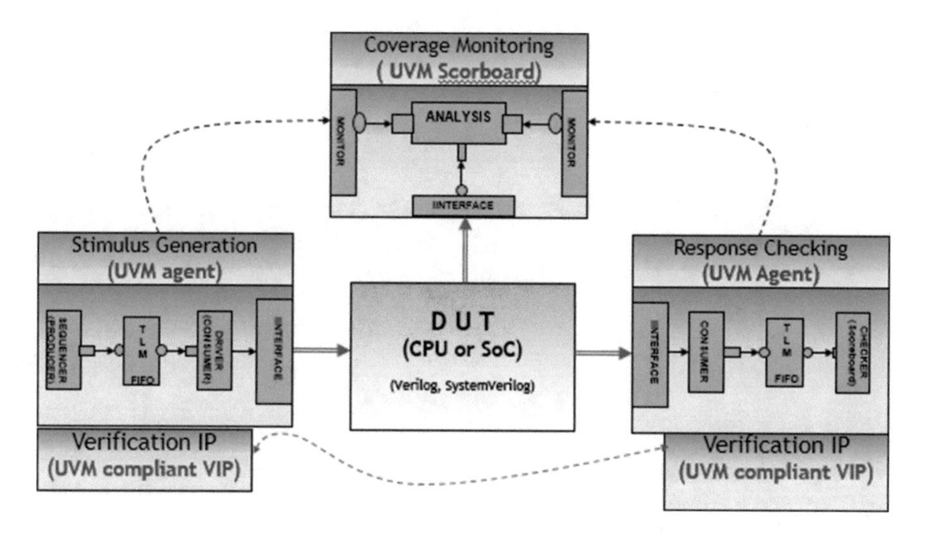

Fig. 4.14 Reusable UVM example

- Agent sequencer
- TLM FIFO
- Consumer
- Entire transaction-level communication mechanism
- All “class” interfaces
- The overall architecture of the entire verification environment

Chapter 5
Constrained Random Verification (CRV)

Chapter Introduction
Constrained Random Verification (CRV) is a methodology that is supported by SystemVerilog which has a built-in constraint solver. This allows you to constraint your stimulus to better target a design function, thereby allowing you to reach your coverage goal faster with accuracy. From that sense, coverage and CRV go hand in hand. You check your coverage and see where the coverage holes are. You then constrain your stimulus to target those holes and improve coverage.

As part of verification strategy, you start with direct testing to target "directly" features that you need to verify. But directed testing runs out of steam very fast. If you jump straight to random, you may or may not hit the corner cases of importance. Fully random can end up wasting a lot of simulation cycles without improving coverage. That's where constrained random comes into picture.

5.1 Productivity Gain with CRV

Figure 5.1 shows a chart used by EDA vendors to highlight the productivity and quality gain with the use of constrained random stimuli. The Y-axis is the coverage achieved and the X-axis is the time it took to achieve that coverage. As you notice, directed testing may take you to that goal but at the expense of lengthy time in test development, debug, and simulation. This is because for every corner case, you will be creating a new test case and hope against hope that you will reach that corner case. A lot of trial and error will take place. The debug time will dramatically increase as the number of tests increase, not to mention the simulation and regression time.

In contrast, if you understand the logic that is not covered and constrain and randomize your stimuli to target that logic, you will not only get to the coverage goal faster but also will find some of those hidden corner cases that you had not even envisioned.

A.B. Mehta, *ASIC/SoC Functional Design Verification*,
DOI 10.1007/978-3-319-59418-7_5

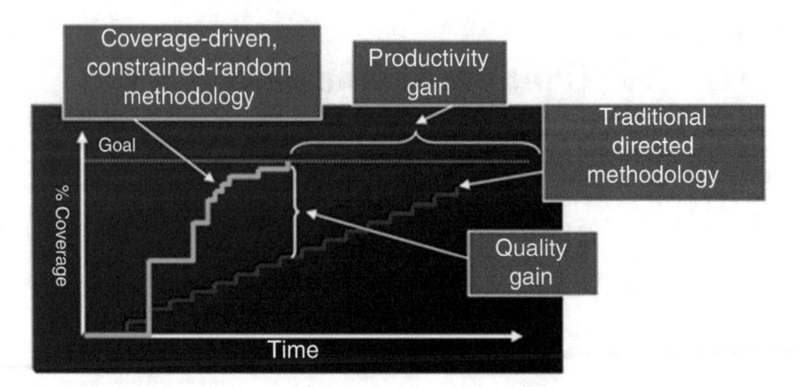

Fig. 5.1 Advantage of coverage-driven constrained random verification methodology

CRV is not new. What is new is that SystemVerilog has incorporated an exhaustive constraint solver that allows you to constraint your stimuli in a logical and organized way. The language semantics are easy to understand and easy to deploy.

Let us look at a simple CRV methodology that shows the need and importance of having coverage as an integral part of your verification methodology and CRV as part of that methodology to cover missing gaps in functional coverage.

5.2 CRV Methodology

Figure 5.2 shows the functional verification methodology in a nut shell. Traditionally, both design and verification teams start running with the specifications written by the architects of the design. The DV (design verification) team starts putting together a verification architecture/environment, test plan, and a testbench. Tests are written per the test plan and the verification cycle begins. The key component missing in this traditional methodology is planning for coverage (and assertions). Without a comprehensive coverage plan, the team has no idea how their tests and testbenches are performing. They use code coverage (if) at the most.

This is where CRV and coverage marry. The first step is to create a functional coverage plan (Chap. 7). Once the coverage plan is ready, you measure the coverage after each simulation (regression) run. This process can also be automated, and the major EDA vendors provide just such mechanisms.

If the coverage is not complete, you identify the holes in your coverage results. This is where CRV comes into picture. Since your directed tests are leaving holes in the coverage results (think corner cases), you now need to move onto constrained random stimuli. Constrained random allows you to narrow down your stimuli to those areas where coverage is lacking. Now, you design your stimulus with constraints. There are many ways to place constraints on your stimulus, some of which are explained in the coming sections.

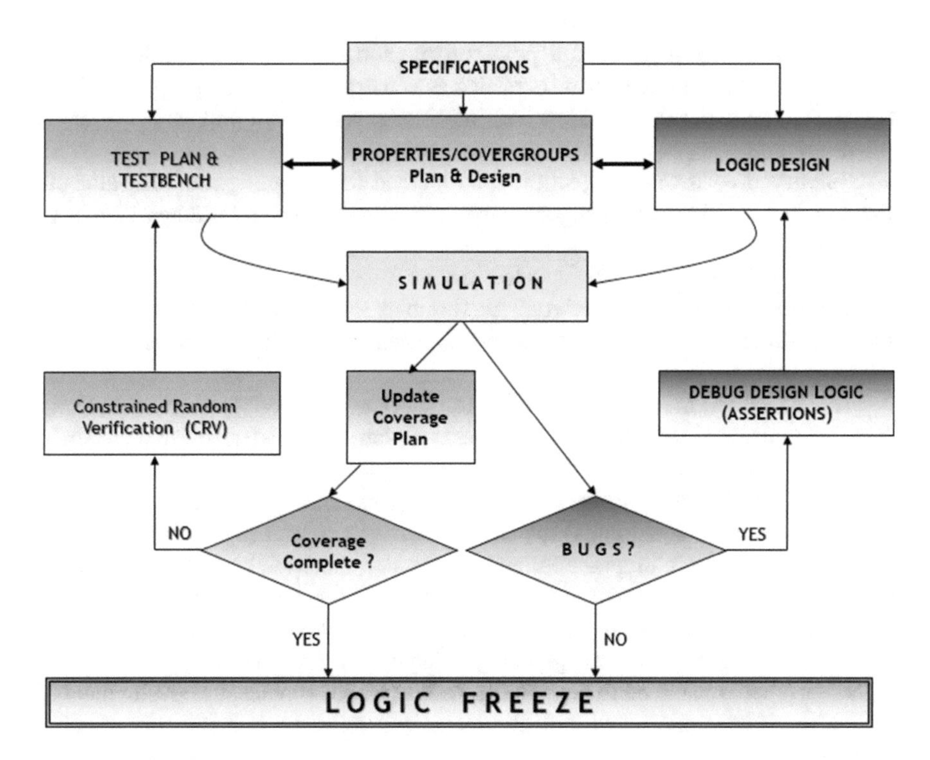

Fig. 5.2 Constrained random verification methodology

After you constrain your stimulus, you go through the simulation cycle and repeat the coverage cycle and further identify remaining coverage holes. You further apply constraints to your stimulus and repeat the entire loop.

CRV is an objective methodology in that you know objectively if you are done verifying your chip (as opposed to subjective measure where as soon as you stop finding bugs, you may stop simulations). A coverage- and assertions-based methodology is discussed in Sect. 7.3.

Now, let us look at some of the CRV basics and features provided by (SystemVerilog_LRM_1800-2012). CRV is a very involved methodology, and I strongly suggest referring to the SystemVerilog LRM to study it in detail. The full scope of CRV discussion is out of the scope of this book.

5.3 Basics of CRV

The random constraints are typically specified on top of an object-oriented data abstraction that models the data to be randomized as objects that contain random variables and user-defined constraints. The constraints determine the legal values that can be assigned to the random variables. Objects are ideal for representing complex aggregate data types and protocols such as network packets.

Constraint programming is a powerful method that lets users build generic, reusable objects that can later be extended or constrained to perform specific functions. The approach differs from both traditional procedural and object-oriented programming.

SystemVerilog uses an object-oriented method for assigning random values to the member variables of an object, subject to user-defined constraints. Here's an example.

Referring to Fig. 5.3, the top-left corner text block shows a class named PacketBase. It uses "rand" keyword for variables src, len, and payload. These are then the variables that can be randomized. Then we provide a constraint on the "payload_size."

We then extend the PacketBase class to EtherPacket class (top right text block). Here we constrain "src" and "len," further constrain "payload_size," and constrain the "payld" to "h aa"

The class "stim_gen" instantiates EhterPacket and uses the method "randomize" to randomize (with constraints) the "rand" variables of PacketBase (and extended class EtherPacket). Calling randomize () causes new values to be selected for all the random variables in an object so that all the constraints are true (satisfied). We don't have any "rand" variables in this example which are unconstrained. But if we did, unconstrained variables are assigned any value in their declared range.

Using inheritance to build layered constraint systems enables the development of general-purpose models that can be constrained to perform application-specific functions

The simulation log is shown in the right bottom block of Fig. 5.3.

Objects can also be further constrained using the "randomize () with" construct which declares additional constraints in-line with the call to randomize(). For example,

```
int randomVal;
randomVal = PacketBase.randomize with {payload_size > 15; pay-
load_size < 30;};
```

The above examples illustrate several important properties of constraints, as follows ((SystemVerilog_LRM_1800-2012):

- Constraints can be any SystemVerilog expression with variables and constants of integral type (e.g., bit, reg, logic, integer, enum, packed struct).
- The constraint solver will be able to handle a wide spectrum of equations, such as algebraic factoring, complex Boolean expressions, and mixed integer and bit expressions.
- If a solution exists, the constraint solver will find it. The solver can fail only when the problem is over-constrained and there is no combination of random values that satisfy the constraints.
- Constraints support only 2-state values. The 4-state values (X or Z) or 4-state operators (e.g., ===, !==) are illegal and will result in an error.
- The solver can randomize singular variables of any integral type.

```
class PacketBase;

  byte header;
  rand bit[7:0] src;
  rand bit[5:0] len;
  rand bit [7:0] payld [];
  constraint payload_size {payld.size >0 ; payld.size < 15;}

  virtual function show;
   $display("PacketBase=%h %h %h",src,len,payld);
 endfunction:show

endclass
```

```
class stim_gen;

  task run;
   EtherPacket e1;
   e1 = new();

    e1.randomize;

  endtask

endclass: stim_gen
```

```
class EtherPacket extends PacketBase;

  rand bit[15:0] len;
  constraint c1 { src inside {[8'h2A:8'h2F]};}
  constraint c2 { len inside{ [64:1518] };}
  constraint payload_size {payld.size >15 ; payld.size < 20;}
  constraint c3 { foreach(payld[i]) { (payld[i])  == 'haa; }}

  function show;
   $display("EtherPacket=%h %h %h",src,len,payld);
  endfunction:show

endclass : EtherPacket
```

```
# EtherPacket=2a 013b aa aa aa aa ...
# EtherPacket=2f 008d aa aa aa aa ...
# EtherPacket=2b 0116 aa aa aa aa aa ...
# EtherPacket=2f 042b aa aa aa ...
# EtherPacket=2d 018d aa aa aa ...
# EtherPacket=2c 0236 aa aa aa aa aa aa ...
```

Fig. 5.3 Constrained random verification example

Note that every class contains pre_randomize() and post_randomize() methods which are automatically called by randomize() before and after computing new random values.

pre_randomize() method is as follows:

```
function void pre_randomize()
```

Same applies to post_randomize()

When obj.randomize() is invoked, it first invokes pre_randomize() on obj and also all of its random object members that are enabled. After the new random values are computed and assigned, randomize() invokes post_randomize() on obj and also all its random object members that are enabled.

Note the following rules that apply to randomize():

Random variables declared as static are shared by all instances of the class in which they are declared. Each time the randomize() method is called, the variable is changed in every class instance.

- If randomize() fails, the constraints are infeasible, and the random variables retain their previous values.
- If randomize() fails, post_randomize() is not called.
- The randomize() method is built-in and cannot be overridden.
- The built-in methods pre_randomize() and post_randomize() are functions and cannot block.

5.3.1 Random Variables: Basics

There are two types of random type–modifier keywords: "rand" and "randc."

Variable declared with "rand" keyword is standard random variables. Their values are *uniformly* distributed over their range. For example:

```
rand bit [3:0] length;
```

This is a 4-bit unsigned variable with a range from 0 to 15. If unconstrained, this variable will be assigned any value in the range from 0 to 15 with *equal probability*. So, in this example, the probability of the same value repeating on successive calls to randomize is 1/16.

In contrast, "randc" are random-cyclic variables that cycle through all the value in a random permutation of their declared range. For example:

```
randc bit [1:0] length;
```

The variable "length" can take on values 0,1,2, and 3. Randomize computes an initial random permutation of the range values of "length" and then returns those values in order on successive calls. After it returns the last element of a permutation, it repeats the process by computing a new random permutation. Here's how these permutations will work.

```
0 -> 3 -> 2 -> 1 (initial)
2 -> 1 -> 3 -> 0 (next permutation)
2 -> 0 -> 1 -> 3 (next permutation)
…. .
```

Note that the semantics of random-cyclic (randc) variables will be solved *before* other random variables. A set of constraints that includes both rand and randc variables will be solved so that the randc variables are solved first.

If a random variable is declared as static, the randc state of the variable will also be static. Thus, randomize chooses the next cyclic value (from a single sequence) when the variable is randomized through any instance of the base class.

"rand" and "randc" on Arrays

- Arrays can be declared "*rand*" or "*randc*" in which case all the array's member elements will be considered '*rand*' or "*randc.*"
- Individual array elements can be constrained, in which case the index expression may include iterative constraint loop variables, constants, and state variables.
- Dynamic arrays, associative arrays, and queues can be declared *rand* or *randc.* All the elements in the array are randomized, overwriting any previous data. Please refer to (SystemVerilog_LRM_1800-2012) for further usage and restrictions on "*rand*" arrays.
- An object handle can be declared rand, in which case all of that object's variables and constraints are solved concurrently with the variables and constraints of the object that contain the handle. Object handles cannot be declared randc.
- An unpacked structure can be declared *rand.* Unpacked structures cannot be declared *randc.* A member of an unpacked structure can be made random by having a rand or randc modifier in the declaration of its type. Members of unpacked structures containing a union as well as members of packed structures cannot be allowed to have a random modifier.

The (SystemVerilog_LRM_1800-2012) provides plenty of examples to further solidify the randomize() method around "rand" and "randc." Complete discussion of those features is out of the scope of this book.

5.3.2 Random Number System Functions and Methods

SystemVerilog provides the following system functions and methods to further augment the constrained random verification methodology.

```
$urandom()
$urandom_range()

srandom()
get_randstate()
set_randstate()
```

Let us look at each in detail.

$urandom() system function provides a mechanism for generating pseudorandom numbers. The functional returns a new 32-bit random number each time it is called. The return number is unsigned.

For example:

```
bit [63:0] cache_line_addr;
bit [7:0] tag_address;

cache_line_addr[31:0] = $urandom (1234);// get a 32 bit unsigned
number with the seed '1234'
cache_line_addr = ($urandom, $urandom) //Get 64 bit unsigned
address for the cache line

tag_address = $urandom & 32'h 0000_00ff; //mask MSB 31:8 and get a
random 8 bit number //for the tag address
```

Note that the "seed" is an optional argument that determines the sequence of random number generated. This is for predictability of random number generation. In other words, the same sequence of random numbers will be generated every time the same seed is used. "seed" is very important for regression runs where each run needs to work with the same sequence of random numbers.

$urandom_range() returns an unsigned integer within a specified range. For example:

```
Value = $urandom_range(15,0);
```

Where 15 is the maxval and 0 is the minval (both are int unsigned). The function will return a value in the range of 15 and 0.

```
Value = $urandom_range(0,15);
```

Here the maxval is greater than minval!! In such a case, the arguments are automatically reversed so that the first argument is larger than the second argument. So, in this case also, value will be in the range from 0 and 15 inclusive.

srandom() is a *method*. It allows manually seeding the random number generation (RNG) of objects or threads. Its syntax is

```
function void srandom (int seed);
```

The *srandom()* method initializes an object's RNG (random number generation) using the value of the given seed.

get_randstate() method retrieves the current state of an object's RNG.

Here's the syntax:

```
function string get_randstate();
```

The *get_randstate()* method returns a copy of the internal state of the RNG associated with the given object.

The RNG state is a string of unspecified length and format. The length and contents of the string are implementation dependent. (SystemVerilog_LRM_1800-2012).

set_randstate() is a method that sets the state of an object's RNG. The syntax of this method is

```
function void set_randstate
```

The RNG state is a string of unspecified length and format. Calling *set_randstate()* with a string value that was not obtained from get_randstate(), or from a different implementation of *get_randstate()*, is undefined. (SystemVerilog_LRM_1800-2012)

5.3.3 *Random Weighted Case:* Randcase

So, what if you want to randomize among a case of statements? Just as in the "case" statement of SystemVerilog, there is the "*randcase*" statement available for CRV.

The keyword "*randcase*" introduces a case statement that randomly selects one of its branches. The randcase_item expressions are nonnegative integral values that constitute the branch weights. An item's weight divided by the sum of all weights gives the probability of taking that branch.

For example:

```
randcase
10: X=X+1;
 5: Y=Y+1;
3: Z=Z+1;
endcase
```

The total weight of all three case statements is 18. The probability of X=X+1 being executed is 10/18. Similarly, Y=Y+1 has the probability of 5/18 and Z=Z=1 has the probability of 3/18. This is a great feature in that you not only get the randomized effect but you can also (in some sense) constrain and get the randomization effects in your favor.

If a branch specifies a zero weight, then that branch is not taken. If all randcase_items specify zero weights, then no branch is taken and a warning will be issued. The *randcase* weights can be arbitrary expressions, not just constants. For example:

```
byte a,b;
randcase
a+b : X=X+1;
a ^ b : Y=Y+1;
a - b : Z=Z=1;
endcase
```

In this example, the sum of the weights is computed using standard addition semantics (maximum precision of all weights), where each sum is unassigned. The weights in this example use 8-bit precision (a,b are declared as byte). The resulting weights are added as 8-bit unsigned values. Each call to randcase retrieves one random number in the range of 0 to the sum of the weights. The weights are then selected in declaration order: smaller random numbers correspond to the first (top) weight statements.

To summarize, this section gives a high-level view of the constrained random features provided by the SystemVerilog LRM. The full description of each of the feature is beyond the scope of this book. Please refer to (SystemVerilog_LRM_1800-2012).

Chapter 6
SystemVerilog Assertions (SVA)

Chapter Introduction
SystemVerilog Assertions (SVA) is one of the most important components of SystemVerilog when it comes to design verification. SVA is instrumental in finding corner cases, ease of debug, and coverage of design's sequential logic. We will discuss high-level SVA methodology, SVA and functional coverage-driven methodology, and plenty of applications to solidify the concepts.

6.1 Evolution of SystemVerilog Assertions

SVA is one of the most important technologies to deploy toward robust design verification. Note that SystemVerilog language syntax and semantics are totally different from SVA. If you know SystemVerilog/UVM only, you will not know SVA language. SystemVerilog and SVA, even though under SystemVerilog umbrella, are totally orthogonal languages (even though they both use the same simulation kernel). Hence, one needs to learn SVA on its own.

First, what's an assertion? Very simple definition. An assertion is simply a check against the specification of your design that you want to make sure never violates. If the specs are violated, you want to see a failure. Let us see how SVA came about. What contributed to its evolution?

Referring to Fig. 6.1 SystemVerilog Assertion evolution, we can see that SystemVerilog Assertion language is derived from many different languages. Features from these languages either influenced the language or were directly used as part of the language syntax/semantic.

Sugar from IBM led to PSL. Both contributed to SVA. The other languages that contributed are Vera, "e," CBV from Motorola, and ForSpec from Intel.

In short, when we use SystemVerilog Assertion language, we have the benefit of using the latest evolution of an assertion language that benefited from many other robust assertion languages.

A.B. Mehta, *ASIC/SoC Functional Design Verification*,
DOI 10.1007/978-3-319-59418-7_6

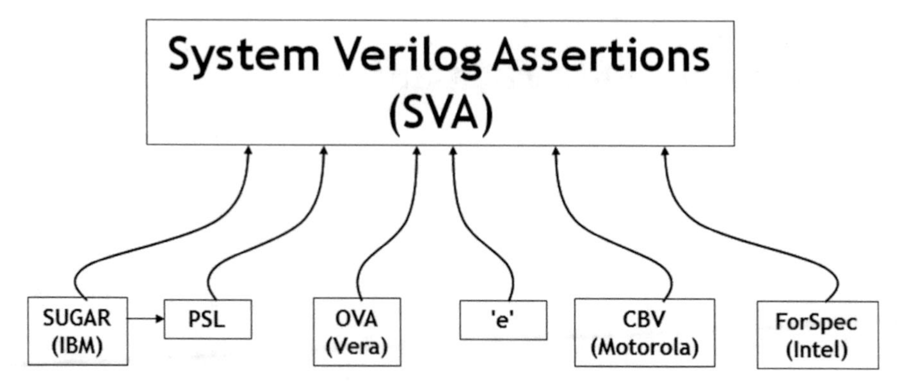

Fig. 6.1 SystemVerilog Assertion evolution

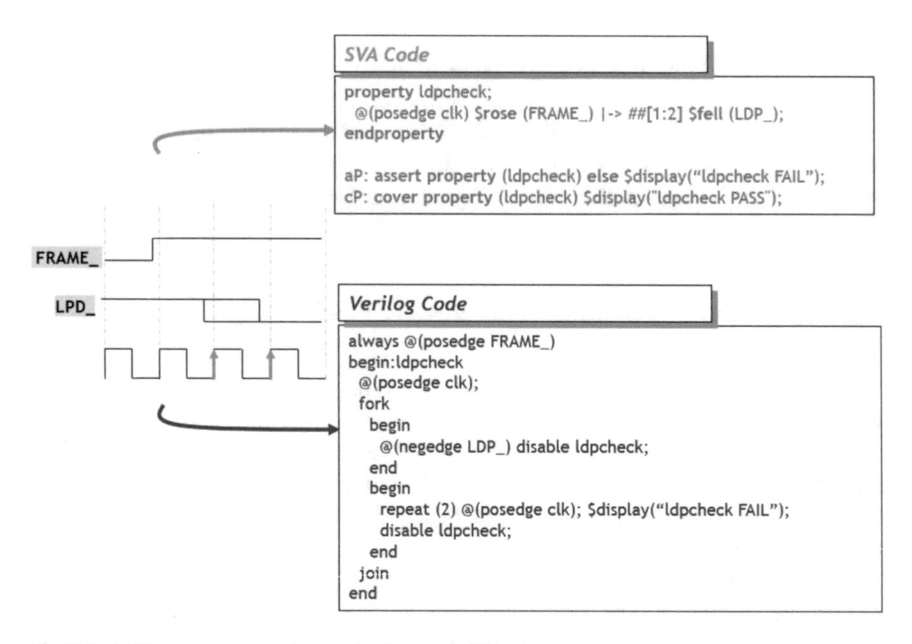

Fig. 6.2 Difference between SystemVerilog and SVA

6.2 SystemVerilog Assertion Advantages

6.2.1 Assertions Shorten Time to Develop

A very simple example of an assertion is shown in Fig. 6.2. This example contrasts SystemVerilog Assertion code with that of behavioral SystemVerilog. As you notice they are completely orthogonal. SVA code is much more readable than SystemVerilog code. This dramatically reduces time to develop complex checkers for your design.

The SVA code is very self-explanatory. There is the property "ldpcheck" that says "at posedge clock, if FRAME_ rises, it implies that within the next two clocks LDP_ falls." This is almost like writing the checker in English. We then "assert" this property, which will check for the required condition to meet at every posedge clk. We also "cover" this property to see that we have indeed exercised (i.e., 'cover'ed) the required condition.

Now examine the SystemVerilog code for the same check. There are many ways to write this code. One of the ways at behavioral level is shown. Here you "fork" out two procedural blocks, one that monitors LDP_ and another that waits for two clocks. You then disable the entire block ("ldpcheck") when either of the two procedural blocks completes. As you can see, not only is the checker very hard to read/interpret but also very prone to errors. You may end up spending more time debugging your checker than the logic under verification.

6.2.2 Assertions Improve Observability

One of the most important advantages of assertions is that they fire at the source of the problem; assertions are located local to sequential conditions in your design. In other words, you don't have to back trace a bug all the way from primary output to *somewhere* internal to the design where the bug originates. Assertions are written such that they are close to logic (e.g., @ (posedge clk) state0 |-> Read); such an assertion is *sitting* close to the state machine, and if the assertion fails, we know that when the state machine was in state0 that Read did not take place. Some of the most useful places to place assertions are FIFOs (especially the asynchronous variety), counters, block-to-block interface, block-to-IO interface, state machines, etc. These constructs are where many of the bugs originate. Placing an assertion that checks for local condition will fire when that local condition fails, thereby directly pointing to the source of the bug. This is shown in Fig. 6.3.

Traditional verification can be called black box verification with black box observability, meaning, you apply vectors/transactions at the primary input of the "block" without caring for what's in the block (black box verification), and you observe the behavior of the block only at the primary outputs (black box observability). Assertions on the other hand allow you to do black box verification with white box (internal to the block) observability.

6.2.3 Assertions Shorten Time to Cover

Referring to Fig. 6.4, assertions not only help you find bugs but also help you determine if you have covered (i.e., exercised) design logic, mainly temporal domain conditions. They are very useful in finding temporal domain coverage of your testbench. Here is the reason why this is so important.

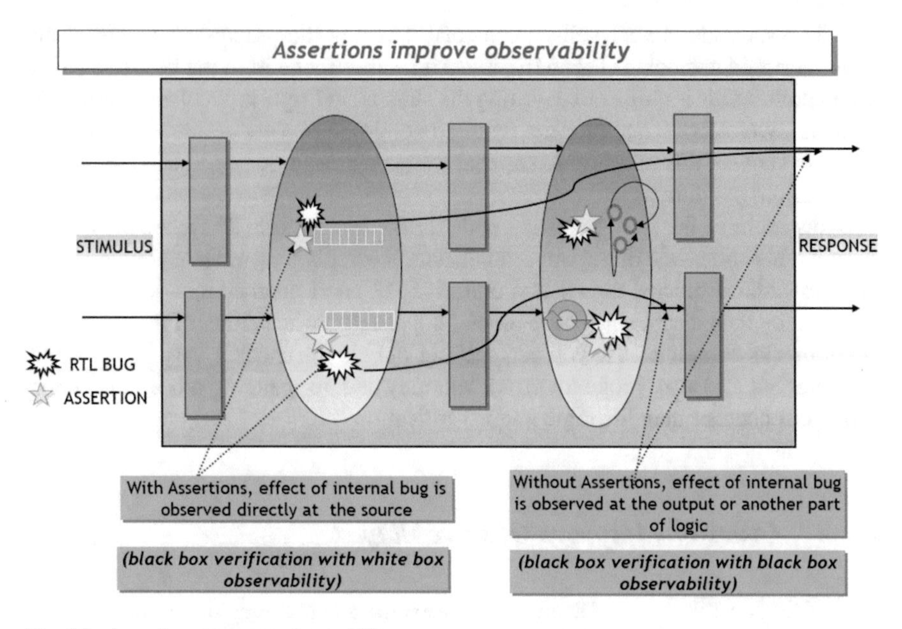

Fig. 6.3 Assertions improve observability

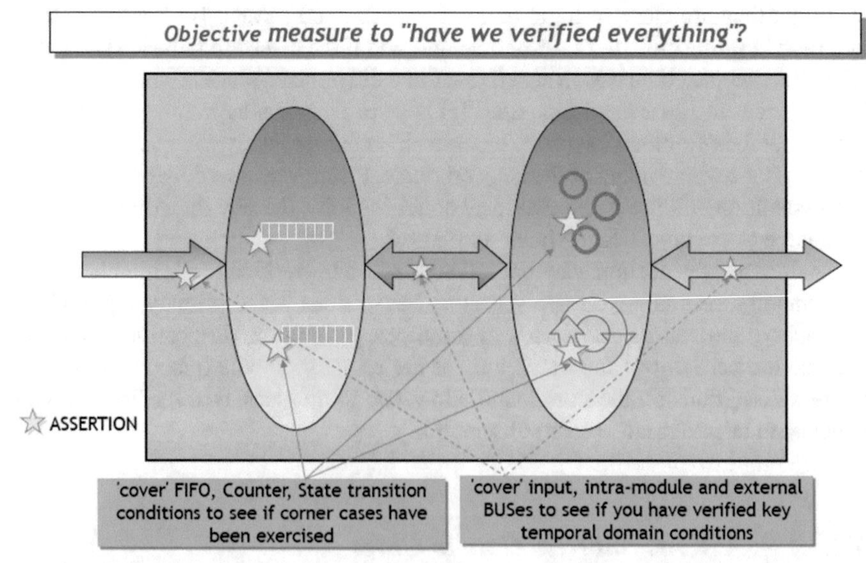

Fig. 6.4 Assertions shorten time to full coverage

Let us say, you have been running regressions 24*7 and have stopped finding bugs in your design. Does that mean you are done with verification? No. Not finding a bug could mean one of two things: (1) there is indeed no bug left in the design or (2) you have not exercised (or covered) the conditions that exercise the bugs. You could be continually hitting the same piece of logic in which no further bugs remain. In other words, you could be reaching a wrong conclusion that all the bugs have been found.

In brief, coverage includes three components (we will discuss this in detail in the chapter on functional coverage): (1) code coverage (which is structural) which needs to be 100%, (2) functional coverage that needs to be designed to cover *functionality* (i.e., intent) of the entire design and must completely cover the design specification, and (3) sequential domain coverage (using SVA "cover" feature) which needs to be carefully designed to fully cover all required sequential domain conditions of the design.

Ok, let us go back to the simple bus protocol assertion that we saw in the previous section. Let us see how the "cover" statement in that SVA assertion works. The code is repeated here for readability.

```
property ldpcheck;

@(posedge clk) $rose (FRAME_) |-> ##[1:2] $fell (LDP_);
endproperty
aP: assert property (ldpcheck) else $display("ldpcheck FAIL");
cP: cover property (ldpcheck) $display("ldpcheck PASS");
```

In this code, you see that there is a "cover" statement. What it tells you is "did you exercise this condition" or "did you cover this property." In other words, and as discussed above, if the assertion never fails, that could be because of two reasons: (1) you don't have a bug or (2) you never exercised the condition to start with! With the cover statement, if the condition gets exercised but does not fail, you get that indication through the "pass" action block associated with the "cover" statement. Since we haven't yet discussed the assertions in any detail, you may not completely understand this concept, but *determination of sequential domain coverage of your design is an extremely important aspect of verification and must be made part of your coverage plan.*

To reiterate, SVA supports the "cover" construct that tells you if the assertion has been exercised (covered). Without this indication and in the absence of a failure, you have no idea if you indeed exercised the required condition. In our example, if FRAME_ never rises, the assertion won't fire, and obviously there won't be any bug reported. So, at the end of simulation, if you do not see a bug or you do not even see

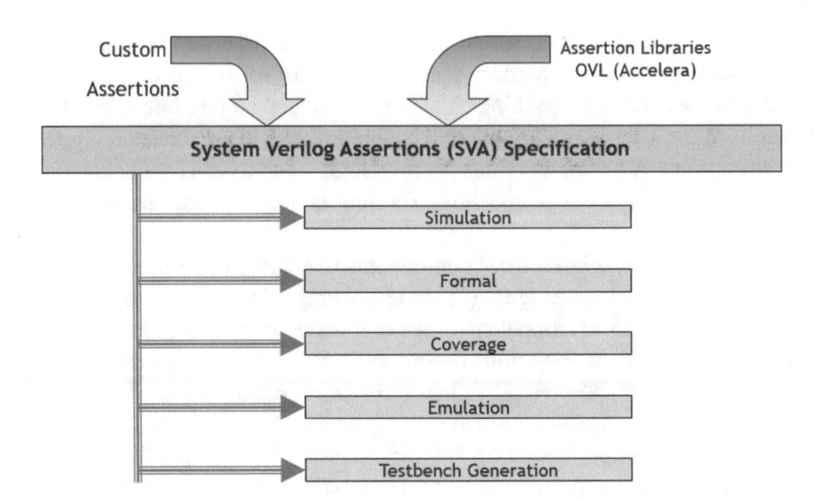

Fig. 6.5 Multiple uses of SVA

the "ldpcheck PASS" display, you know that the assertion never fired. In other words, you must see the "cover property" statement executed in order to know that the condition did get exercised.

6.2.4 One-Time Effort: Many Benefits

Assertions written once have multiple uses as shown in Fig. 6.5. Note that Accelera has produced a library of commonly used assertions which can be an effective way to start learning assertions as well as examine the code written by experts at Accelera and learn SVA coding techniques.

OVL Library Open Verification Library. This library of predefined checkers was written in Verilog before PSL and SVA became mainstream. Currently the library includes SVA (and PSL)-based assertions as well. The OVL library of assertion checkers is intended for use by design, integration, and verification engineers to check for good/bad behavior in simulation, emulation, and formal verification. OVL contains popular assertions such as FIFO assertions, among others. OVL is still in use, and you can download the entire standard library from Accelera website http://www.accellera.org/downloads/standards/ovl.

We won't go into OVL detail since there is plenty of information available on OVL on net. OVL code itself is quite clear to understand. It is also a good place to see how assertions are written for "popular" checks, once you have better understanding of assertion semantics.

6.3 Creating an Assertion Test Plan: PCI Read Example

So, who writes the assertions? This has been a contentious issue between design and verification teams. But the answer is very simple. Writing assertions is the responsibility of *both* design and verification teams.

Design team:

- Microarchitectural-level decisions/assumptions are *not* visible to DV engineers. So, designers are best suited to guarantee uArch-level logic correctness.
- Every assumption is an assertion. If you assume that the "request" you send to the other block will always get an 'ack' in two clocks, that's an assumption. So, design an assertion for it.
- Add assertions as you design your logic, not as an afterthought.

Verification (DV) team:

- Add assertions to check macro functions and chip/SoC-level functionality:
  - Once the packet has been processed for L4 layer, it will indeed show up in the DMA queue.
  - A machine check exception indeed sets PC to the exception handler address.
- Add assertions to check interface IO logic:
  - After reset is de-asserted, none of the signals ever go "X."
  - If the processor is in Wait Mode and no instructions are pending, it must assert a SleepReq to memory subsystem within 100 clocks.
  - On Critical Interrupt, the external clock/control logic block must assert CPU_ wakeup within ten clocks.

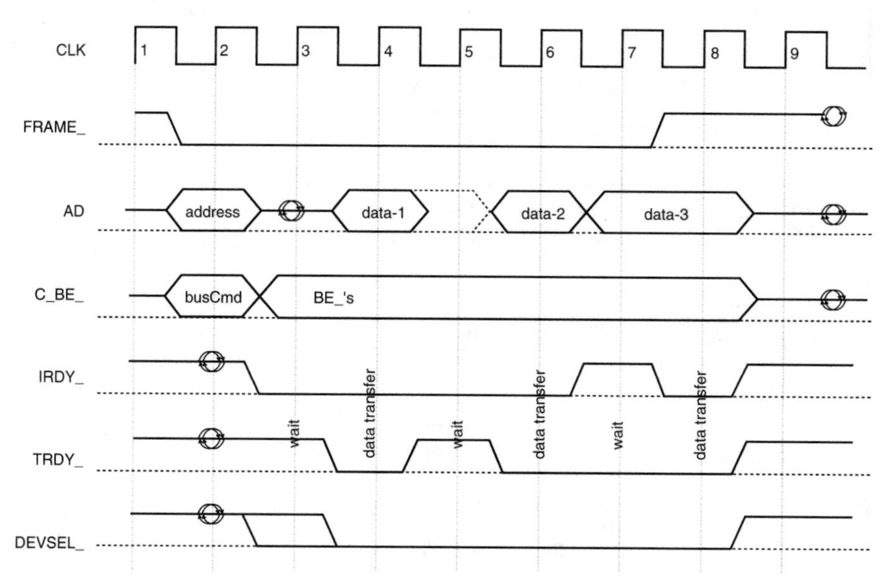

Fig. 6.6 PCI read cycle

Let us consider an example of PCI read protocol. Given the specification in Fig. 6.6, what type of assertions would the design team add, and what type would the verification team add? The tables below describe the difference. I have only given few of the assertions that could be written. There are many more assertions that need to be written by verification and design engineers. However, this example will act as a basis for differentiation.

Designers add assertions at microarchitecture level, while verification engineers concentrate at system level, specifically the interface level in this example.

The PCI protocol is for a simple READ. With FRAME_ assertion, AD address and C_BE_ have valid values. Then IRDY_ is asserted to indicate that the master is ready to receive data. Target transfers data with intermittent wait states. Last data transfer takes place a clock after FRAME_ is de-asserted.

Let us see what type of assertions need to be written by design and verification engineers.

6.3.1 PCI: Read Protocol Assertion Test Plan (Verification Team)

Property Name	Description	Property FAIL?	Property Covered?
Protocol Interface Assertions			
checkPCI_AD_CBE (check1)	On falling edge of FRAME_AD and C_BE_ bus cannot be unknown		
checkPCI_Dataphase (check2)	When both IRDY_and TRDY_are asserted, AD or C_BE_bus cannot be unknown		
checkPCI_Frame_Irdy (check3)	FRAME can be de-asseryted only if IRDY_is asserted		
checkPCI_trdyDevsel (check4)	TRDY_can be asserted only if DEVSEL_is asserted		
checkPCI_CBE_during_trx (check5)	Once the cycle starts (i.e at FRAME_ assertion) C_BE_ cannot float until FRAME_is de-asserted.		

Fig. 6.7 PCI: basic read protocol test plan (verification team)

6.3.2 PCI: Read Protocol Assertions Test Plan (Design Team)

Note the last two columns in the table of Figs. 6.7 and 6.8. (1) Did the property FAIL? (2) Did the property get covered? There is no column for the property PASS, that is, because of "cover" in an assertion that triggers only when a property is exercised but does not fail; in other words, it passes. Hence, there is no need for a PASS column. This "cover" column tells you that you indeed covered (exercised) the assertion and that it did not fail. When the assertion FAILs, it tells you that the assertion was exercised and that it failed during the exercise.

6.4 SVA Assertion Methodology Components

6.4.1 What Type of Assertions Should I Add?

It is important to understand and plan for the types of assertions you need to add. Make this part of your verification plan. It will also help you partition work among your team.

Note the "performance implication" assertions. Many miss on this point. Coming from processor background, I have seen that these assertions turn out to be some of the most useful assertions. These assertions would let us know of the, e.g., cache read latency upfront and would allow us enough time to make architectural changes to improve it.

- RTL assertions (design intent)
  - Intra-module

    Illegal state transitions, deadlocks, and livelocks
    FIFOs, onehot, etc.
- Block interface assertions (block interface intent)
  - Inter-module protocol verification, illegal combinations (ack cannot be "1" if req is "0"), and steady-state requirements (when slave asserts write_queue_full, master cannot assert write_req)
- Chip functionality assertions (chip/SoC functional intent)
  - A PCI transaction that results in Target Retry will indeed end up in the Retry Queue.

Property Name	Description	Property FAIL?	Property Covered?
Microarchitectural Assertions			
check_pci_adrcbe_St	PCI state machine is in 'adr_cbe' state the first clock edge when FRAME_is found asserted		
check_pci_data_St	PCI state machine is in 'data_transfer' state when both IRDY_and TRDY_ are asserted		
check_pci_idle_St	PCI state machine is in 'idle' state when both FRAME_and IRDY_are de-asserted		
check_pci_wait_St	PCI state machine is in 'wait' state if either IRDY_or TRDY_is de-asserted		

Fig. 6.8 Basic read protocol test plan: design team

- Chip interface assertions (chip interface intent)
  - Commercially available standard bus assertion VIPs can be useful in comprehensive check of your design's adherence to std. protocol such as PCIe, AXI, etc.
  - Every design assumption on IO functionality is an assertion.
- Performance implication assertions (performance intent)
  - Cache latency for read; packet processing latency, etc. are to catch performance issues before it's too late. This assertion works like any other. For example, if the "Read Cache Latency" is greater than two clocks, fire the assertion. This is an easy-to-write assertion with very useful return.

6.4.2 *Protocol for Adding Assertions*

- Do *not* duplicate RTL.
  - White box observability does not mean adding an assertion for each line of RTL code. This is a very important point, in that if RTL says "req" means "grant," don't write an assertion that says the same thing!! Read on.
  - Capture the intent.

  For example, a Write that follows a Read to the same address in the request pipe will always be allowed to finish before the Read. This is the intent of the design. How the designer implements reordering logic is not of much interest. So, from verification point of view, you need to write assertions that verify the chip design intent.

  A note here that the above does not mean you do not add low-level assertions. Classic example here is FIFO assertions. Write FIFO assertions for all FIFOs in your design. FIFO is low-level logic, but many of the critical bugs hang around FIFO logic, and adding these assertions will provide maximum bang for your buck.

- Add assertions throughout the RTL design process.
  - They are hard to add as an afterthought.
  - Will help you catch bugs even with your simple block-level testbench.
- If an assertion did not catch a failure...
  - If the test failed and none of the assertions fired, see if there are assertions that need to be added which would fire for the failing case.
  - The newly added assertion is now active for any other test that may trigger it.

Note: This point is very important toward making a decision if you have added enough assertions. In other words, if the test failed and none of the assertions fired, there is a good chance you still have more assertions to add.

- Reuse.
  - Create libraries of common "generic" properties with formal arguments that can be instantiated (reused) with "actual" arguments. We will cover this further in the chapter.
  - Reuse for the next project.

6.4.3 *How Do I know I Have Enough Assertions?*

- It's the "Test plan, test plan, test plan…"
  - Review and re-review your test plan against the design specs.
  - Make sure you have added assertions for every "critical" function that you must guarantee works.

- If tests keep failing but assertions do not fire, you do not have enough assertions.
  - In other words, if you had to trace a bug from primary outputs (of a block or SoC) without any assertions firing, that means that you did not put enough assertions to cover that path.
- "Formal" (aka static formal aka static functional verification) tool's ability to handle assertions.
  - What this means is that if you don't have enough "assertion density" (meaning if a register value does not propagate to an assertion within three to five clocks—resulting in assertions sparsely populated within design), the formal analysis tool may give up on the state/space explosion problem. In other words, a static functional formal tool may not be able to handle a large sequential domain. If the assertion density is high, the tool has to deal with smaller cone of logic. If the assertion density is sparse, the tool has to deal with larger cone of logic in both sequential and combinatorial space, and it may run out of steam.

6.4.4 Use Assertions for Specification and Review

- Use assertions (properties/sequences) for specification.
  - DV (design verification) team:

    Document as much of the 'response checking' part of your test plan as practical directly into executable assertions. Assertions are much easier to read than lengthy description of the response checking code.
    Since assertions are quite readable, use them for verification plan review and update
  - Design team:

    Document micro-arch. level assertions directly into executable assertions.
    Use it for design reviews.
- Assertions cross review.
  - Review:

    DV team reviews macro-, chip-, and interface-level assertions with the design team.
  - Cross review:

    Block A designer reviews Block B interface assertions
    Block B designer reviews Block A interface assertions
  - Incorrect assumptions among teams are detected early on.

6.5 Immediate Assertions

Immediate assertions are simple nontemporal domain assertions that are executed like statements in a procedural block. Interpret them as an expression in the condition of a procedural "if" statement. Immediate assertions can be specified only where a procedural statement is specified. The evaluation is performed immediately with the values taken at that moment for the assertion condition variables. The assertion condition is nontemporal, which means its execution computes and reports the assertion results at the *same* time.

Figure 6.9 describes the basics of an immediate assertion. It is so called because it executes immediately at the time it is encountered in the procedural code. It does not wait for any temporal time (e.g., "next clock edge") to fire itself. The assertion can be *preceded* by a level-sensitive or an edge-sensitive statement. As we will see, concurrent assertions can only work on a "sampling/clock edge"-sensitive logic and not level-sensitive logic.

We see in Fig. 6.9 that there is an immediate assertion embedded in the procedural block that is triggered by @ (posedge clk). The immediate assertion is triggered after @ (posedge d) and checks to see that (b || c) is true.

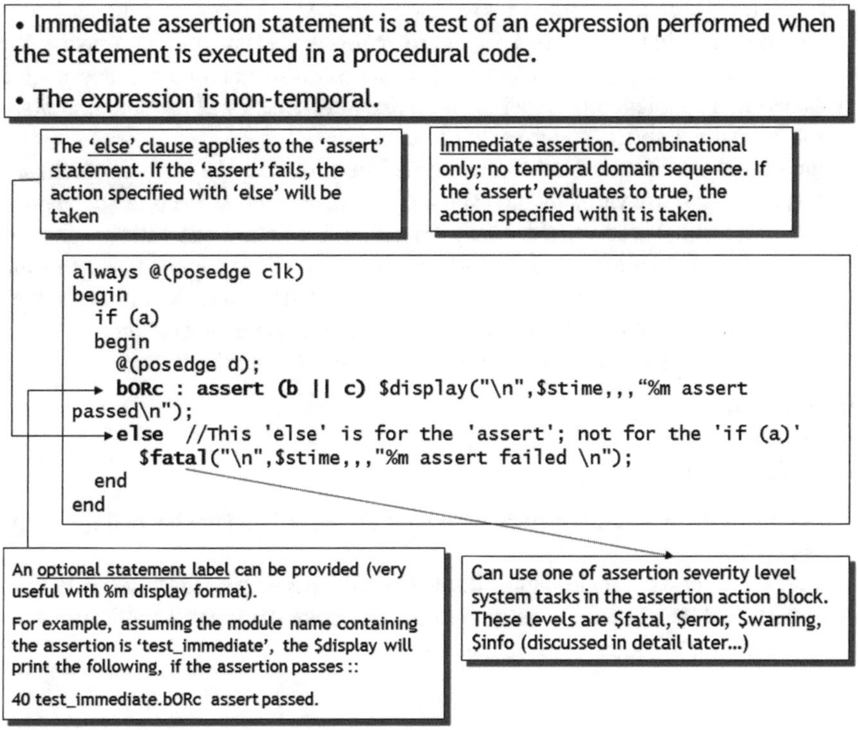

- Immediate assertion statement is a test of an expression performed when the statement is executed in a procedural code.
- The expression is non-temporal.

The 'else' clause applies to the 'assert' statement. If the 'assert' fails, the action specified with 'else' will be taken

Immediate assertion. Combinational only; no temporal domain sequence. If the 'assert' evaluates to true, the action specified with it is taken.

```
always @(posedge clk)
begin
  if (a)
  begin
    @(posedge d);
    bORc : assert (b || c) $display("\n",$stime,,,"%m assert
passed\n");
    else //This 'else' is for the 'assert'; not for the 'if (a)'
      $fatal("\n",$stime,,,"%m assert failed \n");
  end
end
```

An optional statement label can be provided (very useful with %m display format).

For example, assuming the module name containing the assertion is 'test_immediate', the $display will print the following, if the assertion passes ::

40 test_immediate.bORc assert passed.

Can use one of assertion severity level system tasks in the assertion action block. These levels are $fatal, $error, $warning, $info (discussed in detail later...)

Fig. 6.9 Immediate assertion example

We need to note a couple of points here. First, the very preceding statement in this example is @ (posedge d), an edge-sensitive statement. However, it does not have to be. It can be a level-sensitive statement also or any other procedural statement. The reason I am pointing this out is that concurrent assertions can work only off a sampling "edge" and not off a level-sensitive control. Keep this in your back pocket because it will be very useful to distinguish immediate assertions from concurrent assertions when we cover the latter. Second, the assertion itself cannot have temporal domain sequences. In other words, an immediate assertion cannot consume "time." It can only be combinatorial which can be executed in zero time. In other words, the assertion will be computed, and results will be available at the *same* time that the assertion was fired. If the "assert" statement evaluates to 0, X, and Z, then the assertion will be considered to FAIL else it will be considered to PASS.

We also see in the figure that there is (what is known as) an action block associated with FAIL or PASS of the assertion. This is no different than the PASS/FAIL logic we design for an "if…else" statement.

From syntax point of view, an immediate assertion uses only "assert" as the keyword in contrast to a concurrent assertion that requires "assert property."

One key difference between immediate and concurrent assertions is that concurrent assertions always work off the sampled value in preponed region (see Sect. 7.1) of a simulation tick, while immediate assertions work immediately when they are executed (as any combinatorial expression in a procedural block) and do not evaluate its expression in the preponed region. Keep this thought in your back pocket for now since we haven't yet discussed concurrent assertions and how assertions get evaluated in a simulation time tick. But this key difference will become important as you learn more about concurrent assertions.

Finally, as we discussed above, the immediate assertion works on a combinatorial expression whose variables are evaluated "immediately" at the time the expression is evaluated. These variables may transition from one logic value to another (e.g., 1 to 0 to 1) within a given simulation time tick, and the immediate assertion may get evaluated multiple times before the expression variable values "settle" down. Therefore, immediate assertions are also known to be "glitch" prone.

To complete the story, there are three types of immediate assertions:

Immediate **assert**
Immediate **assume**
Immediate **cover**

Details of these assertions (including deferred assertions) are beyond the scope of this book.

Suggested book for in-depth discussion of assertions is "SystemVerilog Assertions and Functional Coverage – a guide to methodology and applications"—Second Edition, Springer 2016, Ashok Mehta.

6.6 Concurrent Assertions

Concurrent assertions are temporal domain assertions that allow creation of complex sequences which are based on *clock* (*sampling*) *edge* semantics. This contrasts with the immediate assertions which are purely combinatorial and do not allow any time consumption.

Concurrent assertions are the gist of SVA language. They are called concurrent because they execute in parallel with the rest of the design logic and are multi-threaded. Let us start with basics and move onto the complex concepts of concurrent assertions.

In Fig. 6.10 we have declared a property "pr1" and asserted it with a label "reqGnt" (label is optional but highly recommended). The figure explains various parts of a concurrent assertion including a property, a sequence and assertion of the property.

The "assert property (pr1)" statement triggers property "pr1." "pr1" in turn waits for the antecedent "cStart" to be true at a (posedge clk), and on it being true implies (fires) a sequence called "sr1." "sr1" checks to see that "req" is high when it is fired and that two "clocks" later "gnt" is true. If this temporal domain condition is satisfied, then the sequence "sr1" will PASS and so will property "pr1" and the "assert property" will be a PASS as well. Let us continue with this example and study other key semantics:

1. "assert"—you must assert a property, i.e., invoke or trigger it.

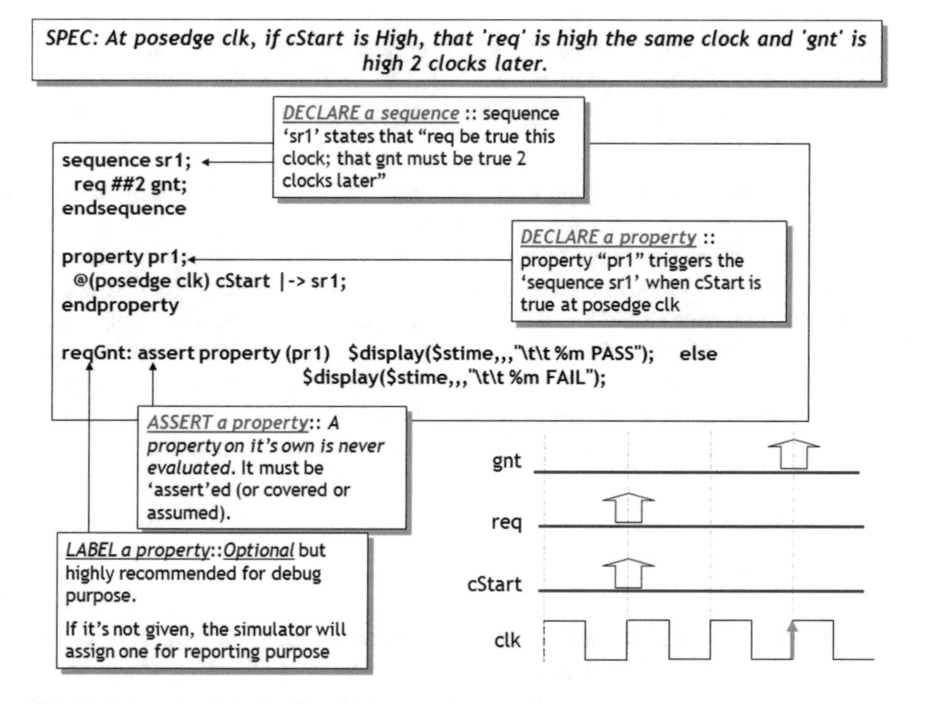

Fig. 6.10 Concurrent assertion example

2. There is an action block associated with either the pass or fail of the assertion.
3. "property pr1" is edge triggered on posedge of clk (more on the fact that you *must* have a sampling edge for trigger is explained further on).
4. "property pr1" has an *antecedent* which is a signal called cStart, which if sampled high (in the preponed region) on the posedge clk, will imply that the *consequent* (sequence sr1) be executed.
5. Sequence sr1 samples "req" to see if it is sampled highly the same as posedge of clk when the sequence was triggered because of the *overlapping implication* (|->) operator, and then wait for two clocks and see if "gnt" is high.
6. Note that each of "cStart," "req," and "gnt" is *sampled* at the edge specified in the property which is the posedge of "clk." In other words, even though there is no edge specified in the sequence, the edge is inherited from property pr1.

Note also that we are using the notion of sampling the values at posedge clk which means that the "posedge clk" is the *sampling edge*. In other words, the sampling edge can be anything (as long as it's an edge and is not level sensitive), meaning it does not necessarily have to be a synchronous edge such as a clock. It can be an asynchronous edge as well. However, *be very careful about using an asynchronous edge* unless you are sure what you want to achieve.

6.6.1 Overlapping and Nonoverlapping Operators

Figure 6.11 further shows the equivalence between overlapping and nonoverlapping operators. "|=>" is equivalent to "|-> ##1." Note that ##1 is not the same as Verilog's #1 delay. ##1 means one clock edge (sampling edge). Hence "|-> ##1" means the same as "|=>."

Suggestion To make debugging easier and have project-wide uniformity, use the overlapping operator in your assertions. Reason? Overlapping is the common denominator of the two types of operator. You can always model nonoverlapping from overlapping, but you cannot do vice versa. What this means is that during debugging, everyone would know that all the properties are modeled using overlapping and that the number of clocks are exactly the same as specified in the property. You do not have to add or subtract from the # of clocks specified in the chip specification. More important, if everyone uses his or her favorite operator, debugging would be very messy not knowing which property uses which operator.

6.7 Clocking Basics

A concurrent assertion is evaluated only on the occurrence of an "edge," known as the "sampling edge." The reason for continually mentioning this "edge" as "clk" is because it is best to have this "edge" synchronous to either posedge or negedge for a signal. You can indeed have an asynchronous edge as well. In Fig. 6.12, we are

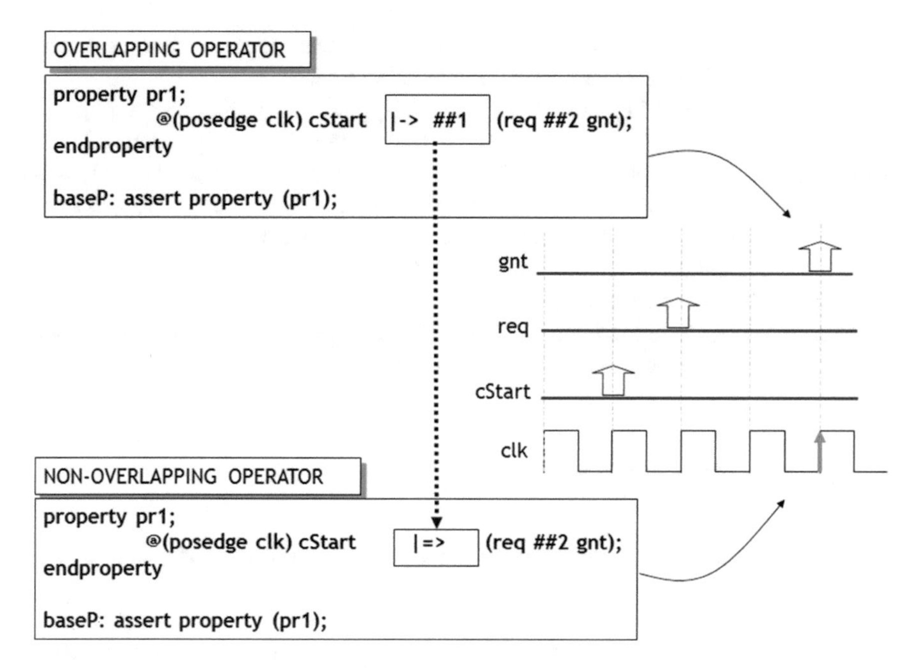

Fig. 6.11 Overlapping and nonoverlapping operators

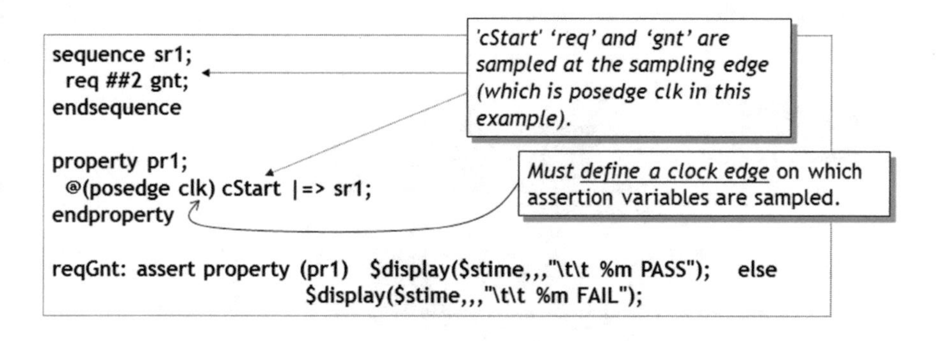

:: CLOCKING BASICS ::

- A concurrent assertion is evaluated only at the occurrence of a clock tick.
- The definition of a clock is explicitly specified by the user.
- *Assertion without a clock (or a sampling edge) will result in a compile Error.*
- The clock expression can be more complex than just a single signal name. E.g., you can have (CLK && Gating_signal).

Fig. 6.12 Clocking basics—assertions

using a nonoverlapping implication operator, which means that at a posedge of clk, if cStart is high, then one clock later sr1 should be executed.

Let us revisit "sampling" of variables. The expression variables cStart, req, and gnt are all sampled in the *preponed region* of posedge clk. In other words, if, for example, cStart=1 and posedge clk changed at the same time, the sampled value of cStart in the "preponed region" will be equal to "zero" and *not* "one." We will soon discuss what "preponed region" really means in a simulation time tick and how it affects the evaluation of an assertion, especially when the sampling edge and the sampled variable change at the same time.

Note again that "sequence sr1" does not have a clock in its expression. The clock for "sequence sr1" is inherited from the "property pr1."

6.7.1 Sampling Edge (Clock Edge)

Now let's look at one of the most important aspects of clocking mechanism of concurrent assertions.

How does the so-called sampling edge sample the variables in a property or a sequence is one of the most important concepts you need to understand when designing assertions. As shown in Fig. 6.13, the important thing to note is that the variables used in assertions (property/sequence/expression) are sampled in the *preponed* region. What does that mean? It means, for example, if a sampled variable

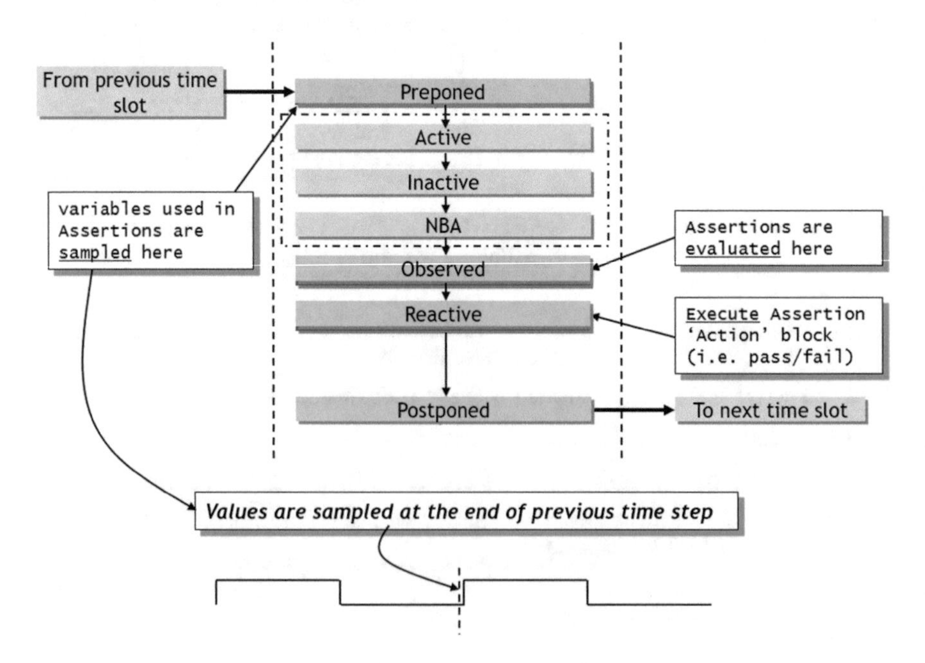

Fig. 6.13 Assertion sampling edge and simulation time tick phases

changes the same time as the sampling edge (e.g., clk), then the value of the variable will be the value it held—*before*—the clock edge.

```
@ (posedge clk) a |=> !a;
```

In the above sequence, let us say that variable "a" changes to "1," the same time that the sampling edge clock goes posedge (and assume "a" was "0" before it went to a "1"). Will there be a match of the antecedent "a"? No! Since "a" went from "0" to "1" the same time that clock went posedge clk, the sampled value of "a" at posedge clk will be "0" (in the preponed region) and not "1." This will not cause the property to trigger because the antecedent is not evaluated to be true. This will confuse you during debug. You would expect "1" to be sampled and the property triggered at posedge clk. However, you will get just the opposite result.

This is a very important point to understand because in a simulation waveform (or for that matter with Verilog $monitor or $strobe), you will see a "1" on "a" with posedge clk and would not understand why the property did not fire or why it failed (or passed for that matter). Always remember that at the sampling edge, the "previous" value (i.e., a delta before the sampling edge in the preponed region) of the sampled variable is used. To reiterate, preponed region is a precursor to the time slot, where only sampling of the data values take place. No value changes or events occur in this region. Effectively, sampled values of signals do not change through the time slot.

6.8 Concurrent Assertions: Binding Properties

"bind" allows us to keep design logic separate from the assertion logic. Design managers do not like to see anything in RTL that is not going to be synthesized. "bind" helps in that direction.

There are three modules (Figs. 6.14 and 6.15). The "designModule" contains the design. The "propertyModule" contains the assertions/properties that operate on the logic in "designModule." And the "test_bindProperty" module binds the propertyModule to the designModule. By doing so, we have kept the properties of the "propertyModule" separate from the "designModule." That is the idea behind "bind." You do not have to place properties in the same module as the design module. As mentioned before, you should keep your design void of all constructs that are non-synthesizable. In addition, keeping assertions and design in separate modules allow both the design and the DV engineers to work in parallel without restrictions of a database management system where a file cannot be modified by two engineers at the same time.

As shown in Fig. 6.15, for "bind" to work, you must declare either the instance name or the module name of the designModule in the "bind" statement. You need the design module/instance name, property module name, and the "bind" instance name for "bind" to work. In our case the design module name is designModule, its instance name is "dM," and the property module name is propertyModule.

module containing the design (designModule)

```
module designModule (da,db,dclk);
input da,dclk;
output logic db;
  always @(posedge dclk) db <= da;
endmodule
```

module containing properties (propertyModule)

```
module propertyModule (pa,pb,pclk);
input pa, pb, pclk;

property rc1;
  pa |-> pb;
endproperty

baseP: assert property (@(posedge pclk) (rc1)) else $display($stime,,,"\tproperty
FAIL");
endmodule
```

Fig. 6.14 Binding properties—assertions

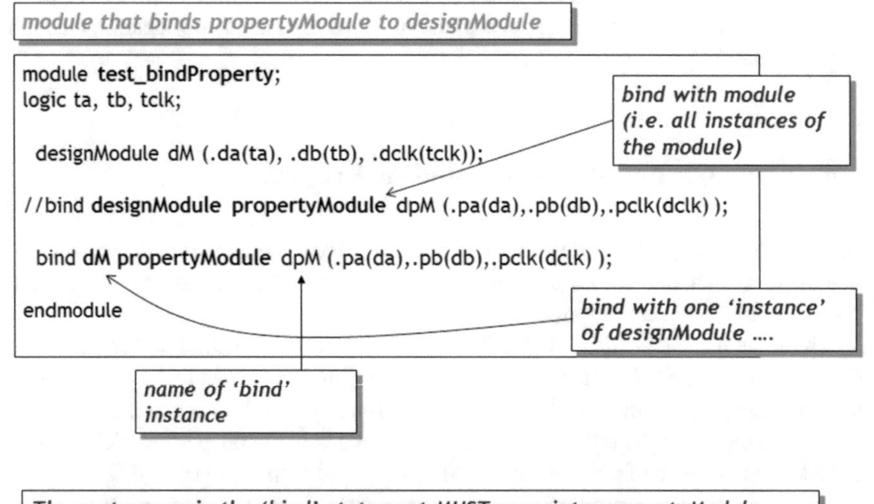

The port names in the 'bind' statement MUST associate propertyModule port names with those of the designModule port names

Fig. 6.15 Binding properties 2—assertions

The (uncommented) "bind" statement uses the module instance "dM" and binds it to the property module "propertyModule" and gives this "bind" an instance name "dpM." It connects the ports of propertyModule with those of the designModule. With this the "property rc1" in propertyModule will act on designModule ports as connected.

The commented "bind" statement uses the module name "designModule" to bind to the "propertyModule," whereby all instances of the "designModule" will be bound to the "propertyModule."

In essence, we have kept the properties/assertions of the design and the logic of the design separate. This is the recommended methodology. You could achieve the same results by putting properties in the same module as the design module but that is highly non-modular and intrusive methodology. In addition, as noted above, keeping them separate allows both the DV and the design engineer to work in parallel.

6.8.1 Binding Properties (Scope Visibility)

But what if you want to bind the assertions of the propertyModule to internal signals of the designModule? That is quite doable.

As shown in Fig. 6.16, "rda" and "rdb" are signals internal to designModule. These are the signals that you want to use in your assertions in the "propertyModule." Hence, you need to make "rda" and "rdb" visible to the "propertyModule." However, you do not want to bring "designModule" internal variables to external ports in order to make them visible to the "propertyModule." You want to keep the "designModule" completely untouched. To do that, you need to add input ports to the "propertyModule" and bind those to the internal signals of the "designModule" as shown in Fig. 6.16. Note that in our example we bind the propertyModule ports "pa" and "pb" to the designModule internal registers "rda" and "rdb." In other words, you can directly refer to the internal signals of designModule during "bind." "bind" has complete scope visibility into the bound module "designModule." Note that with this method you do not have to provide the entire hierarchical instance name when binding to "propertyModule" input ports.

6.9 Operators

Figure 6.17 shows the operators afforded by the SVA language. These operators are the gist of the language.

6.9.1 ##m: Clock Delay

Clock delay is about the most basic of all the operators and probably the one you will use the most! First of all, note that ##m means a delay of "m" number of sampling edges. In this example, the sampling edge is a "posedge clk"; hence, ##m means m number of posedge clks.

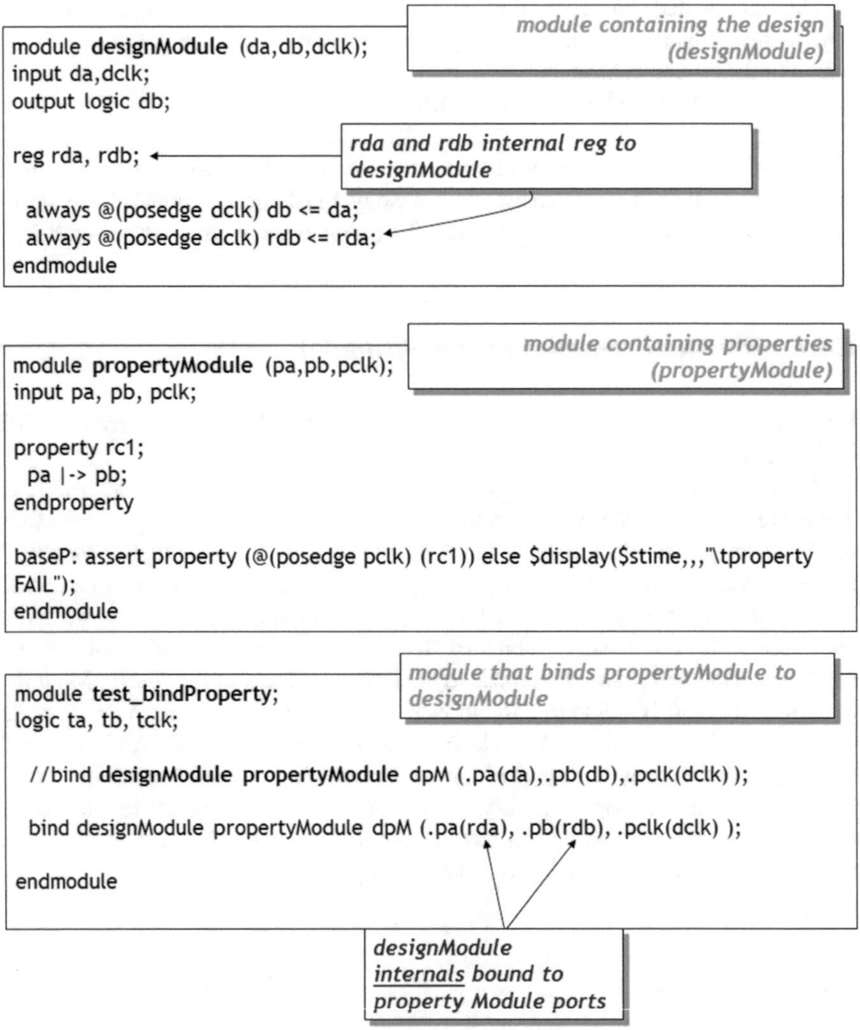
module designModule (da,db,dclk);
input da,dclk;
output logic db;
reg rda, rdb;
always @(posedge dclk) db <= da;
always @(posedge dclk) rdb <= rda;
endmodule
module containing the design (designModule)
rda and rdb internal reg to designModule
module propertyModule (pa,pb,pclk);
input pa, pb, pclk;
property rc1;
pa |-> pb;
endproperty
baseP: assert property (@(posedge pclk) (rc1)) else $display($stime,,,"\tproperty FAIL");
endmodule
module containing properties (propertyModule)
module test_bindProperty;
logic ta, tb, tclk;
//bind designModule propertyModule dpM (.pa(da),.pb(db),.pclk(dclk));
bind designModule propertyModule dpM (.pa(rda), .pb(rdb), .pclk(dclk));
endmodule
module that binds propertyModule to designModule
designModule internals bound to property Module ports

Fig. 6.16 "BIND"ing properties. Scope visibility

As shown in Fig. 6.18, the property evaluates antecedent "z" to be true at posedge clk and implies the sequence "Sab." "Sab" looks for "a" to be true at that same clock edge (because of the overlapping operator used in the property) and, if that is true, waits for two posedge clks and then looks for "b" to be true.

In the simulation log, we see that at time 10, posedge of clk, z==1 and a==1. Hence, the sequence evaluation continues. Two clks later (at time 30), it checks to see if b==1, which it finds to be true and the property passes.

Similar scenario unfolds starting time 40. But this time, b is not equal to 1 at time 60 (two clks after time 40), and the property fails.

Operator	Description
##m **##[m:n]**	Clock delay
[*m] **[*m:n]**	Repetition – Consecutive
[=m] **[=m:n]**	Repetition – Non-consecutive
[->m] **[-> m:n]**	GoTo Repetition – Non-consecutive
sig1 throughout seq1	Signal sig1 must be true throughout sequence seq1
seq1 within seq2	sequence seq1 must be contained within sequence s2
seq1 intersect seq2	'intersect' of two sequences; same as 'and' but both sequences must also 'end' at the same time.
seq1 and seq2	'and' of two sequences. Both sequences must start at the same time but may end at different times
seq1 or seq2	'or' of two sequences. It succeeds if either sequence succeeds.
first_match complex_seq1	matches only the first of possibly multiple matches
not <property_expr>	If <property_expr> evaluates to true, then not <property_expr> evaluates to false; and vice-versa.
if (expression) property_expr1 else property_expr2	If...else within a property
|->	Overlapping implication operator
|=>	Non-overlapping implication operator

Fig. 6.17 Operators—concurrent assertions

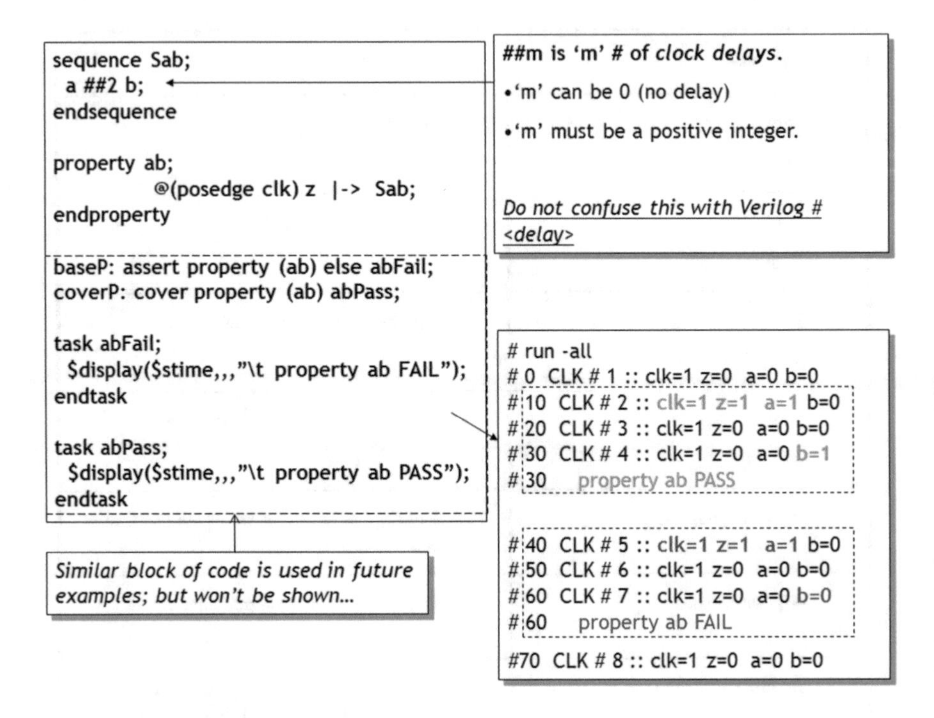

Fig. 6.18 Clock delay operator ##m

6.9.2 ##[m:n]: Clock Delay Range

Since it is quite necessary for a signal or expression to be true in a given *range* of clocks (as opposed to fix number of clocks), we need an operator which does just that.

##[m:n] allows a range of sampling edges (clock edges) in which to check for the expression that follows it. Figure 6.19 explains the rules governing the operator. Note that here also, m and n need to be constants. They cannot be variables.

The property "ab" in the figure says that if at the first posedge of clk the "z" is true, sequence "Sab" will be triggered. Sequence "Sab" evaluates "a" to be true to the same clock that "z" is true and then looks for "b" to be true delayed by either 1 clk or 2 clks or 3 clks. The very first instance that "b" is found to be true within the 3 clocks, the property will pass. If "b" is not asserted within 3 clks, the property will fail.

Note that in the figure, you see three passes. That simply means that whenever b is true the first time within 3 clks, the property will pass. It does not mean that the property will be evaluated and pass three times. To reiterate, the property will pass as soon as (i.e., the first time) that b is true.

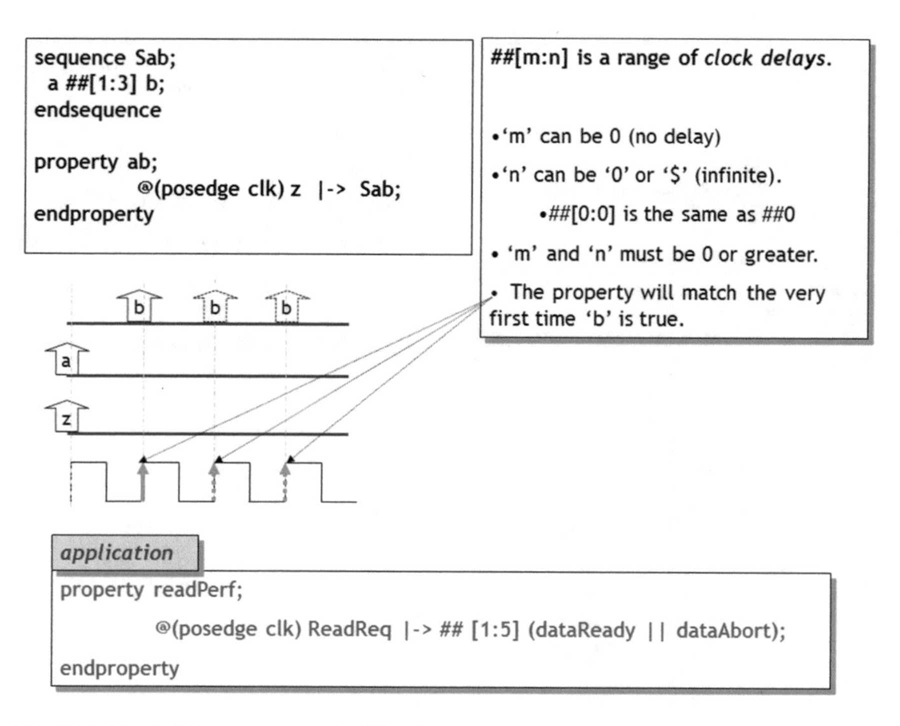

Fig. 6.19 Clock delay range operator ##[m:n]

6.9.3 [*m]: Consecutive Repetition Operator

As depicted in Fig. 6.20, the consecutive repetition operator [*m] sees that the signal/expression associated with the operator stays true for "m" consecutive clocks. Note that "m" *cannot* be $ (infinite # of consecutive repetition).

The important thing to note for this operator is that it will match at the *end* of the last iterative match of the signal or expression.

The example in Fig. 6.20 shows that when "z" is true that at the next clock, sequence "Sc1" should start its evaluation. "Sc1" looks for "a" to be true and then waits for one clock before looking for two consecutive matches on "b." This is depicted in the simulation log. At time 10 "z" is high; at 20 "a" is high as expected (because of nonoverlapping operator in property); at time 30 and 40, "b" remains high matching the requirement b[*2]. At the end of the second high on "b," the property meets all its requirements and passes.

The very next part of the log shows that the property fails because "b" does not remain high for two consecutive clocks. Again, the comparison ends at the *last* clock where the consecutive repetition is supposed to end, and then the property fails.

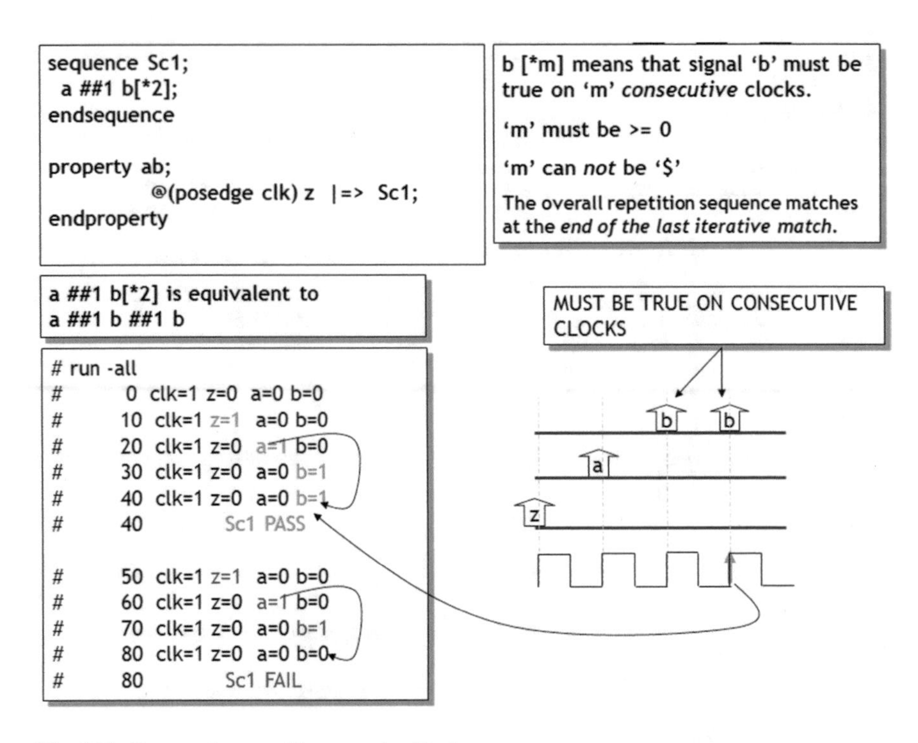

Fig. 6.20 Consecutive repetition operator [*m]

*6.9.4 [*m:n]: Consecutive Repetition Range*

Now let's look at the simulation log (Fig. 6.21). Time 30–90 is straightforward. At time 30, z=1 and a=1, the next clock "b" = 1 and remains "1" for two consecutive clocks and then 1 clock later c=1 as required, and the property passes. But what if "c" was not equal to "1" at time 90? That is what the second set of events shows.

Z=1 and a=1 at time 110 and the sequence Sc1 continues. OK. b=1 the next two clocks. Correct. But why doesn't the property end here? Isn't it supposed to end at the first match? Well, the reason the property does not end at 150 is because it needs to wait for c=1 the next clock. OK, so it waits for C=1 at 170. But it does not see a c=1. Shouldn't the property now fail? NO. This is where the max range :5 comes into picture. Since there is a range [*2:5], if the property does not see a c=1 after the first two consecutive repetitions of "b," it waits for the next consecutive "b" (total 3 now) and then looks for "c=1." If it does not see c=1, it waits for the next consecutive b=1 (total 4 now) and then looks for c=1. Still no "c?" It finally waits for max range fifth b=1 and then the next clock looks for c=1. If it finds one, the property ends and passes. If not, the property fails.

```
sequence Sc1;
  a ##1 b[*2:5] ##1 c;
endsequence

property ab;
        @(posedge clk) z  |-> Sc1;
endproperty
```

```
a ##1 b[*2:5] is equivalent to

a ##1 b ##1 b          ##1 c      ||
a ##1 b ##1 b ##1 b ##1 c      ||
a ##1 b ##1 b ##1 b ##1 b ##1 c ||
a ##1 b ##1 b ##1 b ##1 b ##1 b ##1 c
```

b [*m:n] means that signal 'b' must be true on

minimum 'm' *consecutive* clocks and *maximum* 'n' *consecutive* cycles.

'm' must be >= 0;

'n' can be >=0 or $

The overall repetition sequence matches at the first match of the sequence that meets the required condition.

```
# run -all
#        10  clk=1 z=0  a=0 b=0 c=0
#        30  clk=1 z=1  a=1 b=0 c=0
#        50  clk=1 z=0  a=0 b=1 c=0
#        70  clk=1 z=0  a=0 b=1 c=0
#        90  clk=1 z=0  a=0 b=0 c=1
#        90    Sc1 PASS
#        110  clk=1 z=1  a=1 b=0 c=0
#        130  clk=1 z=0  a=0 b=1 c=0
#        150  clk=1 z=0  a=0 b=1 c=0
#        170  clk=1 z=0  a=0 b=1 c=0
#        190  clk=1 z=0  a=0 b=1 c=0
#        210  clk=1 z=0  a=0 b=1 c=0
#        230  clk=1 z=0  a=0 b=0 c=1
#        230    Sc1 PASS
#        250  clk=1 z=1  a=1 b=0 c=0
#        270  clk=1 z=0  a=0 b=1 c=0
#        290  clk=1 z=0  a=0 b=1 c=1
#        310  clk=1 z=0  a=0 b=0 c=0
#        310  Sc1 FAIL
```

Requirement: After at least 2 consecutive High on 'b', if 'b' goes Low, that 'c' must go High the next clock.

But 'c' is low at #310 and the property fails.

Fig. 6.21 Consecutive range operator

Continuing with the simulation log, the last part shows how the property would fail. One way it would fail is what I have described above. The other way is shown in the log file. I have repeated the log file here to help us concentrate only on that part of the log file:

```
#          250   clk=1 z=1   a=1 b=0 c=0
#          270   clk=1 z=0   a=0 b=1 c=0
#          290   clk=1 z=0   a=0 b=1 c=1
#          310   clk=1 z=0   a=0 b=0 c=0
#          310   Sc1 FAIL
```

At time 250, z=1 and a=1 so the sequence evaluation continues to consecutive operator. “b” is equal to 1 for the next two consecutive clocks. Good. But at time 310, b=0 and—also—c=0. Hence, the property fails. After two consecutive “b,” there should be either a third “b” or a “c=1.” Neither of them is present and the property fails. If C=1 at time 310, the property would pass. If b=1 and c=0 at time 310, the property would continue to evaluate until it sees 5 consecutive “b” or a c=1 before 5 consecutive “b” are encountered. Or after five consecutive “b,” there is a c=1 as shown in the previous part of the simulation log file.

6.9.5 [=m]: Repetition Non-consecutive

Non-consecutive repetition is another useful operator (as the consecutive operator) and used very frequently. In many applications, we want to check that a signal remains asserted or de-asserted several times and that we need *not* know when exactly these transitions take place. For example, if there is a non-burst READ of length 8, that you expect 8 RDACK. These RDACK may come in a consecutive sequence *or not* (based on read latency). But you must have eight RDACK before read is done.

Now, just as in the consecutive operator, the qualifying event (shown as ##1 C in Fig. 6.22) plays a significant role. “##1 c” tells the property that after the last “b,” “c” must occur once and then it can occur *any time* after one clock after the last “b.” Note again that even though we have “##1 c,” “c” does *not* necessarily need to occur one clock after the last “b.” It can occur after *any* # of clks after one clock after the last “b”—as long as—no other “b” occurs while we are waiting for “c.” Confusing! Not really. Let us look at the simulation log in Fig. 6.22. That will clarify things.

In the log, a=1 at time 5; b=1 at time 25 and then at 45. So far so good. We are marching along just as the property expects. Then comes in c=1 at time 75. That also meets the property requirement that “c” occurs any time after last b=1. BUT note that *before* c=1 arrived at time 75, “b” did not go to a “1” after its last occurrence at time 45. The property passes. Let us leave this at that for the moment. Now let us look at the second part of the log.

a=1 at time 95; then b=1 at 105 and 125; we are doing great. Now we wait for c=1 to occur any time after last “b.” C=1 occurs at time 175. But the property fails before that!! What is going on? Note b=1 at time 145. That is not allowed in this property. The property expects a c=1 after the *last* occurrence of “b” but *before* any other b=1 occurs. If another b=1 occurs *before* c=1 (as at time 145), then all bets are off. Property does not wait for the occurrence of c=1 and fails as soon as it sees this extra b=1. In other words, (what I call) the qualifying event “##1 c” encapsulates the property and strictly checks that b[=2] allows only two occurrences of “b” before “c” arrives.

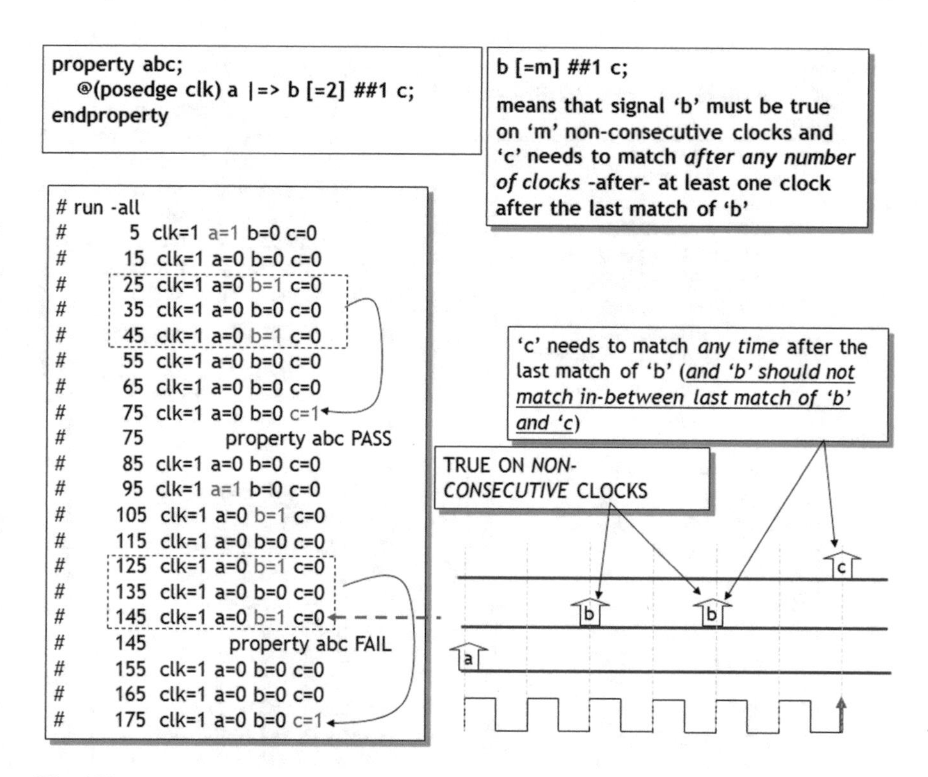

Fig. 6.22 Non-consecutive repetition operator [=m]

6.9.6 [=m:n]: Repetition Non-consecutive Range

Property in Fig. 6.23 is analogous to the non-consecutive (non-range) property, except that this has a range. The range says that "b" must occur minimum two times or maximum five times after which "c" can occur one clock later any time and that no more than maximum of five occurrences of "b" occur between the last occurrence of b=1 and c=1.

Referring to Fig. 6.23, first simulation log (Top left) shows that after a=1 at time 5, b occurs twice (the minimum # of times) at time 15 and 45, and then c=1 at time 75. Why didn't the property wait for five occurrences of b=1? That is because after the second b=1 at time 45, c=1 arrives at time 75, and this c=1 satisfies the property requirement of minimum of two b=1 followed by a c=1. The property passes and does not need to wait for any further b=1. In other words, the property starts looking for "c=1" after the minimum required (2) "b==1." Since it did find a "c=1" after two "b=1," the property ends there and passes.

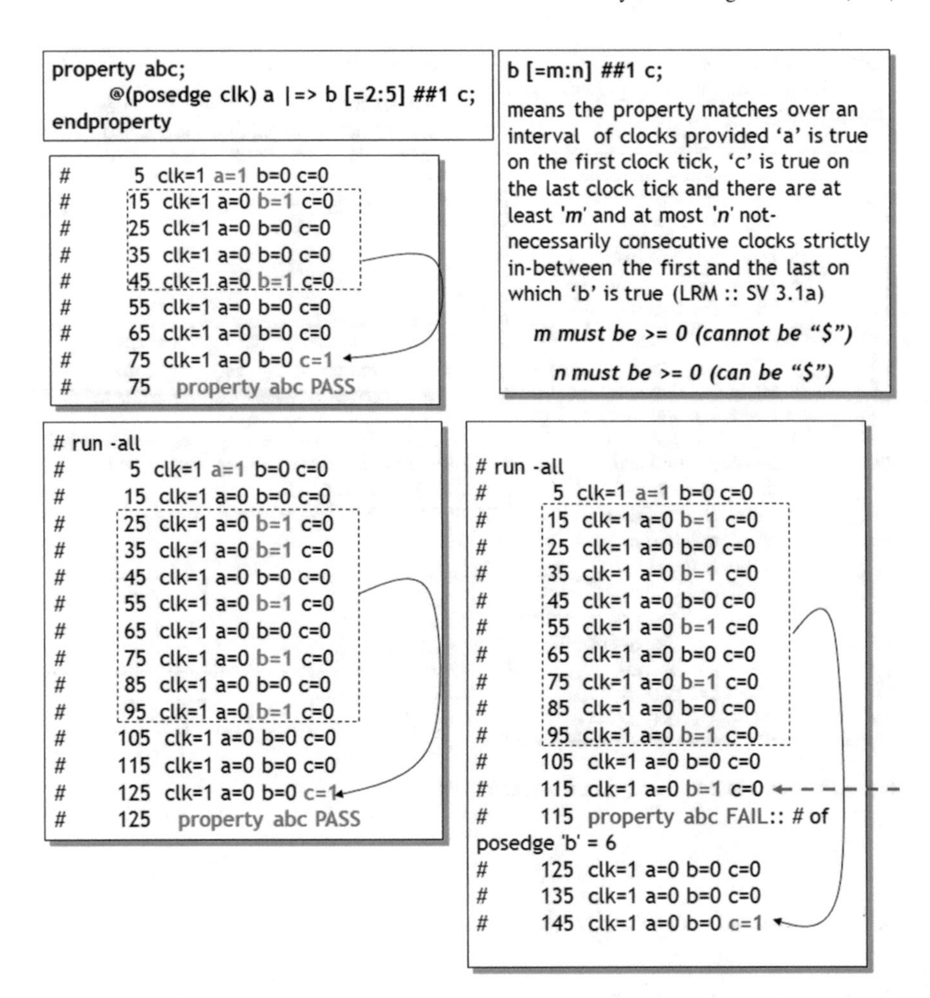

Fig. 6.23 Non-consecutive repetition range operator [=m:n]

Similarly, the simulation log on bottom left shows that "b" occurs five (max) times, and then "c" occurs without any occurrence of b. The property passes. This is how that works. As explained above, after two "b=1," the property started looking for "c==1." But before the property detects "c==1," it sees another "b==1." That's OK because "b" can occur maximum of five times. So, after the third "b==1", the property continues to look for either "c==1" or "b==1" until it has reached maximum of five "b==1." This entire process continues until five "b"s are encountered. Then the property simply waits for a "c." While waiting for a "c" at this stage, if a sixth "b" occurs, the property fails. This failure behavior is shown in simulation log in the bottom right corner of Fig. 6.23.

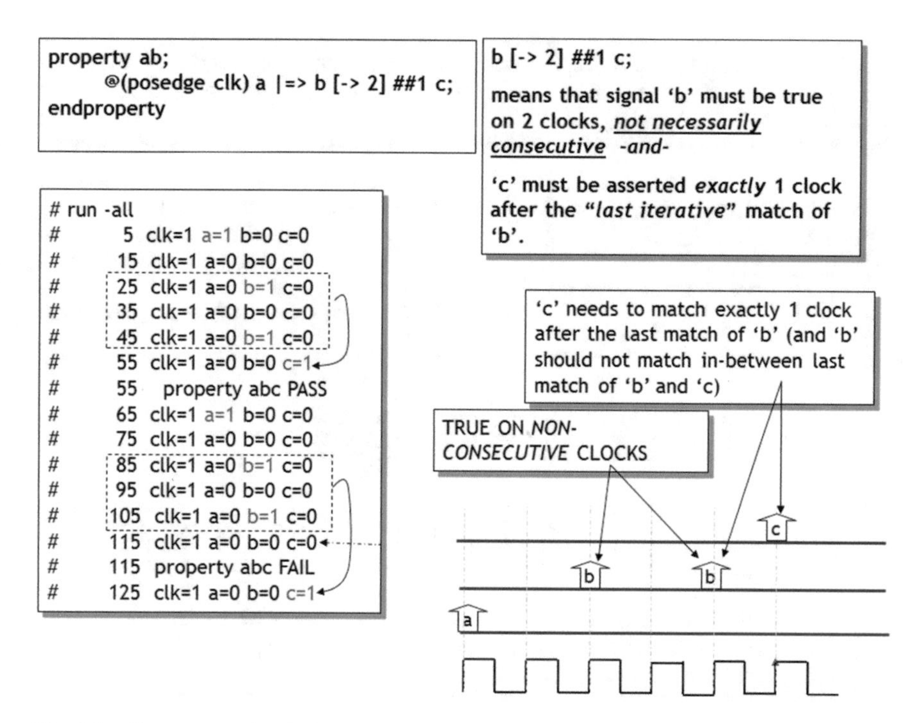

Fig. 6.24 Non-consecutive GoTo repetition operator

6.9.7 [->m] Non-consecutive GoTo Repetition Operator

This is the so-called non-consecutive *goto* operator! Very similar to [=m] non-consecutive operator. Note the symbol difference. The goto operator is [->2].

b[->2] acts exactly the same as b[=2]. So, why bother with another operator with the same behavior? It is the *qualifying event* that makes the difference. Recall that the qualifying event is the one that comes *after* the non-consecutive or the "goto" non-consecutive operator. I call it qualifying because it is *the end event* that qualifies the sequence that precedes for final sequence matching.

The simulation log in Fig. 6.24 shows a PASS and a FAIL scenario. PASS scenario is quite clear. At time 5, a==1, then two non-consecutive "b" occur, and then *exactly* one clock after the last "b=1," "c=1" occurs. Hence, the property passes. The FAIL scenario shows that after two occurrences of b==1, c==1 does *not* arrive *exactly* one clock after the last occurrence of b=1. That is the reason the b[->2] ##1 c check fails.

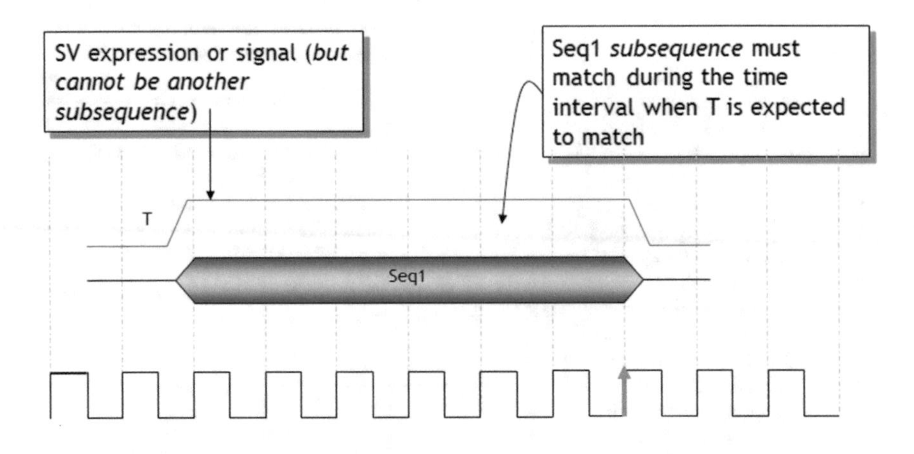

Fig. 6.25 *throughout* operator

6.9.8 sig1 throughout seq1

The *throughout* operator (Fig. 6.25) makes it that much easier to test for condition (signal or expression) to be true *throughout* a sequence. Note that the LHS of *throughout* operator can only be a signal or an expression, but it cannot be a sequence (or subsequence). The RHS of *throughout* operator can be a sequence. So, what if you want a sequence on the LHS as well? That is accomplished with the *within* operator, discussed right after *throughout* operator.

Let us examine the application in Fig. 6.26 which will help us understand the *throughout* operator.

In Fig. 6.26 the antecedent in property pbrule1 requires bMode (burst mode) signal to fall. Once that is true, it requires checkbMode to execute.

Read the property bottom up. checkbMode makes sure that the bMode stays low *throughout* the data_transfer sequence. If bMode goes high before data_transfer is over, the assertion will fail. The data_transfer sequence requires both dack_ and oe_ to be asserted (active low) and to remain asserted for four consecutive cycles. Throughout the data_transfer, burst mode (bMode) should remain low.

There are two simulation logs presented in Fig. 6.27. Both are for FAIL cases! FAIL cases are more interesting than the PASS cases, in this example! The first simulation log (left hand side) shows $fell(bMode) at time 20. Two clocks later at

application

1. When Burst Mode (bMode) is asserted, oe_ and dack_ must be found asserted after 2 clocks.
2. oe_ and dack_ must remain asserted for minimum of 4 clocks after both are found asserted.
3. bMode must remain asserted *throughout* the duration of oe_ && dack_ assertion.

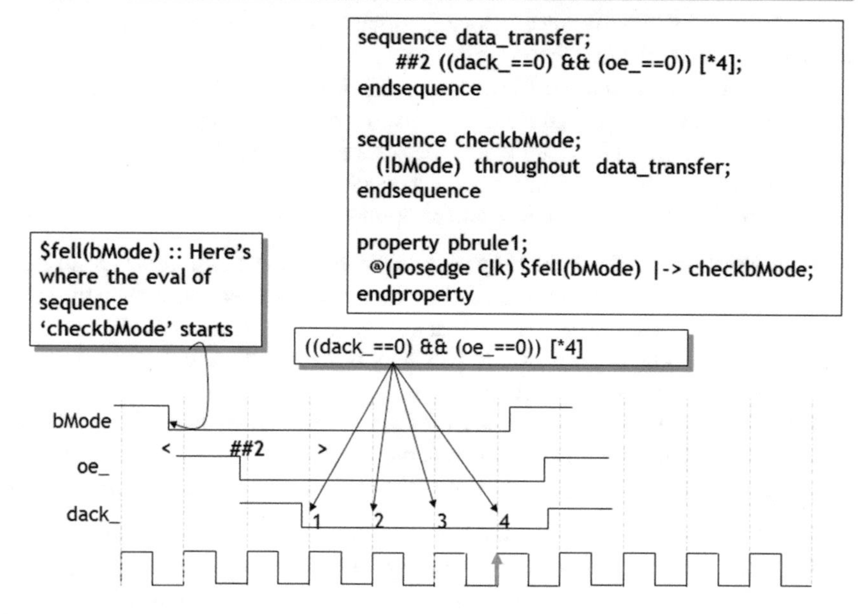

Fig. 6.26 Application: sig1 *throughout* seq1

bMode goes High a clock too early.

```
# 0  CLK #1 :: clk=1 bMode=1 oe_=1 dack_=1
#10  CLK #2 :: clk=1 bMode=1 oe_=1 dack_=1
#20  CLK #3 :: clk=1 bMode=0 oe_=1 dack_=1
#30  CLK #4 :: clk=1 bMode=0 oe_=0 dack_=1
#40  CLK #5 :: clk=1 bMode=0 oe_=0 dack_=0
#50  CLK #6 :: clk=1 bMode=0 oe_=0 dack_=0
#60  CLK #7 :: clk=1 bMode=0 oe_=0 dack_=0
#70    property pbrule1 FAIL
#70 CLK #8 :: clk=1 bMode=1 oe_=0 dack_=0
```

(dack_==0 && oe_==0) does not hold for 4 clocks

```
#100  CLK #11 :: clk=1 bMode=1 oe_=1 dack_=1
#110  CLK #12 :: clk=1 bMode=0 oe_=1 dack_=1
#120  CLK #13 :: clk=1 bMode=0 oe_=0 dack_=1
#130  CLK #14 :: clk=1 bMode=0 oe_=0 dack_=0
#140  CLK #15 :: clk=1 bMode=0 oe_=0 dack_=0
#150  CLK #16 :: clk=1 bMode=0 oe_=0 dack_=0
#160    property pbrule1 FAIL
#160  CLK #17 :: clk=1 bMode=0 oe_=0 dack_=1
#170  CLK #18 :: clk=1 bMode=0 oe_=0 dack_=1
#180  CLK #19 :: clk=1 bMode=0 oe_=0 dack_=1
#190  CLK #20 :: clk=1 bMode=1 oe_=1 dack_=1
```

Fig. 6.27 sig1 *throughout* seq1—application simulation log

time 40, oe_=0 and dack_=0 are detected. So far so good. oe_ and dack_ retain their state for three clocks. That's good too. But in the fourth cycle (time 70), bMode goes high. That's a violation because bMode is supposed to stay low throughout the data transfer sequence, which is four clocks long.

The second simulation log (right hand side) also follows the same sequence as above, but after three consecutive clocks that the oe_ and dack_ remain low, dack_ goes high at time 160. That is a violation because data_transfer (oe_=0 and dack_=0) is supposed to stay low for four consecutive cycles.

This also highlights a couple of other important points:

1. Both sides of the *throughout* operator must meet their requirements. In other words, if either the LHS or the RHS of the throughout sequence fails, the assertion will fail. Many folks assume that since bMode is being checked to see that it stays low (in this case), bMode fails only if the assertion will fail. Not true as we see from the two failure logs.
2. Important point: To make it easier for the reader to understand this burst mode application, I broke it down into two distinct subsequences. But what if someone just gave you the timing diagram and asked you to write assertions for it?

Break down any complex assertion requirement into smaller chunks. This is probably the most important advice I can part to the reader. If you look at the entire AC protocol (the timing diagram) as one monolithic sequence, you will indeed make mistakes and spend more time debugging your own assertion then debugging the design under test.

6.9.9 seq1 within seq2

Analogous to *throughout*, the *within* operator (Fig. 6.28) sees if one sequence is contained *within* or of the same length as another sequence. Note that the *throughout* operator allowed only a signal or an expression on the LHS of the operator. *within* operator allows a sequence on both the LHS and RHS of the operator.

The property "within" ends when the larger of the two sequences end, as shown in Fig. 6.28.

6.9.10 seq1 and seq2

As the name suggests, *and* operator (Fig. 6.29) expects both the LHS and RHS side of the operator *and* to evaluate to true. It does not matter which sequence ends first as long as both sequences meet their requirements. The property ends when the longer of the two sequences ends. *But note that both the sequences must start at the same time.*

'Seq1 within Seq2' matches along a finite interval of consecutive clocks ticks provided that Seq2 matches along the interval and Seq1 matches along some sub-interval of consecutive clock ticks.
Note that both Seq1 and Seq2 can be sequences.

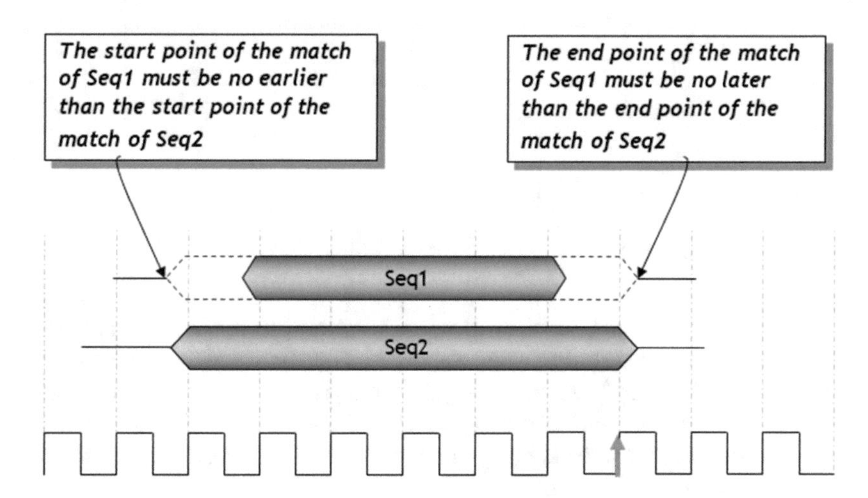

Fig. 6.28 *within* operator

'Seq1 *and* Seq2' match if

- Both sequences start at the same time
- Both sequences match
- The end time of each sequence can be different.

The end time is the end time of either Seq1 or Seq2, *whichever matches last*.

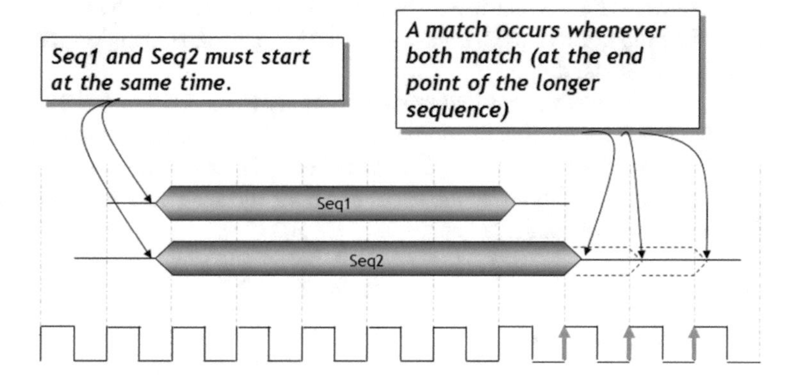

Fig. 6.29 "and" operator

'Seq1 *or* Seq2' match if

• operand 'or' is used when at least one of the two operand sequences is expected to match.

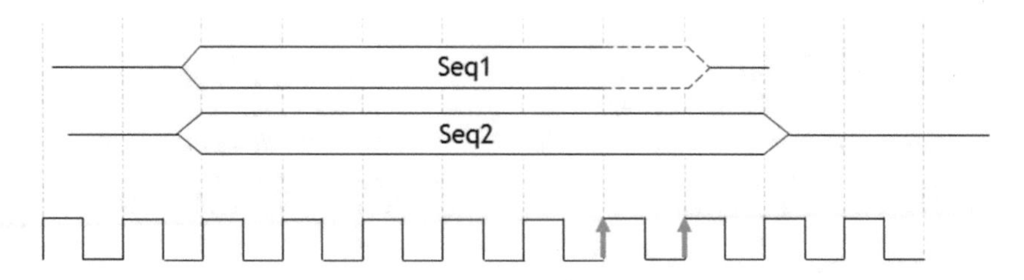

Fig. 6.30 "or" operator

The *and* operator is very useful, when you want to make sure that certain concurrent operations in your design start at the same time and that they both complete/match satisfactorily. As an example, in the processor world, when a Read is issued to L2 cache, L2 will start a tag match and issue a DRAM Read at the same time, in anticipation that the tag may not match. If there is a match, it will abort the DRAM Read. So, one sequence is to start tag compare, while another is to start a DRAM Read (ending in DRAM Read Complete or Abort). The DRAM Read sequence is designed such that it will abort as soon as there is a tag match. This way we have made sure that both sequences start at the same time and that they both end.

6.9.11 *seq1 or seq2*

or of two sequences means that when either of the two sequences match its requirements that the property will pass. Please refer to Fig. 6.30 and examples that follow to get a better understanding.

The feature to note with *or* is that as soon as either of the LHS *or* RHS sequence meets, its requirements are that the property will end. This contrasts with *and* where only after the longest sequence ends that the property is evaluated.

Note also that if the shorter of the two sides fails, the sequence will continue to look for a match on the longer sequence.

6.9.12 *seq1 intersect seq2*

So, with *throughout*, *within*, *and,* and *or* operators, who needs another operator that also seems to verify that sequences match?

'Seq1 *intersect* Seq2' match if

- **Both sequences start at the same time**
- **Both sequences must match**
- ***The lengths of the two matches of the operand sequences must be the same.***

The end time is when both sequences match and end at the same time.

The main difference between 'and and 'intersect' is the requirement on the length of the two sequences. For 'and' each sequence can be of any length. For 'intersect' they must be of the same length.

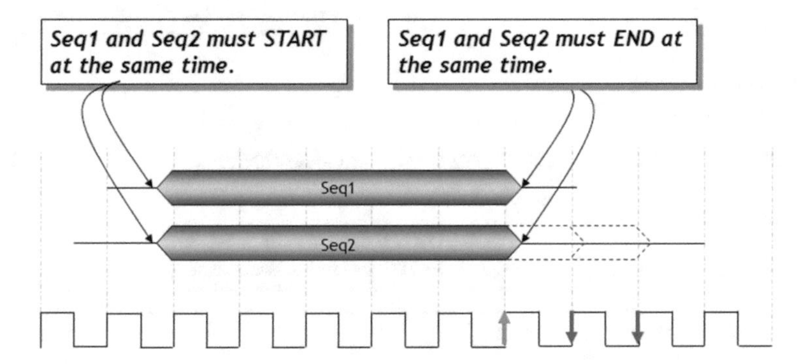

Fig. 6.31 *intersect* operator

throughout or *within* or *and* or *or* does *not* make sure that both the LHS and RHS sequences of the operator are exactly the same. They can be of the same length, but the operators do not care as long as the signal/expression or sequence meets their requirements. That's where *intersection* operator (Fig. 6.31) comes into picture. It makes sure that the two sequences indeed start at the same time and end at the same time and satisfy their requirements. In other words, they intersect.

As you can see, the difference between *and* and *intersect* is that *intersect* requires both sequences to be of the same length and that they both start at the same time and end at the same time, while *and* can have the two sequences of different lengths.

6.10 Local Variables

This is a very brief introduction to local variables. Please refer to (Mehta 2016) for complete detail and examples on this topic. Without the dynamic multi-threaded semantics and features of local variables, many of the assertions would be impossible to write.

Local variables are dynamic variables.

They are dynamically created when needed within an instance of a sequence and removed when the end of the sequence is reached.

'local vars' is one of the most powerful features of SVA language because it allows checking of complex pipelined behavior of the design.

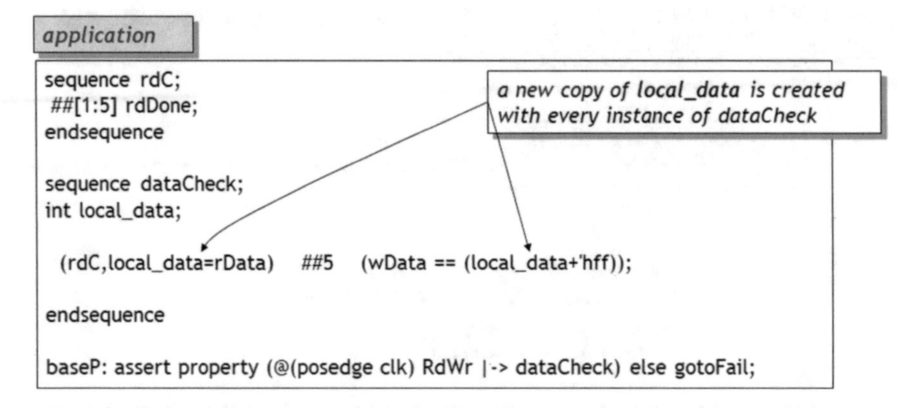

sequence datraCheck reads as ::

on matching 'rdC', store rData in the local var called local_data and ##5 clocks later wData must match local_data+'hff

Note that dataCheck is triggered when 'RdWr' is true. 'RdWr' can be true every clock and dataCheck would be triggered every clock. For every trigger of dataCheck, a new copy of local_data is created which will store rData and check for wData 5 clocks later.

Fig. 6.32 Local variables—basics

Local variable is a feature you are likely to use very often. They can be used both in a sequence and a property. They are called *local* because they are indeed local to a sequence and are not visible or available to other sequences or properties. Figure 6.32 points out key elements of a local var. The most important and useful aspect of a local variable is that it *allows multi-threaded application and creates a new copy of the local variable with every instance of the sequence in which it is used*. User does not need to worry about creating copies of local variables with each invocation of the sequence. Above application says that whenever "RdWr" is sampled high at a posedge clk, "rData" is compared with "wData" five clocks later. The example shows how to accomplish this specification. Local variable "int local_data" stores the "rData" at posedge of clk and then compares it with wData five clocks later. Note that "RdWr" can be sampled true at every posedge clk. Sequence "data_check" will enter every clock, create a new copy of local_data, and create a new pipelined thread that will check for local_data+'hff with "wData" five clocks later.

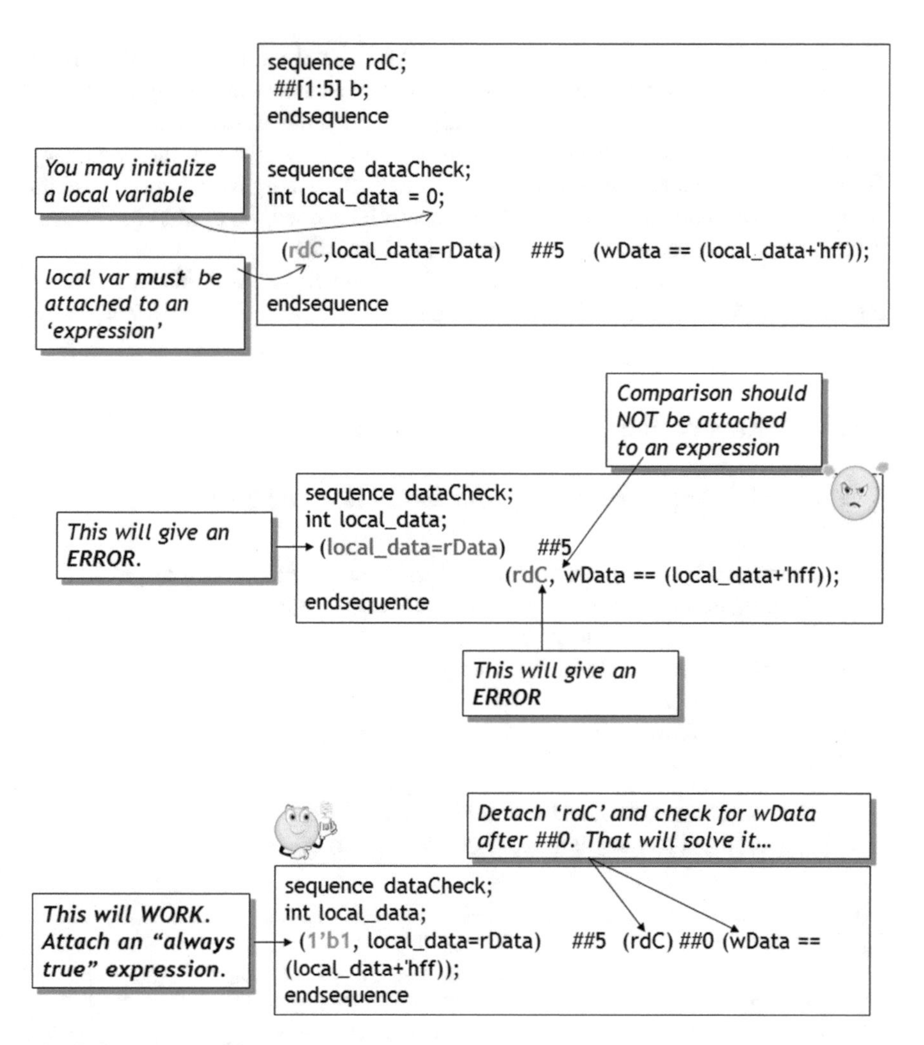

Fig. 6.33 Local variables—do's and don'ts

Note that the sampled value of a local variable is defined as the current value (not the value in the preponed region).

Moving along, Fig. 6.33 shows other semantics of local variables. *Pay close attention to the rule that local variable must be attached to an expression, while comparison cannot be attached to an expression*!!

In this example, "local_data=rData" is attached to the sequence "rdC." In other words, assignment "local_data=rData" will take place only on completion of sequence "rdC." But what if you don't have an expression to attach to the local variable when you are storing a value? Use 1'b1 (always true) as an expression. That will mean whenever you enter a sequence, the expression is always true, and you should store the value in the local variable. Simple!

Note that local variables do not have default initial values. A local variable without an initialization assignment will be unassigned at the beginning of the evaluation attempt. The expression of an initialization assignment to a given local variable may refer to a previously declared local variable. In this case the previously declared local variable must itself have an initialization assignment, and the initial value assigned to the previously declared local variable will be used in the evaluation of the expression assigned to the given local variable.

The detailed discussion of local variables is beyond the scope of this book. Please refer to (Mehta 2016) for in-depth discussion.

6.11 SystemVerilog Assertions: Applications

6.11.1 SVA Application: Infinite Delay Range Operator

In this example (Fig. 6.34), we expect "tErrorBit" to rise in a certain range of clock delays. The figure explains how the assertion works. Note also that you could use "&&" in place of ##0 to achieve the same results. Since assertions are mainly temporal domain, I prefer to tie in everything with temporal domain constructs. But that's a matter of preference.

Note also the following two semantically equal statement but with different syntax. These are short forms:

- ##[*] is used as an equivalent representation of ##[0:$].
- ##[+] is used as an equivalent representation of ##[1:$].

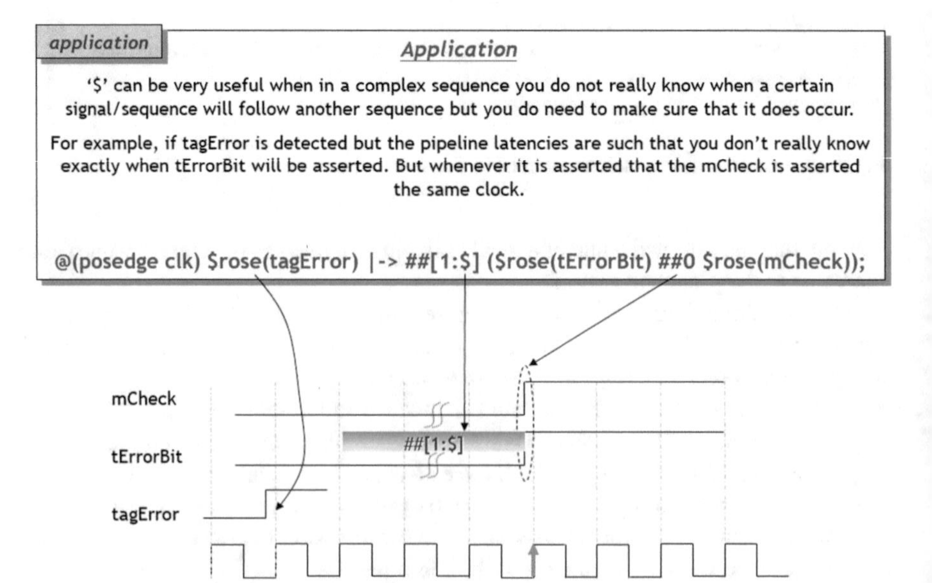

Fig. 6.34 Concurrent assertions—application

application

Specification:

PCI Special cycle requires that DEVSEL_ should remain high (deasserted) during the special cycle.

```
property checkDevSelSpecialCycle;
@(posedge clk)
  $fell(FRAME_) && (CMD == 4'b0001) |-> DEVSEL_ [*1:$] ##0 $rose(FRAME_);
endproperty
```

HINT: You can mix **Edge Sensitive** and **Level Sensitive** expressions in a logic condition.

Fig. 6.35 SVA Application—consecutive delay range operator

6.11.2 SVA Application: Consecutive Delay Range Operator

Figure 6.35 depicts the following application.

A PCI cycle starts when FRAME_ is asserted (goes low) and the CMD is valid. A CMD == 4'b0001 specifies the start of a PCI special cycle. On the start of such a cycle (i.e., the antecedent being true), the consequent looks for DEVSEL_ to be high forever consecutively at every posedge clk until FRAME_ is de-asserted (goes high). This is by far the easiest way to check for an event/expression to remain true (and we do not know for how long) until another condition/expression is true (i.e., until what I call the qualifying event is true).

Note also that you can mix edge-sensitive and level-sensitive expressions in a single logic expression. That is indeed impressive and useful.

6.11.3 SVA Application: Consecutive Delay Range Operator

Property in Fig. 6.36 says that at $rose(tagError), check for tErrorBit to remain asserted until mCheck is asserted. If tErrorBit does not remain asserted until mCheck gets asserted, the property should fail.

So, at $rose(tagError) and one clock later, we check to see that $rose(tErrorBit) occurs. If it does, then we move forward at the same time (##0) with tErrorBit[*1:$]. This says that we check to see that tErrorBit remains asserted consecutively (i.e., at every posedge clk) until the *qualifying event* $rose(mCheck) arrives. In other words,

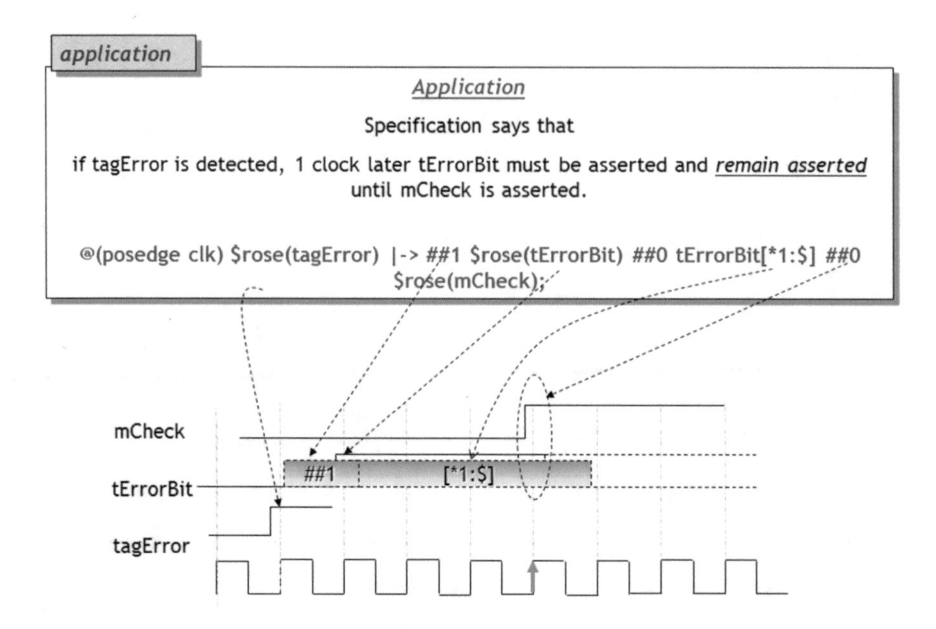

Fig. 6.36 SVA Application—consecutive delay range operator

the qualifying event is what makes consecutive range operator very meaningful as well as useful. Think of the qualifying event as the one that ends the property. This way, you can check for some expression to be true until the qualifying event occurs.

6.11.4 SVA Application: Antecedent as Property Check. Consequent as Hard Failure

Property in Fig. 6.37 states that if the currentState of the state machine is not IDLE and if the currentState remains stable for 32 clocks, the property should fail.

There are a couple of points to observe.

Note that the *entire* expression *((currentState != IDLE) && $stable(currentState))* is checked for consecutive repetition of 32 times because we need to check at every clock for 32 clocks that the currentState is not IDLE, and whatever state that existed in previous clock still remains the same (i.e., $stable). In other words, you must make sure that within these 32 clocks, the currentState does not go back to IDLE. If it does, then the antecedent does not match, and it will start all over again to check for this condition to be true (i.e., the antecedent to be true).

Note that if the antecedent is indeed true, it would mean that the state machine is stuck into the same state for 32 clocks. In such a case, we want the assertion to fire. That is taken care of by a hard failure in the consequent. We simply program consequent to fail without any prerequisite.

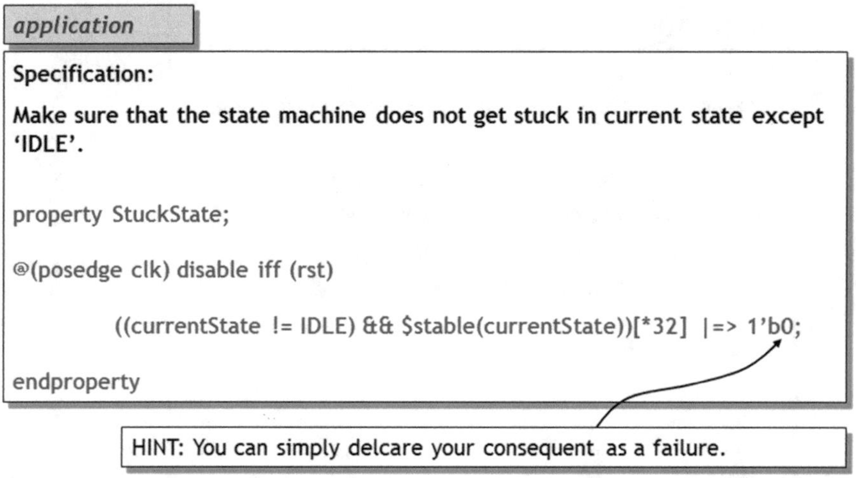

application

Specification:

Make sure that the state machine does not get stuck in current state except 'IDLE'.

```
property StuckState;
@(posedge clk) disable iff (rst)
        ((currentState != IDLE) && $stable(currentState))[*32] |=> 1'b0;
endproperty
```

HINT: You can simply delcare your consequent as a failure.

Fig. 6.37 Concurrent assertions—application

As you notice, this property is unique in that the condition is checked for in the antecedent. The consequent is simply used to declare a failure.

6.11.5 SVA Application: State Transition Check of a State Machine

This application (Fig. 6.38) states that the state machine matches the state transition specification. If we are in `readStart state after one clock, the state machine should be in `readID state and stay in that state until the state machine reaches `readData state. It then is expected to stay in `readData state until `readEnd arrives. In short, we have made sure that the state machine does not stray and do an illegal transition until it reaches `readEnd.

Note the use of `define to establish temporal relationship between signals and states. This makes the code very readable.

Here's the source code showing a simple testbench, the *checkReadStates* assertion, and simulation log:

```
module state_transition;
int readStartState, readIDState, readDataState, readEndState;
logic clk, read_enb;

`define readStart (read_enb ##1 readStartState)
`define readID (readStartState ##1 readIDState)
```

```
application

Specification:
Make sure that the state machine follows the specified transitions

`define readStart (read_enb ##1 readStartState)
`define readID (readStartState ##1 readIDState)
`define readData (readIDState ##1 readDataState)
`define readEnd (readDataState ##1 readEndState)

sequence checkReadStates;

  @(posedge clk)
  `readStart          ##1
  `readID     [*1:$]  ##1
  `readData   [*1:$]  ##1
  `readEnd    ;

endsequence
```

Fig. 6.38 Concurrent assertions—application

```
`define readData (readIDState ##1 readDataState)
`define readEnd (readDataState ##1 readEndState)

property checkReadStates;
  @(posedge clk)
    `readStart     ##1
    `readID    [*1:$]     ##1
    `readData[*1:$]     ##1
    `readEnd    ;
endproperty

sCheck: assert property (checkReadStates) else $display
($stime,,,"FAIL");
cCheck: cover property (checkReadStates) $display ($stime,,,"PASS");

initial
begin
    read_enb=1; clk=0;
    @(posedge clk) readStartState=1;
    @(posedge clk) @(posedge clk); readIDState=1;
    @(posedge clk) @(posedge clk); readDataState=1;
    @(posedge clk) @(posedge clk); readEndState=1;

end
```

```
initial $monitor($stime,,,"clk=",clk,
        "read_enb=%0b",read_enb,,,
        "readStartState=%0b",readStartState,,
        "readIDState=%0b",readIDState,,
        "readDataState=%0b",readDataState,,
        "readEndState=%0b",readEndState);

always #10 clk=!clk;


endmodule


/*
#          0    clk=0  read_enb=1readStartState=0  readIDState=0
readDataState=0 readEndState=0
#          10   clk=1 read_enb=1   readStartState=1 readIDState=0
readDataState=0 readEndState=0
#          20   clk=0 read_enb=1   readStartState=1 readIDState=0
readDataState=0 readEndState=0
#          30   clk=1 read_enb=1   readStartState=1 readIDState=0
readDataState=0 readEndState=0
#          40   clk=0 read_enb=1   readStartState=1 readIDState=0
readDataState=0 readEndState=0
#          50   clk=1 read_enb=1   readStartState=1 readIDState=1
readDataState=0 readEndState=0
#          60   clk=0 read_enb=1   readStartState=1 readIDState=1
readDataState=0 readEndState=0
#          70   clk=1 read_enb=1   readStartState=1 readIDState=1
readDataState=0 readEndState=0
#          80   clk=0 read_enb=1   readStartState=1 readIDState=1
readDataState=0 readEndState=0
#          90   clk=1 read_enb=1   readStartState=1 readIDState=1
readDataState=1 readEndState=0
#          100  clk=0 read_enb=1   readStartState=1 readIDState=1
readDataState=1 readEndState=0
#          110  clk=1 read_enb=1   readStartState=1 readIDState=1
readDataState=1 readEndState=0
#          120  clk=0 read_enb=1   readStartState=1 readIDState=1
readDataState=1 readEndState=0
#          130  clk=1 read_enb=1   readStartState=1 readIDState=1
readDataState=1 readEndState=1
#          140  clk=0 read_enb=1   readStartState=1 readIDState=1
readDataState=1 readEndState=1
# 150PASS
```

```
#           150  clk=1 read_enb=1   readStartState=1 readIDState=1
readDataState=1 readEndState=1
#           160  clk=0 read_enb=1   readStartState=1 readIDState=1
readDataState=1 readEndState=1
#           170  PASS
#           170  clk=1 read_enb=1   readStartState=1 readIDState=1
readDataState=1 readEndState=1
#           180  clk=0 read_enb=1   readStartState=1 readIDState=1
readDataState=1 readEndState=1
#           190  PASS
#           190  clk=1 read_enb=1   readStartState=1 readIDState=1
readDataState=1 readEndState=1
*/
```

6.11.6 SVA Application: Multi-threaded Operation

This is a very interesting behavior of multi-threaded assertions (Fig. 6.39). This is something you need to understand.

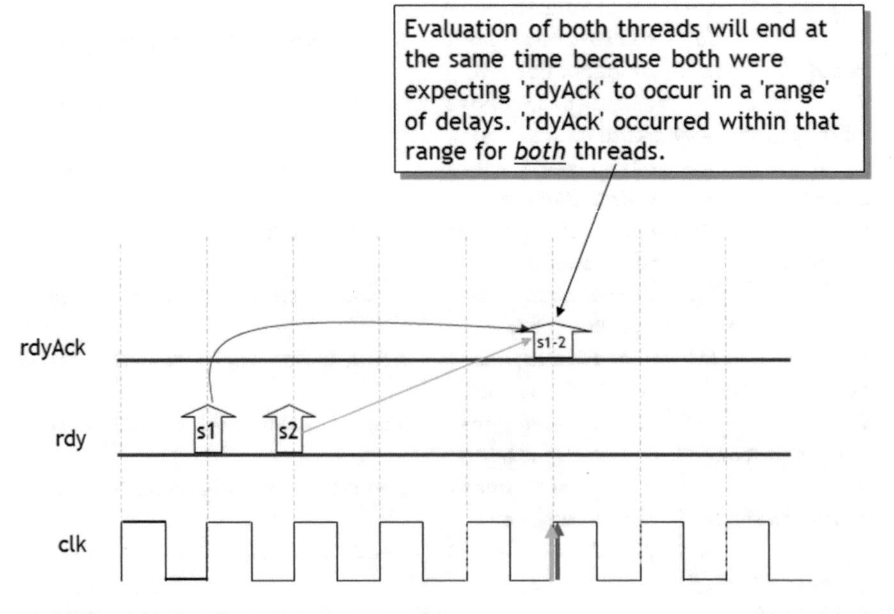

Fig. 6.39 Multi-threading—concurrent assertions

At s1, “rdy” is high and the antecedent is true. That implies that “rdyAck” be true within the next 5 clks. s1 thread starts looking for “rdyAck” to be true. The very next clock, rdyAck, is not yet true, but luck has it that “rdy” is indeed true at this next clk (s2). This will fork off another thread that will also wait for “rdyAck” to be true within the next 5 clks. The “dyAck” comes along within 5 clks from s1, and that thread is satisfied and will pass.

But the second thread will also pass at the same time, because it also got its rdyAck within the 5 clks that it was waiting for.

This is a *–very-* important point to understand. *The range operator can cause multiple threads to complete at the same time, giving you false positive.* This contrasts with what we saw earlier with ##m constant delay where each thread will always complete only after the fixed ##m clock delays. There is a separate end to each separate thread. With the range delay operator, multiple threads can end at the same time.

Important Let us further explore this concept since *it can indeed lead to false positive*. How would you know if rdyAck that satisfied both “rdy”s is for which “rdy?” Also, if you did not receive “rdyAck” for the second “rdy,” you will indeed get a false positive.

One hint is to keep the antecedent an edge-sensitive function. For example, in the above example, instead of “@ (posedge clk) rdy” we could have used “@ (posedge clk) $rose(rdy)” which would have triggered the antecedent only once, and there won’t be any confusion of multiple threads ending at the same time. This is a performance hint as well. Use edge-sensitive-sampled value functions whenever possible. Level-sensitive antecedent can fork off unintended multiple threads affecting simulation performance.

But a better solution is to use local variables to ID each “rdy” and “rdyAck.” This will indeed make sure that you received a “rdyAck” for *each* “rdy” and that each “rdyAck” is associated with the correct “rdy.”

You don’t need to understand local variables (which are scantly covered in this book) in detail to understand the following example. Local variables are *dynamic* variables. Each instance of a sequence forks off another independent thread. Very powerful feature. You don’t need to keep track of the pipelined behavior. The local variable does it for you. Having understood that, you should be able to follow the following example.

Problem statement:

To recap our problem definition, two “rdy” signals are asserted on consecutive clocks with a range of clocks during which a “rdyack” must arrive. “rdyack” arrives that satisfies the time range requirements for *both* “rdy”s and the property passes. We have no idea whether a “rdyack” arrived for each “rdy.” The PASS of the assertion does not guarantee that.

First, let us simulate the problem definition. The following code simulates the property and shows that both instances of "rdy" will indeed PASS with a single "rdyAck":

```
module range_problem;
logic clk, rdy, rdyAck;
initial
begin
  clk=1'b0; rdy=0; rdyAck=0;
  #500 $finish(2);
end
always begin
  #10 clk=!clk;
end

initial
begin
        repeat (5) begin @(posedge clk) rdy=~rdy; end
end

initial $monitor($stime,,,"clk=",clk,,,"rdy=",rdy,,,"rdyAck=",rdy
Ack);
initial
begin
        repeat (4) begin @(posedge clk); end
        rdyAck=1;
end

sequence rdyAckCheck;
   (1'b1, $display($stime,,,"ENTER SEQUENCE rdy ARRIVES")) ##[1:5]
($rose(rdyAck),$display($stime,,,"rdyAck ARRIVES"));
endsequence

gcheck: assert property (@(posedge clk) $rose (rdy) |-> rdyAck-
Check) begin $display($stime,,,"PASS"); end
    else begin $display($stime,,,"FAIL"); end

endmodule


/* Simulation log
0  clk=0  rdy=0  rdyAck=0
#          10  clk=1  rdy=1  rdyAck=0
#          20  clk=0  rdy=1  rdyAck=0
#          30  ENTER SEQUENCE rdy ARRIVES Á First 'rdy' is detected
```

```
#         30  clk=1  rdy=0  rdyAck=0
#         40  clk=0  rdy=0  rdyAck=0
#         50  clk=1  rdy=1  rdyAck=0
#         60  clk=0  rdy=1  rdyAck=0
#         70  ENTER SEQUENCE rdy ARRIVES Á Second 'rdy' is detected
#         70  clk=1  rdy=0  rdyAck=1
#         80  clk=0  rdy=0  rdyAck=1
#         90  clk=1  rdy=1  rdyAck=1
#         90  rdyAck ARRIVES Á 'rdyack' detected for both 'rdy's
and the property PASSes.
#         90  PASS
*/
```

The following code shows the solution using local variables:

```
module range_solution;
logic clk, rdy, rdyAck;
byte rdyNum, rdyAckNum;

initial
begin
  clk=1'b0; rdy=0; rdyNum=0; rdyAck=0; rdyAckNum=0;
  #500 $finish(2);
end
always
begin
  #10 clk=!clk;
end
initial
begin
            repeat (4)
            begin
                @(posedge clk) rdy=0;
                @(posedge clk) rdy=1; rdyNum=rdyNum+1;
            end
end

initial $monitor($stime,,,"clk=",clk,,,"rdy=",rdy,,,"rdyNum=",rdy
Num,,,"rdyAckNum",rdyAckNum,,,"rdyAck=",rdyAck);
always
begin
        repeat (4)
          begin
        @(posedge clk); @(posedge clk); @(posedge clk);
            rdyAck=1; rdyAckNum=rdyAckNum+1;
```

```
            @(posedge clk) rdyAck=0;
          end
end

sequence rdyAckCheck;
byte localData; /*local variable 'localData' declaration. Note
this is a dynamic variable. For every entry into the sequence it
will create a new instance of localData and follow an independent
thread.*/

              (1'b1,localData=rdyNum,    $display($stime,,,"ENTER
SEQUENCE",,,"LOCAL rdyNum=", localData))


##[1:5]


((rdyAck && rdyAckNum==localData),
$display($stime,,,"rdyAck ARRIVES ",,,"LOCAL",,,"rdyNum=",localD
ata,,, "rdyAck=",rdyAckNum));
endsequence

gcheck: assert property (@(posedge clk) $rose (rdy) |-> rdyAck-
Check) begin $display($stime,,,"PASS"); end
else begin $display($stime,,,"FAIL",,,"rdyNum=",rdyNum,,,"rdyAckN
um=",rdyAckNum); end

endmodule


/* Simulation Log
#       0  clk=0  rdy=0  rdyNum=   0  rdyAckNum   0  rdyAck=0
#      10  clk=1  rdy=0  rdyNum=   0  rdyAckNum   0  rdyAck=0
#      20  clk=0  rdy=0  rdyNum=   0  rdyAckNum   0  rdyAck=0
#      30  clk=1  rdy=1  rdyNum=   1  rdyAckNum   0  rdyAck=0
#      40  clk=0  rdy=1  rdyNum=   1  rdyAckNum   0  rdyAck=0
#      50  ENTER SEQUENCELOCAL rdyNum= 1 Á First 'rdy' arrives. A
'rdyNum' (generated in your testbench as shown above) is assigned
to 'localData'. This 'rdyNum' is a unique number for each invoca-
tion of the sequence and arrival of 'rdy'.

#      50  clk=1  rdy=0  rdyNum=  1  rdyAckNum   1  rdyAck=1
#      60  clk=0  rdy=0  rdyNum=  1  rdyAckNum   1  rdyAck=1
#      70  rdyAck ARRIVES LOCALrdyNum= 1rdyAck= 1
```

```
#       70  PASS Á When 'rdyAck' arrives, the sequence checks to
see that its 'rdyAckNum' (again, assigned in the testbench) cor-
responds to the first rdyAck. If the numbers do not match the prop-
erty fails. Here they are indeed the same and the property PASSes.

#       70  clk=1  rdy=1  rdyNum=   2  rdyAckNum   1  rdyAck=0
#       80  clk=0  rdy=1  rdyNum=   2  rdyAckNum   1  rdyAck=0
#       90  ENTER SEQUENCELOCAL rdyNum= 2 Á Second 'rdy' arrives.
localData is assigned the second 'rdyNum'. This redNum will not
overwrite the first rdyNum. Instead a second thread is forked off
and 'localData' will maintain (store) the second 'rdyNum'.

#       90  clk=1  rdy=0  rdyNum=   2  rdyAckNum   1  rdyAck=0
#      100  clk=0  rdy=0  rdyNum=   2  rdyAckNum   1  rdyAck=0
#      110  clk=1  rdy=1  rdyNum=   3  rdyAckNum   1  rdyAck=0
#      120  clk=0  rdy=1  rdyNum=   3  rdyAckNum   1  rdyAck=0
#      130  ENTER SEQUENCELOCAL rdyNum= 3
#      130  clk=1  rdy=0  rdyNum=   3  rdyAckNum   2  rdyAck=1
#      140  clk=0  rdy=0  rdyNum=   3  rdyAckNum   2  rdyAck=1
#      150  rdyAck ARRIVES LOCALrdyNum= 2rdyAck= 2
#      150  PASS Á When 'rdyAck' arrives, the sequence checks to
see that its 'rdyAckNum' (again, assigned in the testbench) cor-
responds to the second rdyAck. If the numbers do not match the
property fails. Here they are indeed the same and the property
PASSes. This is what we mean by pipelined behavior, in that, the
second invocation of the sequence maintains its own copy of 'local-
Data' and compares with the second 'rdyAck'. This way there is no
question of which 'rdy' was followed by which 'rdyAck'. No false
positive. Rest of the simulation log follows the same chain of
thought.

Can you figure out why the property fails at #270? Hint: Start
counting clocks at time #170 when fourth 'rdy' arrives. Did a 'rdy-
Ack' arrive for that 'rdy'?


#      150  clk=1  rdy=1  rdyNum=   4  rdyAckNum   2  rdyAck=0
#      160  clk=0  rdy=1  rdyNum=   4  rdyAckNum   2  rdyAck=0
#      170  ENTER SEQUENCELOCAL rdyNum= 4
#      170  clk=1  rdy=1  rdyNum=   4  rdyAckNum   2  rdyAck=0
#      180  clk=0  rdy=1  rdyNum=   4  rdyAckNum   2  rdyAck=0
#      190  clk=1  rdy=1  rdyNum=   4  rdyAckNum   2  rdyAck=0
#      200  clk=0  rdy=1  rdyNum=   4  rdyAckNum   2  rdyAck=0
#      210  clk=1  rdy=1  rdyNum=   4  rdyAckNum   3  rdyAck=1
#      220  clk=0  rdy=1  rdyNum=   4  rdyAckNum   3  rdyAck=1
```

```
#       230   rdyAck ARRIVES LOCALrdyNum= 3rdyAck= 3
#       230   PASS

#       230   clk=1   rdy=1   rdyNum=    4  rdyAckNum 3rdyAck=0
#       240   clk=0   rdy=1   rdyNum=    4  rdyAckNum 3rdyAck=0
#       250   clk=1   rdy=1   rdyNum=    4  rdyAckNum 3rdyAck=0
#       260   clk=0   rdy=1   rdyNum=    4  rdyAckNum 3rdyAck=0
#       270   FAILrdyNum= 4rdyAckNum= 3

#       270   clk=1   rdy=1   rdyNum=    4  rdyAckNum 3rdyAck=0
#       280   clk=0   rdy=1   rdyNum=    4  rdyAckNum 3rdyAck=0
#       290   clk=1   rdy=1   rdyNum=    4  rdyAckNum 4rdyAck=1
*/
```

Sequence “rdyAckCheck,” in the above code, is explained as follows.

Upon entry in the sequence, a copy of localData is created and a rdyNum is stored into it. While the sequence is waiting for #[1:5] for the rdyAck to arrive, another “rdy” comes in, and sequence “rdyAckCheck” is invoked. Again, the localData is assigned the next rdyNum and stored. This is where dynamic variable concept comes into picture. The second store of rdyNum into localData does not clobber the first store. A second copy of the localData is created, and its thread will also now wait for #[1:5]. This way we make sure that for each “rdy” we will indeed get a unique “rdyAck.” Please carefully examine the simulation log to see how this works. I’ve placed comments in the simulation log to explain the operation.

Detail of local variables is beyond the scope of this book. Please refer to (Mehta 2016) book to fully understand the semantic of local variables.

6.11.7 SVA Application: A Request ⇔ Grant Bus Protocol

Refer to Fig. 6.40

6.11.8 SVA Application: Machine Check Exception

Property in Fig. 6.41 says that at $rose(tagError), check for tErrorBit to remain asserted until mCheck is asserted. If tErrorBit does not remain asserted until mCheck gets asserted, the property should fail.

So, at $rose(tagError) and one clock later, we check to see that $rose(tErrorBit) occurs. If it does, then we move forward at the same time (##0) with tErrorBit[*1:$]. This says that we check to see that tErrorBit remains asserted consecutively (i.e., at every posedge clk) until the *qualifying event* $rose(mCheck) arrives. In other words, the qualifying event is what makes consecutive range operator very meaningful as

<u>Specification</u>:

- When request is asserted that grant is asserted the very next clock.
    - *grant must have been de-asserted prior to it's assertion.*
- grant must remain asserted as long as request is asserted.
- grant must de-assert the very next clock after request is de-asserted.

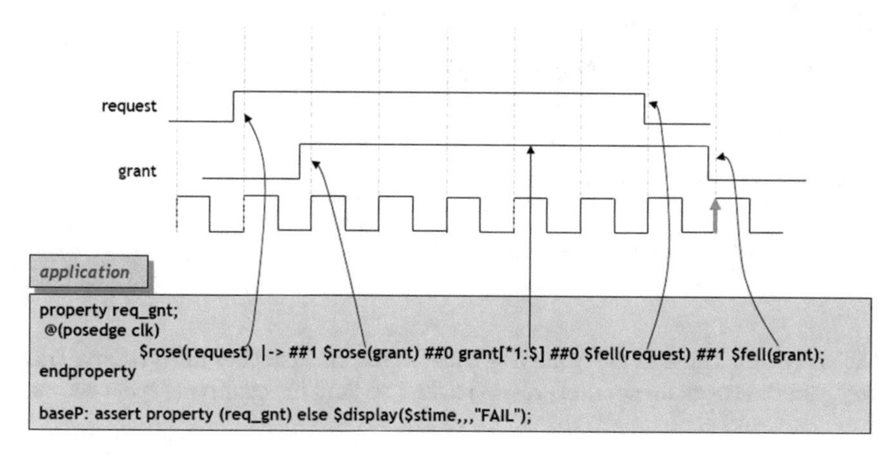

Fig. 6.40 Concurrent assertions—application

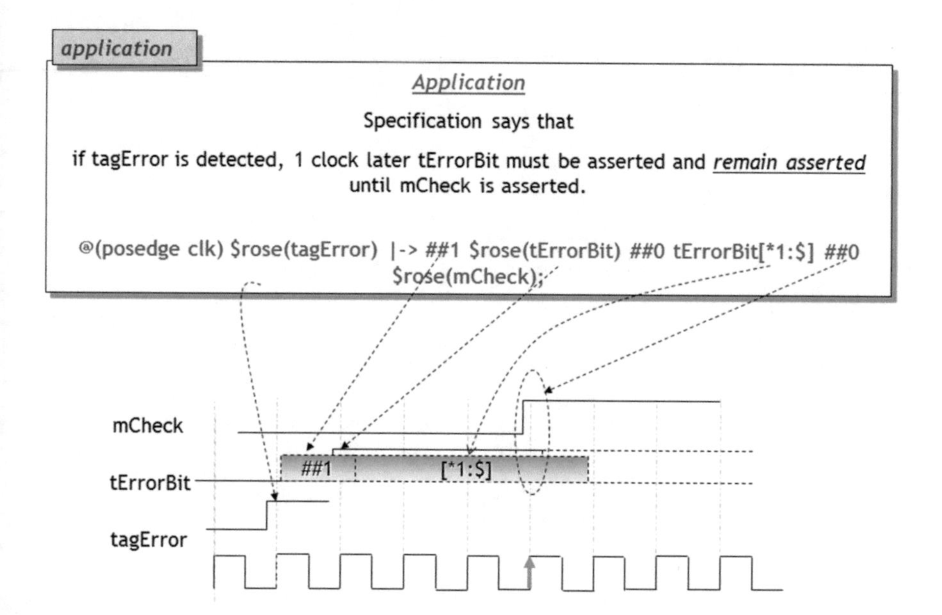

Fig. 6.41 SVA Application: machine check exception

application

Specification:

- For every 'req' you must get *at least* 1 'ack' and 'ack' must clear the next clock.

```
property ReqAckCheck;
        @(posedge clk) $rose(req) |=> ack[->1] ##1 !ack;
endproperty
aP: assert property (reqAckCheck);
```

Fig. 6.42 SVA Application: "req" followed by "ack"

well as useful. Think of the qualifying event as the one that ends the property. This way, you can check for some expression to be true until the qualifying event occurs.

6.11.9 SVA Application: "req" followed by "ack"

Refer to Fig. 6.42

Chapter 7
SystemVerilog Functional Coverage (SFC)

Chapter Introduction
SystemVerilog functional coverage (SFC) is another important component that falls within SystemVerilog. In this chapter, we will discuss the difference between code and functional coverage and SFC fundamentals such as "covergroup," "coverpoint," "cross," "transition," etc. along with complete examples.

7.1 Difference Between Code Coverage and Functional Coverage

Ah, so you have done everything to check the design. But what have you done to check your testbench? How do you know that your testbench has indeed covered everything that needs to be covered? That's where functional coverage comes into picture. But first let us make sure we understand the difference between the good old code coverage and the new functional coverage methodology.

- Code Coverage
  - Measures coverage of design *structure* (branch, expression, state transition, etc.)
    - Tool specific and derived automatically (*not* user specified)
  - But does not cover the *intent* of the design
    - RTL Code with bug :: OUT = CTRL ? R1 : R2;
    - RTL Code without bug :: OUT = CTRL ? R2 : R1;
      - Code coverage won't catch this functional bug
- Functional Coverage

A.B. Mehta, *ASIC/SoC Functional Design Verification*,
DOI 10.1007/978-3-319-59418-7_7

- Functional coverage is based on the design's functional specification (user specified).
- It measures coverage of the design *intent*. For example:
- Control-oriented coverage
  - Have I exercised all possible protocols that read cycle supports (burst, non-burst, etc.)?
  - *Transition* coverage
    - Did we issue transactions that access Byte followed by Qword followed by multiple Qwords (use SystemVerilog *transition* coverage)?
    - A Write to L2 is followed by a read from the same address (and vice versa). Again, the *transition* coverage will help you determine if you have exercised this condition.
  - *Cross* coverage
    - Tag and Data Errors must be injected at the same time (use SystemVerilog *cross* coverage).
- Data-oriented coverage
  - Have we accessed cache lines at all granularity levels (odd bytes, even bytes, word, quadword, full cache line, etc.)?

7.2 SystemVerilog Components for Complete Coverage

First let us examine the components of SystemVerilog language that contribute to functional coverage (Fig. 7.1).

The first component of the complete coverage picture is the "cover" statement associated with an assertion. This "cover" statement allows us to measure sequential domain functional coverage. Recall that "assert" checks for failures in your design, and "cover" sees if the property did get exercised (i.e., got covered). Pure combinatorial coverage is not sufficient. What I call "low-level" sequential conditions such as every req should be followed by a gnt-type sequential assertion. If this assertion does not fail, it could be because the logic is correct or *because you never really asserted "req" to start with.* The "cover" completes this story. We "cover" exactly the same property that we "assert." In the req/gnt example, if "cover" passes we know that the property did get exercised by the testbench, and it did not fail (if "assert" did not fail).

The second component is the functional coverage language which is the gist of this section. Functional coverage allows you to specify the "function" you want to cover via the so-called coverpoints and covergroups. More importantly, it also allows you to measure *transition* as well as *cross* coverage to see that we have indeed covered finer details of our design. This section will clarify all this.

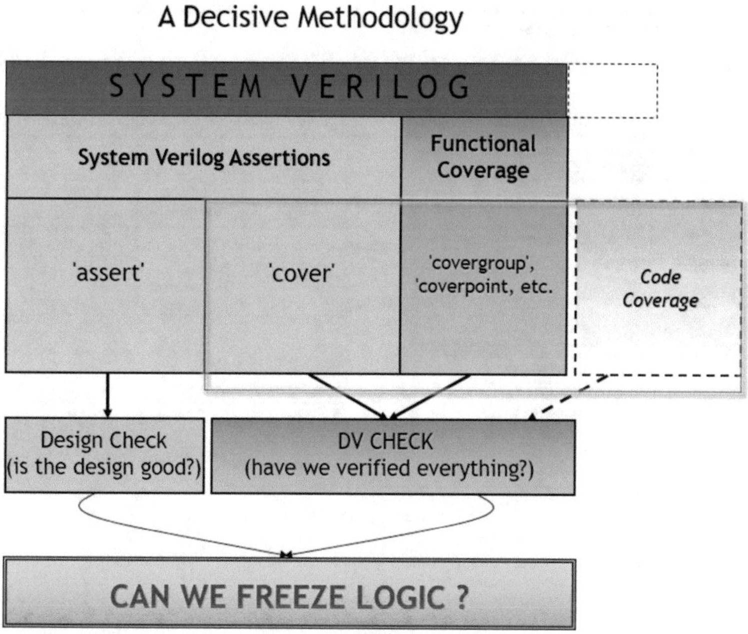

Fig. 7.1 ASIC design functional coverage methodology

Note that code coverage is still important. It will cover structural coverage of the design, for example, states of a state machine, conditional coverage, branch coverage, line coverage, etc. We are all familiar with code coverage; hence I'll leave it at that.

7.3 Assertion (ABV) and Functional Coverage (SFC)-Based Methodology

Here are high-level points for an automated robust project methodology:

- Your test plan is (obviously) based on what functions you want to test (i.e., cover).
- So, create a Functional Cover Matrix based on your test plan that includes each of the functions (control and data) that you want to test.
  - Identify in this matrix all your functional covergroups/coverpoints (more on that coming soon)
  - Measure their coverage during verification/simulation process

You may even automate updating the matrix directly from the coverage reports. That methodology is depicted in Fig. 7.2. Measure the effectiveness of your tests

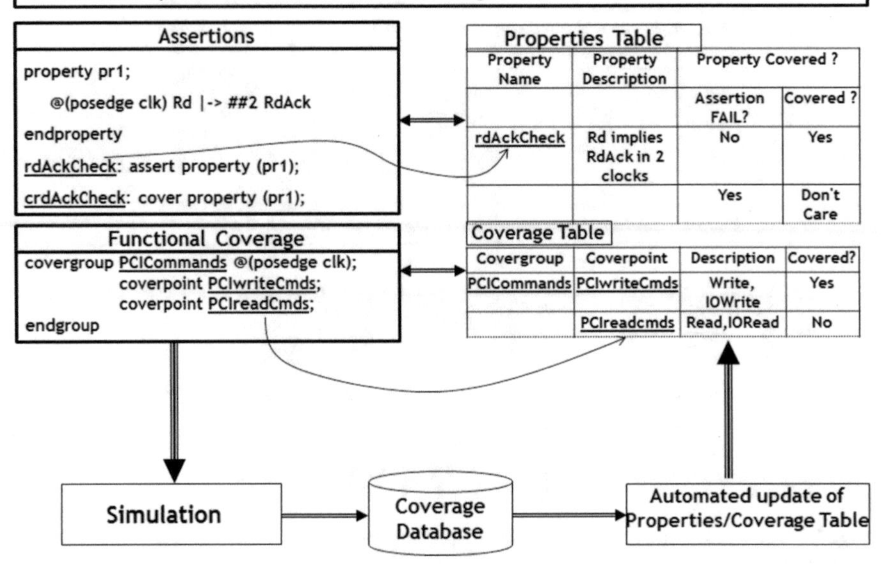

Fig. 7.2 Automated coverage-driven design verification methodology

from the coverage reports. The following points are indeed the gist of what functional coverage allows you to accomplish.

- For example, if your tests are accessing mostly 32-byte granules in your cache line, you will see byte, word, quadword coverage low, or not covered. Change or add new tests to hit bytes/words, etc. Use constrained random methodology to narrow down the target of your tests. Constrained random is a very powerful methodology and goes hand in hand with functional coverage. Complete description of constrained random is beyond the scope of this book.
- Or that the tests do not fire transactions that access Byte followed by Qword followed by multiple Qwords. Check this with *transition* coverage.
- Or that Tag and Data Errors must be injected together at some point in time (*cross* coverage between Tag and Data Errors)
- "Cover" sequential domain assertions.
- And add more *coverpoints* for critical functional paths through design.
  - For example, a Write to L2 is followed by a read from the same address and that this happens from both processors in all possible write/read combinations.
- Remember to update your functional cover plan as verification progresses.

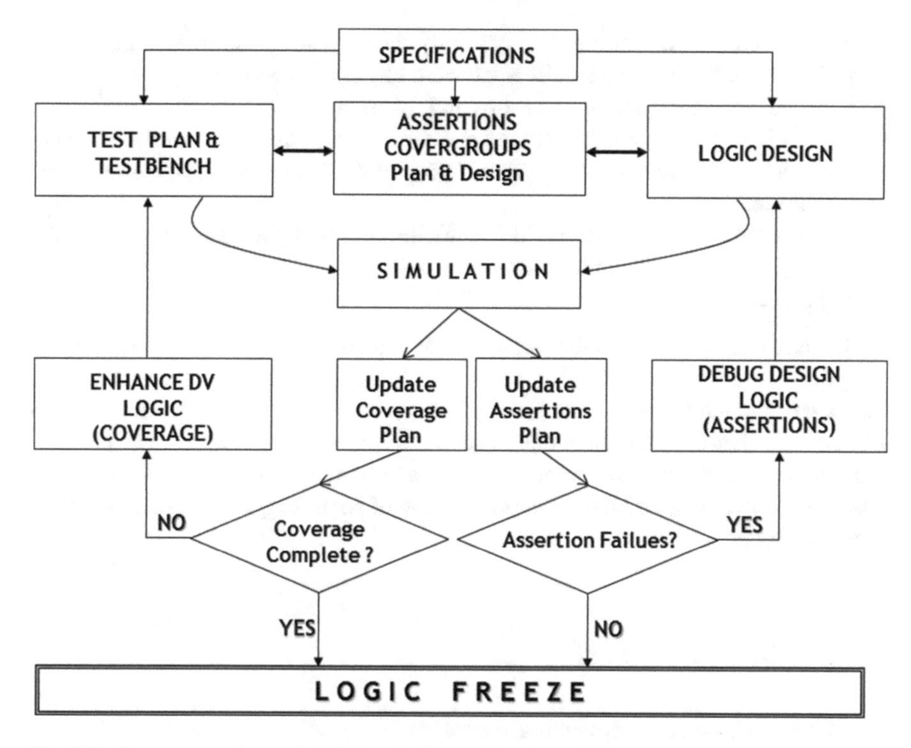

Fig. 7.3 Coverage- and assertion-driven design verification methodology

- Just because you created a plan in the beginning of the project, it does not mean it's an end in all.
- As your knowledge of the design and its corner cases increase, so should the exhaustiveness of your test plan and the functional cover plan.
- Continue to add *coverpoints* for any function that you didn't think of at the onset.

Figures 7.2 and 7.3 show an assertion- and coverage-driven methodology.

1. For every "assert" in a property, have an associated "cover." Give meaningful names to the property and assert labels.
2. Create a properties table which automatically reads in your assertion description and creates a *fail*/covered matrix. If the assertion *fails*, well, fill in the *fail* column. If it not and if it gets covered, fill in the covered column. How do we fill in this matrix? Read on….
3. Create a functional coverage plan with *covergroup* and *coverpoint*. Again, give meaningful names to covergroup and coverpoint(s).
4. Create a coverage table that automatically derives the covergroup/coverpoint names from step 3 and creates a matrix for "covered" results. This matrix is for those functions that are not covered by assertion "cover" nor are they covered by

code coverage. So, you need to carefully design your covergroups and coverpoints. Do not cover what's already covered by code coverage.
5. Simulate your design with assertions and functional cover groups.
6. Simulation will create a "coverage database." This database has all the information about failed assertions, "covered" properties, and covered covergroups and coverpoints.
7. Using EDA vendor provided API, shift through this database and update the properties table and coverage table.
8. Loop back.

The advantage of such methodology is that you continually know if you are spinning the wheel without increasing coverage. Without such continual measure, you may keep simulating; bugs don't get reported; you start feeling comfortable only to realize later that the functional coverage was inadequate. You were basically running the tests that target the same logic repeatedly. If you have a methodology as described above, you will have a correct notion of what functional logic to target to increase bug rate.

7.3.1 Follow the Bugs!

- So, when do you *start* collecting coverage?
  - Code and functional coverage add to simulation overhead.
  - So, don't turn on code/functional coverage at the very "beginning" of the project.
  - But what does "beginning" of the project mean? When does the "beginning" end?
- That's where the bugs come into picture!
  - Create bug report charts.
  - During the "beginning" time, bug rate will (should) be high. All low-hanging fruits are being picked. ☺
  - When the bug rate starts to drop, the "beginning" has come to an "end."
  - That's when your existing test strategy is running out of steam. ☺
  - That's when you start code and functional coverage to determine:
    - If new tests are simply exercising the same logic repeatedly
    - And which part of logic is not yet covered
  - Develop tests for the uncovered functionality. Use constrained random methodology.
  - Your bug rate will again go up (guaranteed! ☺).

7.4 SystemVerilog "Covergroup" Basics

Here is a basic example to give you a flavor of the language syntax/semantics. Let's see what's a covergroup.

- "Covergroup" is a user-defined type that allows you to collectively sample all those variables/transitions/cross that are sampled at the same clock (sampling) edge.
- "The 'covergroup' construct encapsulates the specification of a coverage model."
- A "covergroup" can be defined in a "package," "a module," a "program," an "interface," or a "class."

Figure 7.4 is self-explanatory with its annotations. Key syntax of the covergroup and coverpoint is pointed out. A few points to reiterate are as follows:

1. *Covergroup* without a coverpoint is useless, *and* the compiler won't give an error (at least the simulators that the author has tried).
2. *Covergroup*, as the name suggests, is a group of coverpoints, meaning you can have multiple coverpoints in a covergroup.
3. You must instantiate the covergroup.
4. You may provide (not mandatory) a sampling edge to determine when the *coverpoints* in a *covergroup* get sampled. If the clocking event is omitted, you must procedurally trigger the coverage sample window using a built-in method called *sample*().
5. A "*covergroup*" can be declared in:

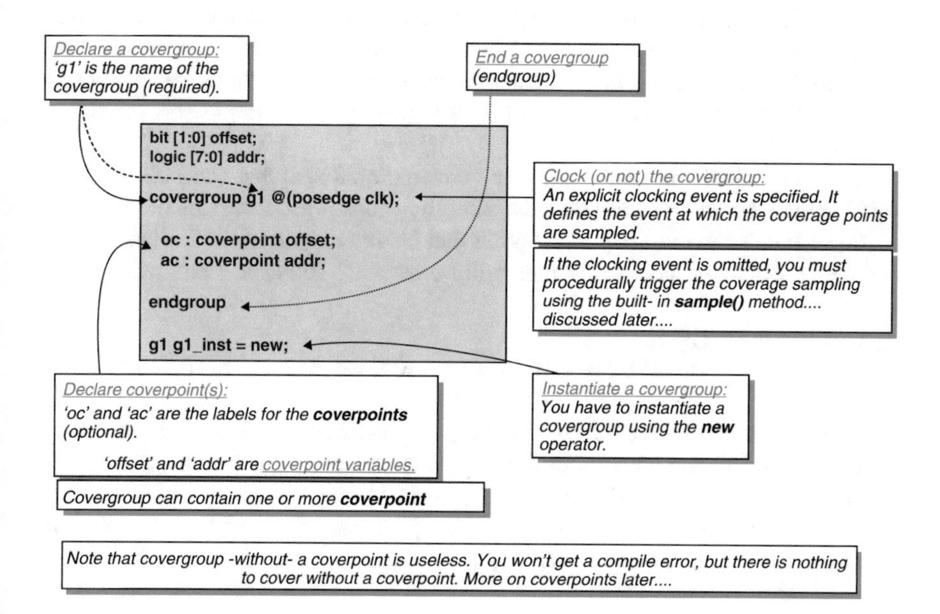

Fig. 7.4 Covergroup basics

a. Package
b. Interface
c. Module
d. Program
e. Class (We'll see an example soon.)

Other points are annotated in Fig. 7.4. Carefully study them so that the rest of the chapter is easier to digest.

7.5 SystemVerilog "Coverpoint" Basics

- A coverpoint is a variable or an expression that functionally covers design parameters (reg, logic, enum, etc.).
- Each coverpoint includes a set of bins associated with its sampled value or its value transition.
- The so-called bins can be defined by the user or created automatically by an EDA tool. A bin tells you the actual coverage measure.

7.6 SystemVerilog "Bins": Basics…

What's a "bin"? A "bin" is something that collects coverage information (collect in a "bin"). Bins are created for coverpoints. If a coverpoint is covering a variable (let's say the 8-bit "adr" as shown in Fig. 7.5) and would like to have different values of that variable be collected in different collecting entities, the "bins" will be those entities. "Bins" allow you to organize the coverpoints' sample (or transition) values.

You can declare bins many ways for a coverpoint. Recall that bins collect coverage. From that point of view, you must carefully choose the declaration of your bins.

Okay, here's the most important point that is *very* easy to misunderstand. In the following statement, how many bins will be created? 16, 4, or 1 and what will it cover?

bins adrbin1 = {[0:3]};

Answer: One bin will be created to cover "adr" values equal to "0," "1," "2," or "3."

Note that "bins adrbin1" is without the [] brackets. In other words, "bins adrbin1" will *not* automatically create four bins for "adr" values {[0:3]}; it will create only one bin to cover "adr" values "0," "1," "2," and "3."

Very important point: Do not confuse {[0:3]} to mean that you are asking the bin to collect coverage for adr0 to adr15. {[0:3]} literally means "adr" value =0, =1, =2, and =3.

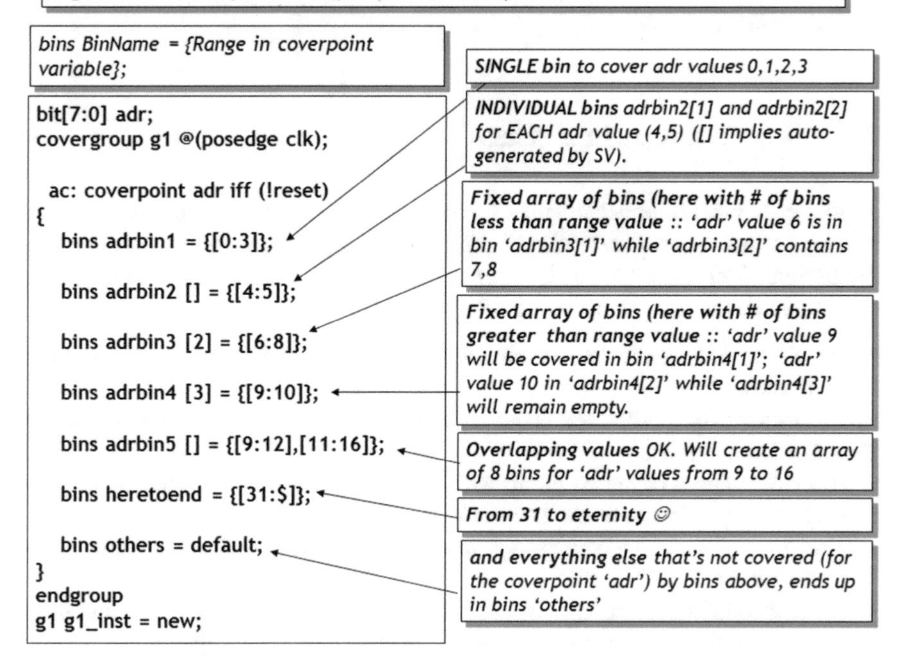

Fig. 7.5 "Bins": Basics

Another important point "bins adrbin1 = {[0:3]};" also is that if we hit *either* of the "adr" value ("0," "1," "2," or "3"), the single bin will be considered *completely* covered. Not very intuitive, I agree. But that's what the language semantics dictate. Again, you don't have to cover all four values to have "bins adrbin1" considered covered. You hit any one of those four values, and the "adrbin1" will be considered 100% covered.

But what if you want each value of the variable "adr" be collected in separate bins so that you can indeed see if each value of "adr" is covered explicitly? That's where "bins adrbin2[] ={[4:5]};" comes into picture. Here "[]" tells the simulator to create two explicit bins called adrbin2[1] and adrbin2[2] covering the two "adr" values =4 and =5, respectively. adrbin2[1] will be considered covered if you exercised adr==4, and adrbin2[2] will be considered covered if adr==5 is exercised.

Other ways of creating bins are described in Fig. 7.5 with annotation to describe the nuances. Note that you can have "less" or "more" number of bins than the "adr" values on the RHS of a bin's assignment. How will "bins" be allocated in such cases is explained in the figure. Note also the case {[31:$]} called "bins heretoend." What does "$" mean in this case? It means [32:255] since "adr" is an 8-bit variable.

The rest of the semantics is well described with annotation in the figure. Do study them carefully, since they will be very helpful when you start designing your strategy to create "bins."

7.7 "Covergroup" in a "Class"

So where do you use or declare this "covergroup"? One of the best places to embed a coverage group is within a "class." Why a class? Here are some reasons. (Note—discussion of "class" is beyond the scope of this book. The author is assuming readers' familiarity with SystemVerilog "class").

- An embedded covergroup defines a coverage model for protected and local properties.
- Class members can be used in coverpoint expressions, coverage constructs, and option initialization.
- By embedding a coverage group within a class definition, the covergroup provides a simple way to cover a subset of the class properties.
- This style of declaring covergroups allows for modular verification environment development.
- An embedded covergroup can define a coverage model for protected and local class properties without any changes to the class data encapsulation.

Okay, let us see what Fig. 7.6 depicts.

```
class xyz;
  bit [3:0] m_x;
  int m_y;
  bit m_z;

  covergroup xyzCover @(m_z);
        coverpoint m_x;
        coverpoint m_y;
  endgroup

  function new();
    xyzCover xyzCovInst = new;
  endfunction
endclass
```

- *By embedding a 'covergroup' within a class definition, the 'covergroup' provides a simple way to cover as part of class definition the class properties (modular development)*
- *A 'class' can have more than one 'covergroup'*

Fig. 7.6 Covergroup in a class

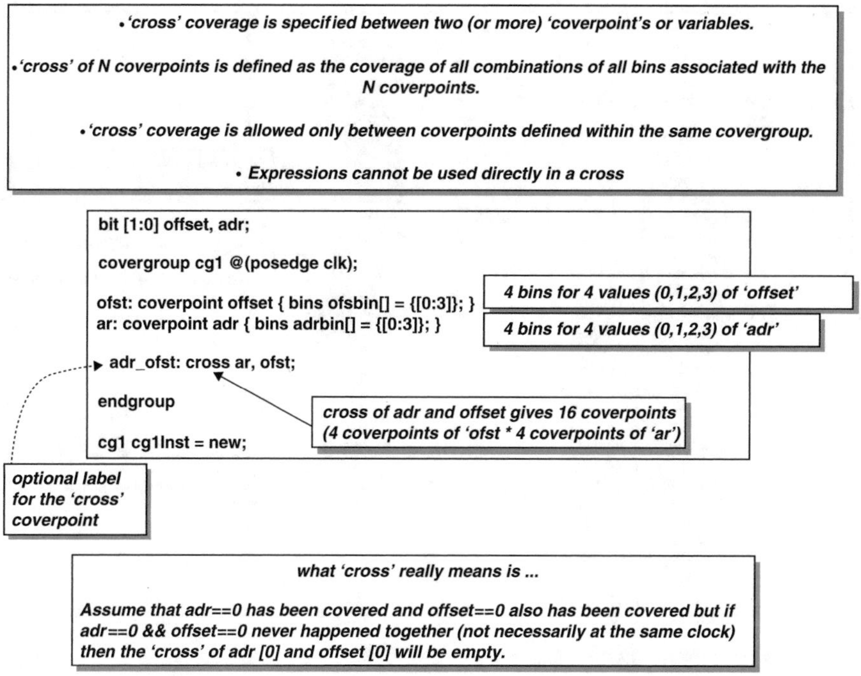

Fig. 7.7 Cross coverage

"Covergroup xyzCover" is sampled on any change on variable "m_z." This covergroup contains two coverpoints, namely, "m_x" and "m_y." Note that there are no explicit bins specified for the coverpoints. How many bins for each coverpoint will be created?

Note that covergroup is instantiated within the "class." That makes sense since the covergroup is embedded within the class. Obviously, if you do not instantiate a covergroup in the "class," it will not be created, and there will not be any sampling of data.

7.8 "Cross" Coverage

"Cross" is a very important feature of functional coverage. This is where code coverage completely fails. Figure 7.7 describes the syntax and semantics.

Syntax:

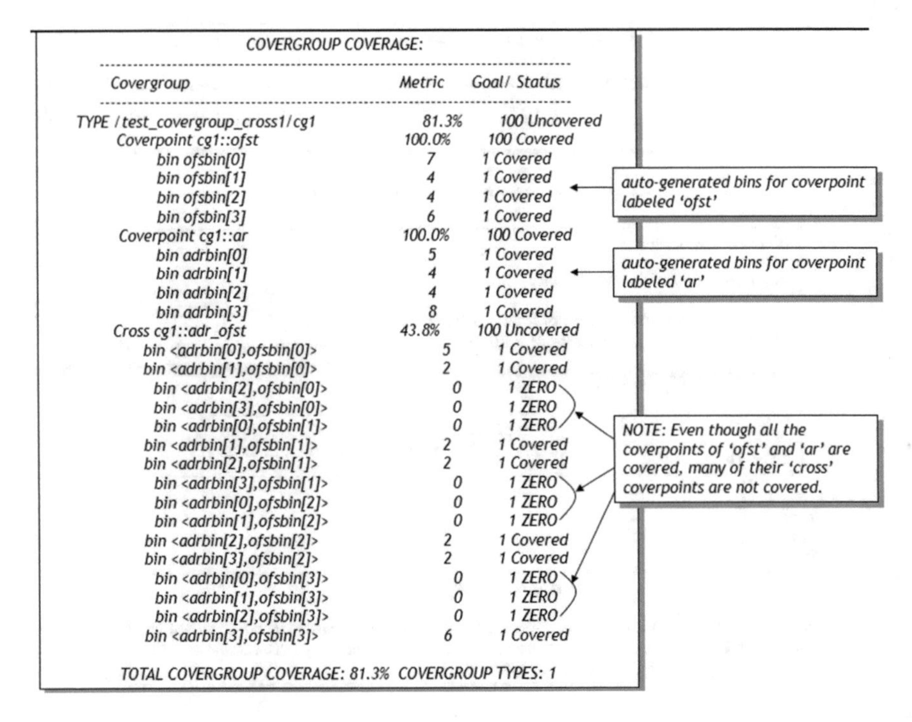

Fig. 7.8 "Cross" coverage: Simulation log

```
cover_cross ::= [ cross_identifier : ] cross list_of_cross_items [
iff ( expression ) ] cross_body
```

Two variables, "offset" and "adr," are declared. Coverpoint for "offset" creates four bins called ofsbin[0] ... ofsbin[3] for the four values of "offset," namely, 0, 1, 2, and 3. Coverpoint "adr" also follows the same logic and creates adrbin[0]... adrbin[3] for the four values of "adr," namely, 0, 1, 2, and 3.

adr_ofst is the label given to the "cross" of ar, ofst. First of all, the "cross" of "ar" (label for coverpoint adr) and "ofst" (label for coverpoint offset) will create another set of 16 bins (four bins of "adr" *, four bins of "offset"). These "cross" bins will keep track of the result of "cross." However, what does "cross" mean?

Four values of "adr" need to be covered (0, 1, 2, 3). Let us assume adr==2 has been covered (i.e., adrbin[2] is covered). Similarly, there are four values of "offset" that need to be covered (0, 1, 2, 3), and that offset==0 has also been covered (i.e., ofsbin[0] has been covered). However, have we covered "cross" of adr=2 (adrbin[2]) and offset=0 (ofsbin[0])? Not necessarily. "Cross" means that adr=2 and offset=0 must be true "together" at some point in time. This does *not* mean that they need to be "covered" at the *same* time. It simply means that, e.g., if adr==2, it should remain at that value until offset==0 (or vice versa). This will make both of them true "together." If that is the case, then the "cross" of adrbin[2] and ofsbin[0] will be considered "covered."

In the simulation log (Fig. 7.8), we see that both adrbin[2] and ofsbin[0] have been individually covered 100%. However, their “cross” has not been covered.

First, you will see the four bins (ofsbin[0] to ofsbin[3]) of coverpoint cg1::ofst. All four bins are covered, and hence coverpoint cg1::ofst is 100% covered. Next, you will see the four bins (adrbin[0] to adrbin[3]) of coverpoint cg1::ar. All bins are covered here as well and so is the coverpoint cg1::ar.

Now let us look at the “cross” of 4bins*4bins=16 bins coverage. Both “ofst” and “ar” are 100% covered, but the three cases that follow (among many others) are not covered because whatever values the testbench drove, these bins never had the same value at any given point in time (e.g., adrbin[2] is “2” at time t, and then ofsbin[0] should be “0” either at time t or any time after that, as long as adrbin[2] = “2”).

Hence,

```
bin <adrbin[2],ofsbin[0]>          0          1 ZERO
```

Similarly, there are other cases of “cross” that are not covered as shown in the simulation log. Such a log will clearly identify the need to enhance your testbench. To reiterate, such “cross” cannot be derived from code coverage.

7.9 “Bins” for Transition Coverage

As noted in Fig. 7.9, this is by far the most useful feature of functional coverage. Transaction level transitions are very important to cover. For example, did the CPU issue a read followed by write-invalid? Did you issue a D$miss followed by a D$hit cycle? Such transitions are where the bugs occur, and we want to make sure that we have indeed exercised such transitions.

Figure 7.9 explains how the semantics work. Note that we are addressing both the “transition” and the “cross” of “transition” coverage.

There are two transitions in the example.

bins ar1 = (8’h00 => 8’hff); which means that adr1 should transition from “0” to “ff” on two consecutive posedge of clk. In other words, the testbench must exercise this condition for it to be covered.

Similarly, there is the “bins ar2” that specifies the transition for adr2 (1 => 0).

The cross of transitions is shown at the bottom of Fig. 7.9. It is very interesting how this works. Take the first values of each transition (viz., adr1=0 && adr2=1). This will be the start points of cross transition at the posedge clk. If at the next (posedge clk) values are adr1=’ff’ && adr2=0, the cross transition is covered.

More on the “bins” of transition is shown in Fig. 7.10. In the figure, different styles of transitions have been shown. “bins adrb2” requires that “adr1” should transition from 1=>2=>3 on successive posedge clk. Of course, this transition sequence can be of arbitrary length. “bins adrb3[]” shows another way to specify multiple transitions. The annotation in the figure explains how we get four transitions.

> ***A very important feature of functional coverage is the ability to see if required 'transitions' in a design have been exercised.***
>
> ***This temporal domain coverage is not possible with code coverage (except for state transition coverage which is restricted strictly to the state machines 'derived' by the code coverage tool).***

```
bit[7:0] adr1;
bit adr2;

covergroup gc @(posedge clk);
  ac: coverpoint adr1
  {
      bins ar1 = (8'h00 => 8'hff);
  }
  dc: coverpoint adr2
  {
      bins ar2 = (1'b1 => 1'b0);
  }

  acdc: cross ac,dc;

endgroup
gc gcInst = new;
```

This means that adr1 is '00' at this posedge clk and it should be 'ff' the next posedge clk (since 'posedge clk' is the sample point).

This means that adr2 is '1' followed by '0' at successive sample points (i.e. successive posedge clk in this example).

This means that if the following condition is met that the cross will be covered;

adr1=00 && adr2=1 at this posedge clock
and at the next posedge clock
adr1=ff && adr2=0

Fig. 7.9 Transition coverage

"bins adrb5" (in some sense) is analogous to the consecutive operator of assertions. Here 'hf [*3] means that adr1='hf should repeat 3 times at successive posedge clk.

Similarly, the nonconsecutive transition ('ha [->3]) means that adr1 should be equal to 'ha, three times and *not* necessarily at consecutive posedge clk. Note that just as in nonconsecutive operator, here also 'ha needs to arrive three times with arbitrary number of clocks in between their arrival and that "adr1" should *not* have any other value in between these three transitions. The simulation log shows the result of a testbench that exercises all the transition conditions.

7.10 Performance Implications of Coverage Methodology

Introduction: This section describes the methodology components of functional verification and what you should cover, when you should cover, performance implications, and applications on how to write properties that combine the power of assertions with power of functional coverage.

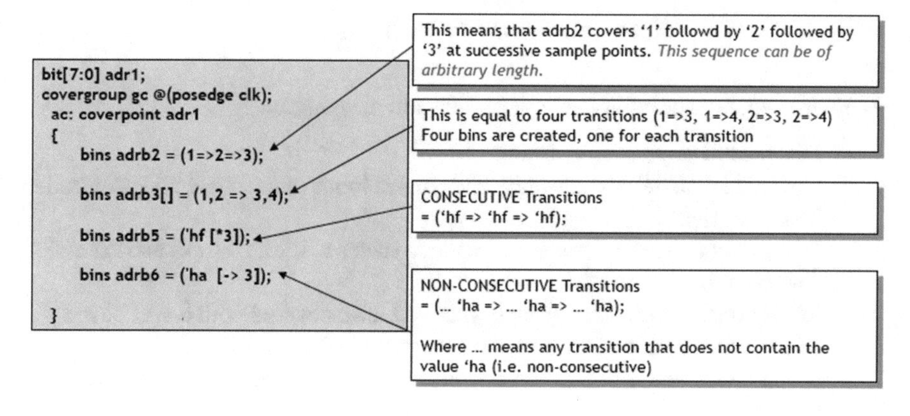

COVERGROUP COVERAGE:

Covergroup	*Metric*	*Goal/ Status*
Coverpoint gc::ac	*100.0%*	*100 Covered*
bin adrb2[1=>2=>3]	*2*	*1 Covered*
bin adrb3[2=>4]	*4*	*1 Covered*
bin adrb3[2=>3]	*3*	*1 Covered*
bin adrb3[1=>4]	*2*	*1 Covered*
bin adrb3[1=>3]	*4*	*1 Covered*
*bin adrb5[15[*3]]*	*1*	*1 Covered*
bin adrb6	*1*	*1 Covered*

Fig. 7.10 "Bins" of transition

7.10.1 Know What *You Should Cover*

- Don't cover the entire 32-bit address bus.
  - Cover only the addresses of interest (e.g., Byte/word/dword aligned, start/end address, bank crossing address, etc.)
- Don't cover the entire counter range.
  - Cover only the rollover counter values (transition from all 1s to all 0s)
- No need to cover the entire 2K FIFO.
  - Cover only FIFO full, FIFO empty, FIFO full crossed with FIFO_push, FIFO empty crossed with FIFO read, etc.
- Autogenerated bins are both a convenience and a nuisance. They may create a lot of clutter that may not be relevant. Be judicious in the usage of autogenerated "bins."
- Use "cross" and "intersect" to weed out unwanted "bins" and also "illegal_bins" and "ignore_bins."

7.10.2 *Know* When *You Should Cover*

- Enable your cover points only when they are meaningful.
  - Disable coverage during "reset."
  - Cover "test mode" signals only when in test mode (e.g., JTAG TAP Controller TMS asserted).
  - Make effective use of coverage methods such as "start," "stop," and "sample" (more on this later).
  - Do not repeat with covergroups what you have covered with SVA "cover."
  - Make effective use of covergroup "trigger" condition.
  - Make effective use of the "action" block associated with "cover" to activate a covergroup.

7.11 When to "Cover" (Performance Implication)

Functional coverage should be carefully collected as discussed above. The language does allow tasks that allow you to control when to start collecting coverage and when to stop. These tasks can be associated with an instance of a covergroup and invoked from procedural block.

Figure 7.11 shows the covergroup "rg" with two coverpoints "pc" and "gc." "pc" covers all the pending requests, and "gc" covers the number of masters on the bus when those requests are made. "my_rg" is the instance of this covergroup.

Since we want to start collecting pending requests at the assertion of req, when the requests are granted, we don't want to cover pending requests and number of masters anymore. "gnt"-related cover can be another covergroup.

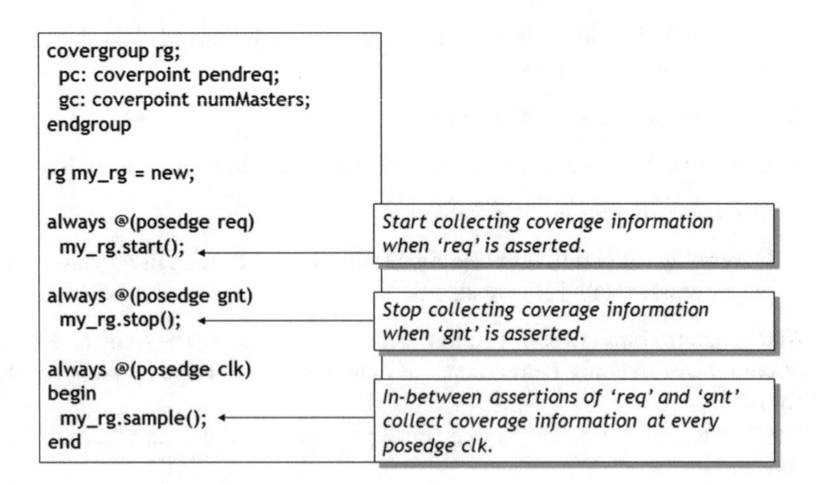

Fig. 7.11 Functional coverage: Performance implication

There is simple control but very good performance improvement. Use it wisely to speed up your simulation and a more meaningful coverage log.

Lastly, there is the sampling edge task sample() which derives its sampling edge from "always @ (posedge clk)" and applies it to "my_rg" as this sampling edge. This also tells us that we can have covergroup-specific sampling edges. It is a very good feature. Note that my_rg.sample() will "start" when my_rg.start() is executed and will stop when my_rg.stop() is executed. This is about as easy as it gets when it comes to controlling collection of coverage information.

Note that optionally, there is also a "strobe" option that can be used to modify the sampling behavior. When the strobe option is not set (the default), a coverage point is sampled the *instant* the clocking event takes place, as if the process triggering the event were to call the built-in sample() method. If a variable toggles multiple times in a time step, the coverage point will also be sampled multiple times. The "strobe" option can be used to specify that coverage points are sampled in the postponed region, thereby filtering multiple clocking events so that only one sample per time slot is taken. The strobe option only applies to the scheduling of samples triggered by a clocking event.

7.12 SystemVerilog Functional Coverage: Applications

7.12.1 PCI Cycles

Figure 7.12 shows an application on how to cover all PCI commands. It also shows how to categorize these commands into different *explicit bins*. Categorizing this way helps a lot during debug and measurement of functional coverage.

Figure 7.13 shows an application of how you can make sure that you have "covered" all possible PCI transactions and all possible *transitions* among them (e.g., IORead to MemWrite, MemRead to MemWrInv, etc.). Without such features, you will deploy ad hoc methods to see that you have covered (i.e., your testbench have covered) all possible *transitions*.

In this application, we are covering all transitions from read to write and write to read. For example, the following transitions will be covered (only a sample of transitions is shown):

For read to write (bins R2W[]):

```
IORead => IOWrite
IORead => MemWrite
IORead => ConfWrite
IORead => MemWrInv
MemRead => IOWrite
<etc.>
```

For write to read (bins W2R[]):

Requirement: Cover all PCI Cycle Types.

```
// PCI C/BE Commands
enum {iack, SpecialC, IORead, IOWrite, MemRead, MemWrite, ConfRead, ConfWrite,
MemRMult, DualAddr, MemReadLine, MemWrInv} pciCommands;

covergroup pciCommands_cover @(negedge FRAME_);

          pciCmdCover : coverpoint pciCommands

          {
          bins pcireads []={IORead, MemRead, ConfRead, MemRMult, MemReadLine};
          bins pciwrites [] = {IOWrite, MemWrite, ConfWrite, MemWrInv};
          bins pcimisc [] = {iack, SpecialC};
          }

endgroup
```

EXPLICIT bins to categorize PCI cycles in different bins. So, for example, when pcireads bins are 100% covered, we know that all PCI Read type cycles have been exercised.

Fig. 7.12 Functional coverage: Application

Requirement: Cover all PCI Cycle Types and transitions among Read and Write cycles.

```
// PCI C/BE Commands
enum {iack, SpecialC, IORead, IOWrite, MemRead, MemWrite, ConfRead, ConfWrite,
MemRMult, DualAddr, MemReadLine, MemWrInv} pciCommands;

covergroup pciCommands_cover @(posedge clk);
          pciCmdCover : coverpoint pciCommands
          {
          bins pcireads [] = {IORead, MemRead, ConfRead, MemRMult, MemReadLine};
          bins pciwrites [] = {IOWrite, MemWrite, ConfWrite, MemWrInv};
          bins pcimisc [] = {iack, SpecialC};

          bins R2W [] = (IORead, MemRead, ConfRead, MemRMult, MemReadLine =>
IOWrite, MemWrite, ConfWrite, MemWrInv);
          bins W2R [] = (IOWrite, MemWrite, ConfWrite, MemWrInv => IORead,
MemRead, ConfRead, MemRMult, MemReadLine);
          }
endgroup
```

Fig. 7.13 Functional coverage: Application

```
IOWrite => IORead
IOWrite => MemRead
IOWrite =>ConfRead
IOWrite +> MemRMult
IOWrite => MemReadLine
MemWrite => IORead
<etc.>
```

7.12.2 Frame Length Coverage

Figure 7.14 shows the application that combines local variables, subroutine calls, covergroups, and interaction with procedural code outside of the assertion. Here's how it works.

Read this example bottom-up.

Property frame length says that when the rising edge of TX_EN is sampled, we should check the length of the transmitted frame using sequence frmLength.

This application exemplifies the use of

- *local variables*
- *subroutine call associated with an expression to update a variable*
- *covergroup triggered from an explicit event.*

```
logic [7:0] FrameLngth = 0;
event measureFrameLength;

covergroup length_cg @(measureFrameLength );
        coverpoint FrameLngth;
endgroup

task store_Frame_Lngth;
input [7:0] x;
        FrameLngth = x;
        -> measureFrameLength;
endtask

sequence frmLength;
int cnt;
        (TX_EN, cnt=1)  ##1  ((TX_EN, cnt++)[*0:$])
        ##1 (!TX_EN, store_Frame_Lngth(cnt))
endsequence

property frameLength;
        @(posedge TX_CLK) $rose(TX_EN) |-> frmLength;
endproperty

fLength: assert property (frameLength);
```

Fig. 7.14 Functional coverage: Application

Sequence frmLength declares a local variable “cnt” and, at TX_EN==1, initializes cnt=1. One clock later (##1), it increments cnt forever ((TX_EN, cnt++)[*0:$]) until TX_EN deasserts (falls). At that time, we call a task (i.e., a subroutine) called store_Frame_Lngth(cnt) and provide it the final count as a parameter. This final count is the length of the frame that started with TX_EN assertion.

The task store_Frame_Lngth takes the “cnt” as input and assigns it to “logic”-type FrameLngth and triggers a named event called measureFrameLength.

Now the covergroup length_cg triggers at “measureFrameLength” edge, which we just triggered explicitly from task store_Frame_Lngth. The coverpoint covers FrameLngth.

In short, we measure the frame length starting assertion of TX_EN until deassertion of it. We measure the frame length between assertion and deassertion of TX_EN and cover it. With every new assertion of TX_EN, we measure the length of a new frame.

Note that “coverpoint FrameLngth” does not specify any explicit bins. That will create 256 explicit bins each containing a frame length. This way we make sure that we have covered all (i.e., 256) different frame lengths.

Chapter 8
Clock Domain Crossing (CDC) Verification

Chapter Introduction
Clock domain crossing (CDC) has become an ever-increasing problem in multi-clock domain designs. One must solve issues not only at RTL level but also consider the physical timing. This chapter will start with understanding of metastability and then dive into different synchronizing techniques. It will also discuss the role of SystemVerilog Assertions in verification of CDC. We will then discuss a complete methodology.

8.1 Design Complexity and CDC

There are hardly any designs today that operate on a single clock. A typical SoC will have three or more clocks, and these will be asynchronous. We have all done CDC checks using lint tools, among others. But the problem is that there is a disconnect between RTL static or simulation-based analysis and what we see in the physical chip. The issue of metastability due to clock domain crossing is not very predictable at RTL or gate level. Therefore, simulation does not accurately predict silicon behavior, and critical bugs may escape the verification process. This results in almost 25% of all respins due to clocking issues, CDC being the chief among them.

Here's an example of typical real-life designs and the number of clocks and CDC signals they have. This is just a representative data point [(PING YEUNG PH.D.)].

Design type	Number of clock domains	Number of CDC signals
Gigabit Ethernet interface	24–28	~ 11,000
Graphics application	36–42	~18,000
Multimedia SoC	4	~54
Wireless	8	~365

This table goes to show the complexity of CDC verification. Both single-bit and multi-bit synchronizations need to take place.

A.B. Mehta, *ASIC/SoC Functional Design Verification*,
DOI 10.1007/978-3-319-59418-7_8

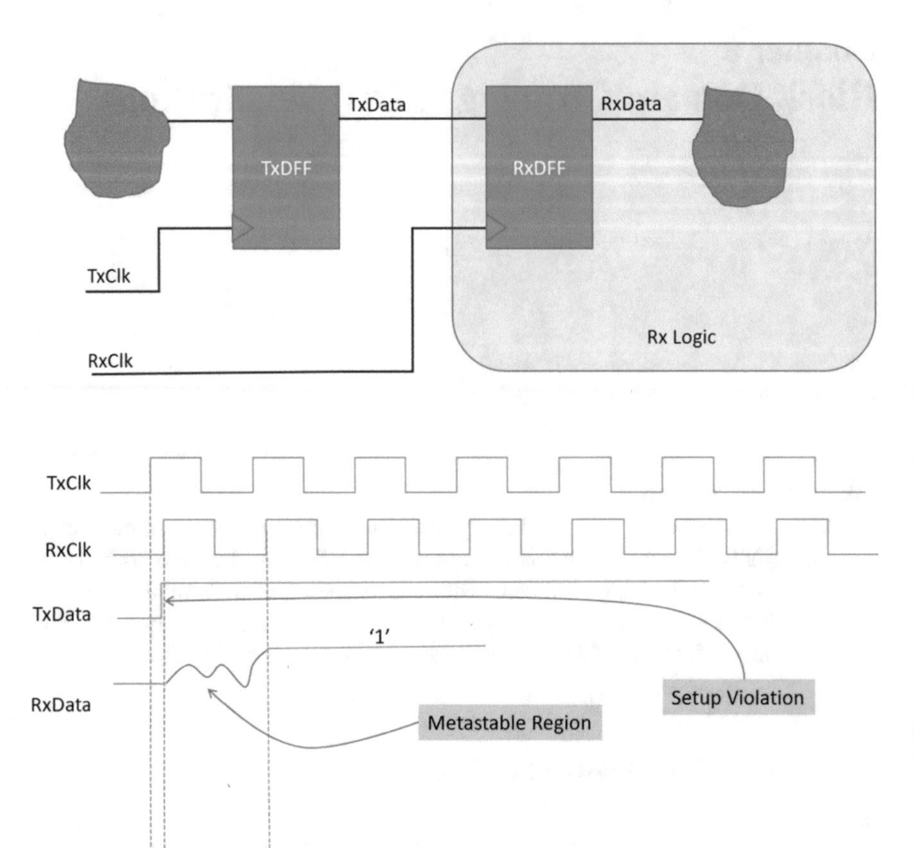

Fig. 8.1 Clock domain crossing—metastability

8.2 Metastability

The main culprit in CDC is the metastability of data that occurs when data crosses from one clock domain to another. The first can be slower or faster compared to the other clock domain. The data that crosses the boundary can end up violating setup/hold requirements of the second clock domain. This is explained via Fig. 8.1. This figure shows a synchronization failure that occurs when a TxData generated in TxClk clock domain is sampled too close (setup violation) to the rising edge of RxClk of the Rx logic domain. Synchronization failure is caused by an output going metastable and not converging to a legal stable state.

When TxData violates setup time of the RxClk, RxData goes metastable, meaning we don't know what state will it settle down to or settle down at all within one clock. If TxData is held "long" enough, RxData will eventually become stable and end up in a correct state. For the sake of simplicity, I've shown the metastable RxData to stabilize in one clock. But that may not necessarily be the case in all instances. If the metastable RxData is fed directly into the forward logic, you do not

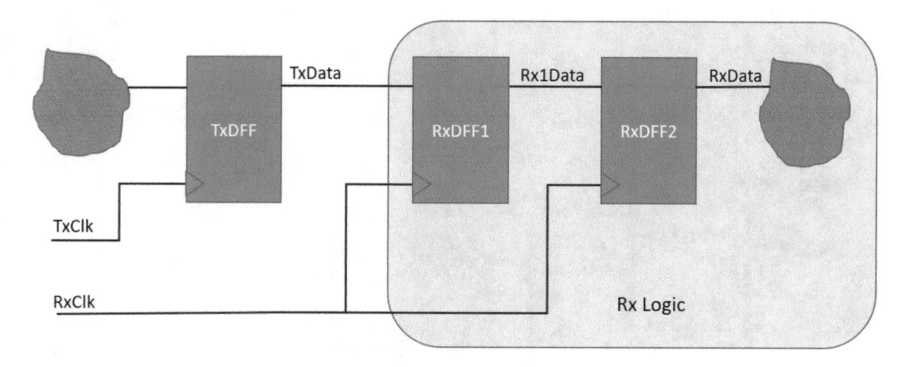

Fig. 8.2 Clock domain crossing—two-flop single-bit synchronizer

know what metastable state got propagated to the forward logic. Since the CDC signal can fluctuate for some period of time, the input logic in the receiving clock domain might recognize the logic level of the fluctuating signal to be different values and hence propagate erroneous signals into the receiving clock domain. In RTL simulation, this metastable state will be regarded as "X" (unknown) state (correctly so), and the logic beyond RxDFF may be rendered useless (i.e., "X" propagation will cause all sorts of issues in the logic).

In short, synchronization failure is caused by an output going metastable and not converging to a legal stable state by the time the output must be sampled again.

8.3 Synchronizer

8.3.1 Two-Flop Synchronizer (Identical Transmit and Receive Clock Frequencies)

A synchronizer is a device that samples an asynchronous signal and outputs a version of the signal that has transitions synchronized to a sample clock.

The simplest synchronizer used in designs is a two-flop synchronizer (Fig. 8.2). The idea is that the first flop on the transmit side samples data input on the first flop's clock (let's call this the transmit clock). The first flop on the receiving clock can be very close to the transmit clock. In this case, the output of the transmit clock flop when captured by the receiving clock will output (at the output of the receive flop) a metastable signal, because the data output of the transmit flop violated the setup/hold requirement of the receive flop. If you let the receive flop output propagate to the design, the results will be unpredictable, because this output can be a "1" or a "0"; you don't know.

But if you insert a second flop in the receiving circuit, the metastable signal output of the first flop of the receive clock will have time (one clock's worth) to stabilize before being latched into the second flop on receive side. Now, the output of this

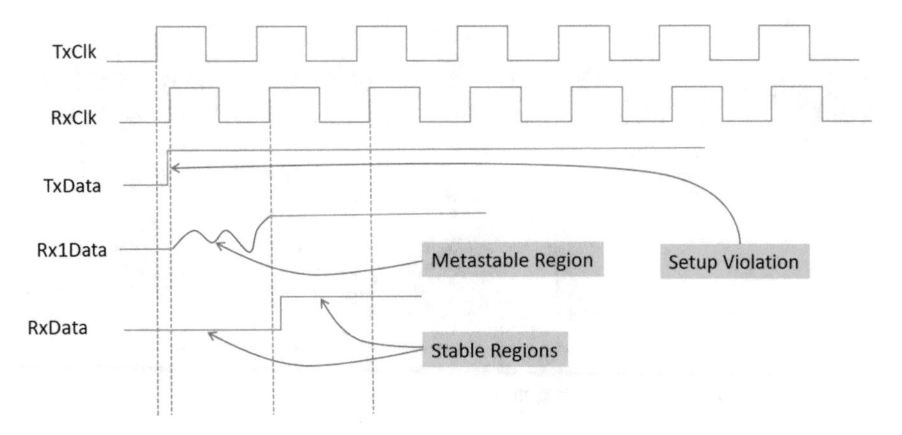

Fig. 8.3 Clock domain crossing—synchronizer—waveform

second flop will have a stable value and can propagate to the rest of the design without unpredictability. Please refer to Fig. 8.3 to understand this scenario. To reiterate, the first flip-flop samples the asynchronous input signal into the new clock domain and waits for a full clock cycle to permit any metastability on the stage-1 output signal to decay, and then the stage-1 signal is sampled by the same clock into a second-stage flip-flop, with the intended goal that the stage-2 signal is now a stable and valid signal synchronized and ready for distribution within the new clock domain.

A couple of implementation guidelines for the two-flop synchronizer:

1. There should not be any combinational logic between the Transmit DFF and the Receive DFF. This allows for maximum metastability resolution time.
2. RxDFF1 and RxDFF2 synchronizer flops should be placed as close as possible during layout. Most companies nowadays offer a predefined, laid out, and verified synchronizer macros which can be hand placed in RTL.

8.3.2 Three-Flop Synchronizer (High-Speed Designs)

For some very high-speed designs, the mean time between failure (MTBF) is too short since the data may change before the second flop synchronizes the TxData. In such cases, you may need three-flop synchronizers to compensate for the high speed. Metastability may not settle down at RxDFF2 (Rx2Data) and hence the need for the third flop (RxDFF3) (Fig. 8.4).

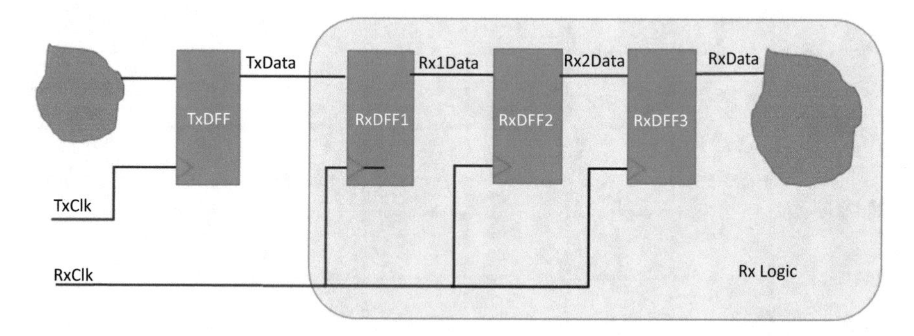

Fig. 8.4 Three-flop single-bit synchronizer

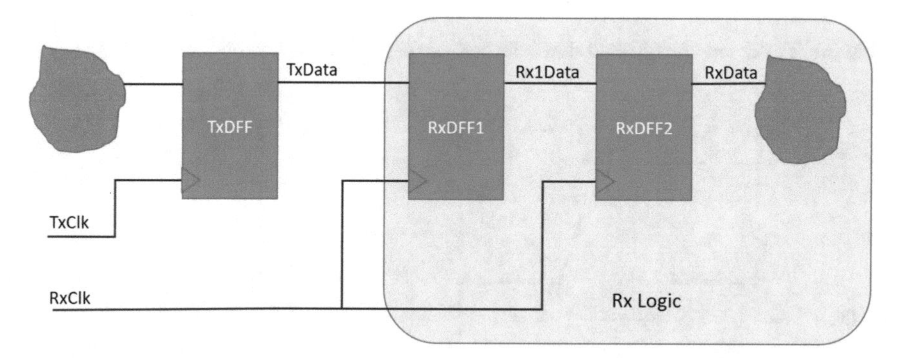

Fig. 8.5 Two-flop single-bit synchronizer

8.3.3 *Synchronizing Fast-Clock (Transmit) into Slow-Clock (Receive) Domains*

So far, we have seen synchronizers that work when both the transmit and the receive clocks are of the same frequency. Note that if the transmit clock is slower than the receive clock, the two (or three) flop synchronizers will work quite well. Recognizing that sampling slower signals into faster-clock domains causes fewer potential problems than sampling faster signals into slower-clock domains, a designer might want to take advantage of this fact by using simple two flip-flop synchronizers to pass single CDC signals between clock domains.

But when the transmit clock is faster than the receive clock, there is the possibility that a signal from the transmit logic may change values twice before it can be sampled or might be too close to the sampling edge of the slower receive clock domain.

For the ensuing discussion, let us call the signal that needs synchronization as the CDC signal. That will make it easier to describe the concept. Here's the two-flop synchronization (Fig. 8.5) for ease of reference.

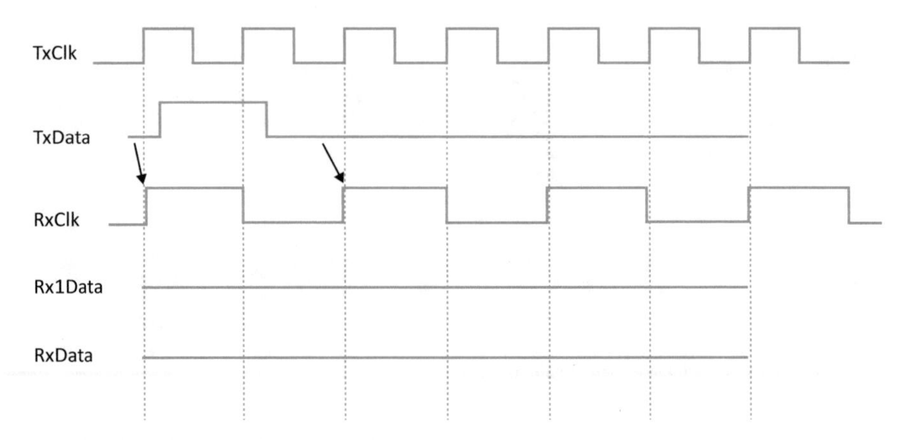

Fig. 8.6 Faster transmit clock to slower receive clock—two-flop synchronizer won't work

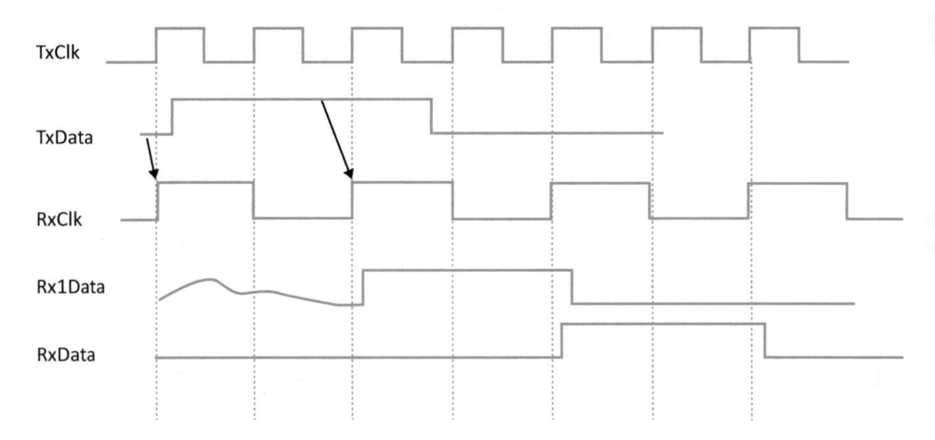

Fig. 8.7 Lengthened transmit pulse for correct capture in receive clock domain

If the CDC signal is only pulsed for one fast-clock cycle, the CDC signal could go high and low between the rising edges of a slower clock and not be captured into the slower-clock domain. This is shown in Fig. 8.6. In this figure, TxData goes high and then goes low (1 high pulse) *in between the RxClk period*. In other words, this high pulse will not be captured by the RxClk. That results into the Rx1Data remaining at the previously captured state of "0" and so does RxData. The high pulse on TxData is dropped by the receive logic which will result in incorrect behavior in the receive logic.

Hence, a two-flop synchronizer won't work when the transmit clock is faster than the receive clock.

One potential solution to this problem is to assert the TxData signal (i.e., the CDC signal) for a period that exceeds the cycle time of the receive clock. This is shown in Fig. 8.7. The general rule of thumb is that the minimum pulse width of the transmit signal be 1.5x the period of the receive clock frequency. The assumption is that the CDC signal will be sampled at least once by the receive clock. The issue with this solution will arise if an engineer mistakes this solution to be a general-purpose solution and miss the transmit (CDC) signal period requirement. This is where SystemVerilog Assertions come into picture. Put an assertion on the CDC signal for its period check when crossing from the high-frequency to the low-frequency domain.

There are other solutions to tackle this problem, which are beyond the scope of this book.

8.3.4 Multi-bit Synchronization

When passing multiple signals between clock domains, simple synchronizers do not guarantee safe delivery of the data. A frequent mistake made by engineers when working on multi-clock designs is passing multiple CDC bits required in the same transaction from one clock domain to another and overlooking the importance of the synchronized sampling of the CDC bits.

The problem is that multiple signals that are synchronized to one clock will experience small data-changing skews that can occasionally be sampled on different rising clock edges in a second clock domain. Even if we could perfectly control and match the trace lengths of the multiple signals, differences in rise and fall times as well as process variations across a die could introduce enough skew to cause sampling failures on otherwise carefully matched traces.

Here are a couple of solutions to solve the multi-bit synchronization problem. In-depth discussion of these solutions is out of scope of this book, but I highly recommend a SNUG paper by Cliff Cummings mentioned in the Bibliography (Clifford E. Cummings).

1. The Gray Code Solution Where Multiple CDC Bits Are Passed Using Gray Codes

The safest counters that can be used in multi-clock designs are Gray Code counters. Gray Codes only allow one bit to change for each clock transition, eliminating the problem associated with trying to synchronize multiple changing CDC bits across a clock domain. Standard Gray Codes have very nice translation properties to convert gray to binary and back again. Using these conversions, it is simple to design efficient Gray Code counters.

I am sure we are familiar with Binary to Gray and Gray to Binary code conversion formulas. But they are presented here for the sake of completeness.

4-bit Gray to Binary conversion:

```
binary [0] = gray[3] ^ gray[2] ^ gray[1] ^ gray[0];
binary [1] = gray[3] ^ gray[2] ^ gray[1];
binary [2] = gray[3] ^ gray[2];
binary [3] = gray[3];
```

This can also be represented as:

```
binary [0] = gray[3] ^ gray[2] ^ gray[1] ^ gray[0] ; // gray>>0
binary [1] = 1'b0 ^ gray[3] ^ gray[2] ^ gray[1] ; // gray>>1
binary [2] = 1'b0 ^ 1'b0 ^ gray[3] ^ gray[2] ; // gray>>2
binary [3] = 1'b0 ^ 1'b0 ^ 1'b0 ^ gray[3] ; // gray>>3
```

And here's the Binary to Gray conversion:

```
gray[0] = binary[0] ^ binary [1];
gray[1] = binary [1] ^ binary [2];
gray[2] = binary [2] ^ binary [3];
gray[3] = binary [3] ^ 1'b0 ;
```

2. Asynchronous FIFO Implementation

Passing multiple bits, whether data bits or control bits, can be done through an asynchronous FIFO. An asynchronous FIFO is a shared memory or register buffer where data is inserted from the write clock domain and data is removed from the read clock domain. Since both sender and receiver operate within their own respective clock domains, using a dual-port buffer, such as a FIFO, is a safe way to pass multi-bit values between clock domains. A standard asynchronous FIFO device allows multiple data or control words to be inserted if the FIFO is not full. The receiver can then extract multiple data or control words if the FIFO is not empty.

8.3.5 Design of an Asynchronous FIFO Using Gray Code Counters

The Gray Code counters are used in this asynchronous FIFO design for the Read_ pointer and the Write_pointer guaranteeing successful transfer of multi-bit data from write clock (aka the transmit clock) to read clock (aka the receive clock). Let us look at an asynchronous FIFO design that uses Gray Code counter.

```
module asynchronous_fifo (
   // Outputs
   fifo_out, full, empty,
   // Inputs
   wclk, wclk_reset_n, write_en,
   rclk, rclk_reset_n, read_en,
   fifo_in
  );

   `define FF_DLY 1'b1
   parameter    D_WIDTH = 20;
   parameter    D_DEPTH = 4;
   parameter    A_WIDTH = 2;
```

```
input                  wclk_reset_n;
input                  rclk_reset_n;
input                  wclk;
input                  rclk;
input                  write_en;
input                  read_en;
input [D_WIDTH-1:0]    fifo_in;

output [D_WIDTH-1:0] fifo_out;
output                      full;
output                      empty;

reg [D_WIDTH-1:0]      reg_mem[0:D_DEPTH-1];
reg [A_WIDTH:0]        wr_ptr;
reg [A_WIDTH:0]        wr_ptr_gray;
reg [A_WIDTH:0]        wr_ptr_gray_rclk_q;
reg [A_WIDTH:0]        wr_ptr_gray_rclk_q2;
reg [A_WIDTH:0]        rd_ptr;
reg [A_WIDTH:0]        rd_ptr_gray;
reg [A_WIDTH:0]        rd_ptr_gray_wclk_q;
reg [A_WIDTH:0]        rd_ptr_gray_wclk_q2;

reg                   full;
reg                   empty;

wire [A_WIDTH:0]      nxt_wr_ptr;
wire [A_WIDTH:0]      nxt_rd_ptr;
wire [A_WIDTH:0]      nxt_wr_ptr_gray;
wire [A_WIDTH:0]      nxt_rd_ptr_gray;
wire [A_WIDTH-1:0]    wr_addr;
wire [A_WIDTH-1:0]    rd_addr;
wire                  full_d;
wire                  empty_d;

assign wr_addr = wr_ptr[A_WIDTH-1:0];
assign rd_addr = rd_ptr[A_WIDTH-1:0];

always @ (posedge wclk)
  if (write_en) reg_mem[wr_addr] <= #`FF_DLY fifo_in;

assign fifo_out = reg_mem[rd_addr];


always @ (posedge wclk or negedge wclk_reset_n)
 if (!wclk_reset_n) begin
    wr_ptr <= #`FF_DLY {A_WIDTH+1{1'b0}};
```

```
      wr_ptr_gray <= #`FF_DLY {A_WIDTH+1{1'b0}};
     end else begin
      wr_ptr <= #`FF_DLY nxt_wr_ptr;
      wr_ptr_gray <= #`FF_DLY nxt_wr_ptr_gray;
     end

   assign nxt_wr_ptr = (write_en) ? wr_ptr+1 : wr_ptr;
   assign nxt_wr_ptr_gray = ((nxt_wr_ptr>>1) ^ nxt_wr_ptr);

   always @ (posedge rclk or negedge rclk_reset_n)
    if (!rclk_reset_n) begin
      rd_ptr <= #`FF_DLY {A_WIDTH+1{1'b0}};
      rd_ptr_gray <= #`FF_DLY {A_WIDTH+1{1'b0}};
     end else begin
      rd_ptr <= #`FF_DLY nxt_rd_ptr;
      rd_ptr_gray <= #`FF_DLY nxt_rd_ptr_gray;
     end

   assign nxt_rd_ptr = (read_en) ? rd_ptr+1 : rd_ptr;
   assign nxt_rd_ptr_gray = (nxt_rd_ptr>>1) ^ nxt_rd_ptr;

   // check full
   always @ (posedge wclk or negedge wclk_reset_n)
    if (!wclk_reset_n)
      {rd_ptr_gray_wclk_q2, rd_ptr_gray_wclk_q} <= #`FF_DLY {{A_
WIDTH+1{1'b0}}, {A_WIDTH+1{1'b0}}};
      else
      {rd_ptr_gray_wclk_q2, rd_ptr_gray_wclk_q} <= #`FF_DLY {rd_
ptr_gray_wclk_q, rd_ptr_gray};

    assign full_d = (nxt_wr_ptr_gray == {~rd_ptr_gray_wclk_q2[A_
WIDTH:A_WIDTH-1], rd_ptr_gray_wclk_q2[A_WIDTH-2:0]});

   always @ (posedge wclk or negedge wclk_reset_n)
    if (!wclk_reset_n)
    full <= #`FF_DLY 1'b0;
    else
    full <= #`FF_DLY full_d;

   // check empty
   always @ (posedge rclk or negedge rclk_reset_n)
    if (!rclk_reset_n)
      {wr_ptr_gray_rclk_q2, wr_ptr_gray_rclk_q} <= #`FF_DLY {{A_
WIDTH+1{1'b0}}, {A_WIDTH+1{1'b0}}};
   else
   {wr_ptr_gray_rclk_q2, wr_ptr_gray_rclk_q} <= #`FF_DLY {wr_ptr_
```

```
gray_rclk_q, wr_ptr_gray};

   assign empty_d = (nxt_rd_ptr_gray == wr_ptr_gray_rclk_q2);

   always @ (posedge rclk or negedge rclk_reset_n)
    if (!rclk_reset_n)
       empty <= #`FF_DLY 1'b1;
      else
       empty <= #`FF_DLY empty_d;

endmodule
```

In the next section, we will see how to use SystemVerilog Assertions to make sure that data are not dropped when write data (on write clock) are transferred through Gray Code counter synchronization logic to read data (on read clock).

8.4 CDC Checks Using SystemVerilog Assertions

As we saw, in Chap. 6, SystemVerilog Assertions (SVA) are a great way to check for sequential domain conditions at clock (or sampling edge) boundaries. The CDC signals crossing from one clock domain to another are perfect candidates to check for using SVA. SVA fully supports multi-clock domain assertions as well as multi-threaded local variables to make full proof checkers to see that your CDC synchronizers (whatever the design style) work as promised. Note that the assertions presented here can be used both for simulation-based checking and formal-based checking (static functional). But I will focus on simulation-based checking since the formal/static functional is still not fully adopted by many engineering groups and requires a complete chapter in itself.

Let us start with the simplest of the design. Later we will see a comprehensive assertion for CDC multi-bit data transfer using the Gray Code counter-based asynchronous FIFO described above.

Here's a wonderful two-flop synchronizer repeated for the sake of convenience.

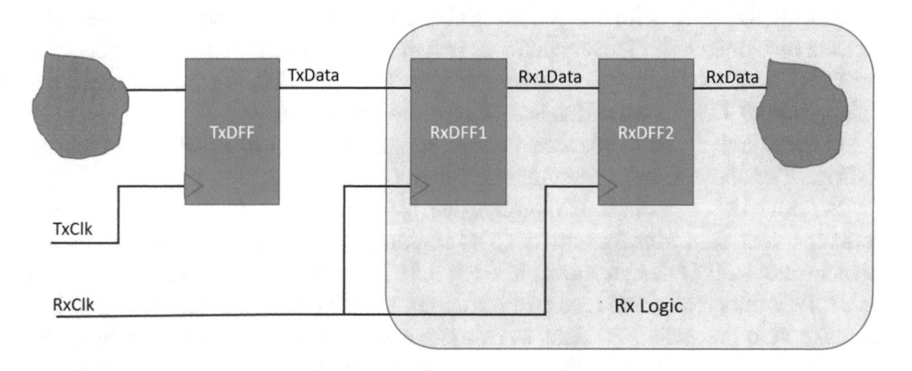

For any synchronizer design, there will be assumptions on TxData stability. Should it remain stable for two clocks? Three clocks? This is to make sure that the CDC signal Rx1Data has enough time to filter the metastability region and pass the correct value to RxData (the output). Let us go with the assumption that TxData should remain stable for two clocks every time it assumes a new value (i.e., it changes). This assumption is required since we assume that TxClk is faster than RxClk. Refer to Fig. 8.7 for the timing diagram of this design.

Here's a simple assertion to check for TxData stability:

```
property TxData_stable;
   @(posedge Txclk) $changed(TxData) |=> $stable(TxData) [*2];
endproperty

assert property (TxData_stable);
```

Let us now see how to make sure that this two-flop single-bit synchronizer correctly transfers data so that RxData === TxData after metastability filter:

```
property Tx_to_Rx_CDC_DataCheck;
local Data;

    @(posedge Txclk) ($changed(TxData)) |=>
     (1'b1, (Data = TxData)) ##1
      @(posedge RxClk) (Rx1Data === Data) ##1 (Rxdata === Data);
endproperty: Tx_to_Rx_CDC_DataCheck

assert property (Tx_to_Rx_CDC_DataCheck);
```

First, the assertion checks that TxData has changed at posedge of TxClk. If it has, we first store the TxData into the multi-threaded local variable Data. 1'b1 is required because local data store must be attached to an expression. Since we don't have any condition, we simply say "always true" is the expression. "Always true" means always store TxData into the data, whenever TxData changes. Then, we check at the CDC boundary clock RxClk that the data has indeed transferred to Rx1Data by comparing Rx1Data with the stored TxData (in the data). One clock later, the RxData must match the TxData that was transmitted on TxClk. This guarantees that the CDC 1-bit two-flop synchronization works as intended. Again, note that the assumption of TxClk faster than RxClk must be adhered to.

As an exercise, see if you can write a simple assertion to check for glitch on TxData. The above solution assumes no glitch on TxData.

Ok, now let us write a comprehensive assertion for a multi-bit Gray Code counter-based data transfer across CDC region. This assertion is written for the asynchronous FIFO design shown in Sect. 3.5. The write data are written to fifo_in on wclk (write clock); and read from fifo_out on rclk (read clk). The assertion has to make sure that whatever data were written into FIFO at the write pointer, the same data is read out from FIFO when read pointer is equal to the write pointer:

```
sequence rd_detect(ptr);
 ##[0:$] (read_en && !empty && (aff1.rd_ptr == ptr));
endsequence

property data_check(wrptr);
integer ptr, data;
  @ (posedge wclk) disable iff (!wclk_reset_n || !rclk_reset_n)
       (write_en && !full, ptr=wrptr, data=fifo_in,
    $display($stime,"\t Assertion Disp wr_ptr=%h data=%h", aff1.
wr_ptr, fifo_in))

|=>
   @ (negedge rclk) first_match(rd_detect(ptr),
    $display($stime,,," Assertion Disp FIRST_MATCH ptr=%h Compare
data=%h fifo_out=%h", ptr, data, fifo_out))
          ##0 (fifo_out === data);
endproperty

dcheck : assert property (data_check(aff1.wr_ptr))
else$display($stime,,,"FAIL: DATA CHECK");
dcheckc : cover property (data_check(aff1.wr_ptr))
$display($stime,,,"PASS: DATA CHECK");
```

In this assertion, data_check property checks to see if FIFO is not full. If so, saves wr_ptr into the local variable "ptr" and the data from FIFO into local variable "data" and display so that we can easily see how the assertion is progressing during simulation.

If the antecedent is true, the consequent says that the first match of rd_ptr being the same as wr_ptr (note wr_ptr was stored in local variable *ptr*) and that the read data is the same as the write data (note write data were stored in local variable *data* in the antecedent).

Sequence rd_detect(ptr) is used as an expression to first_match. It says that wait from now until forever until you detect a read, and its rd_ptr is equal to the wr_ptr (which is stored in the local variable "ptr" in the antecedent).

Many such assertions can be written to see that your synchronizer design works. As an exercise, try writing simple assertions for your synchronizer design.

8.5 CDC Verification Methodology

Metastability from the intermixing of multiple clock signals is not accurately modeled by simulation. Unless you leverage exhaustive, automated clock domain crossing (CDC) analyses to identify and correct problem areas, you will inevitably suffer unpredictable behavior when the chip samples come back from the fab. Bottom line: automated CDC verification solutions are mandatory for multi-clock designs.

Traditional CDC verification methods include manually inspecting RTL code for the presence of synchronizers, running full timing simulations, sweeping clocks against each other, and using special simulation models to randomly vary the delays through synchronizers. These methods find only a subset of errors in a given design.

An effective CDC verification methodology should include structural, protocol, and re-convergence fanout verification [(PING YEUNG PH.D.)].

Structural Verification

Each synchronizer must have the correct structure for the type of signal being sent across clock domains. For example, a 2-DFF synchronizer is usually the best solution for single-bit signals but should not be used for multi-bit signals unless they are gray-coded to ensure that only one bit changes at a time. Multi-bit signals may be synchronized across domains using a separate control signal, an asynchronous FIFO, or other methods. Also, there should be no combinational logic inside or before a synchronizer.

Protocol Verification

Each synchronizer must follow a set of rules, called a transfer protocol, to ensure that the CDC signal is properly transferred across clock domains. For example, even the simplest 2-DFF synchronizer requires that the transmitting signal be held stable long enough to guarantee that it is captured in the receiving domain. This may not occur if the transmitting clock is faster than the receiving clock. Synchronization structures for multi-bit signals require more complex protocol checks. When CDC transfer protocols are violated, an error may not occur in simulation but will eventually occur in real hardware. Protocol analysis should be done using static formal methods. SVA should be deployed to check for correct protocol adherence.

Re-convergence Fanout Verification (Fig. 8.8)

Re-convergence occurs when multiple signals are synchronized separately from one clock domain to another and then used by the same logic in the receiving domain. If that logic assumes a timing relationship between the signals, the design is not tolerant of metastability and will eventually fail. This is because the purpose of synchronizers is to "filter out" metastability to ensure that unpredictable values are not seen by the receiving logic.

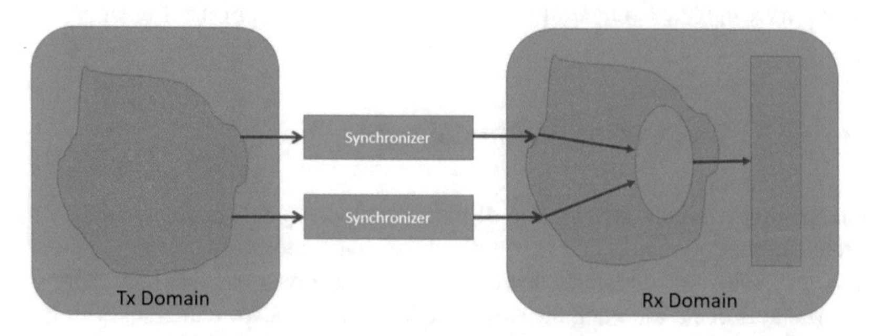

Fig. 8.8 Clock domain crossing—re-convergent fanout and CDC

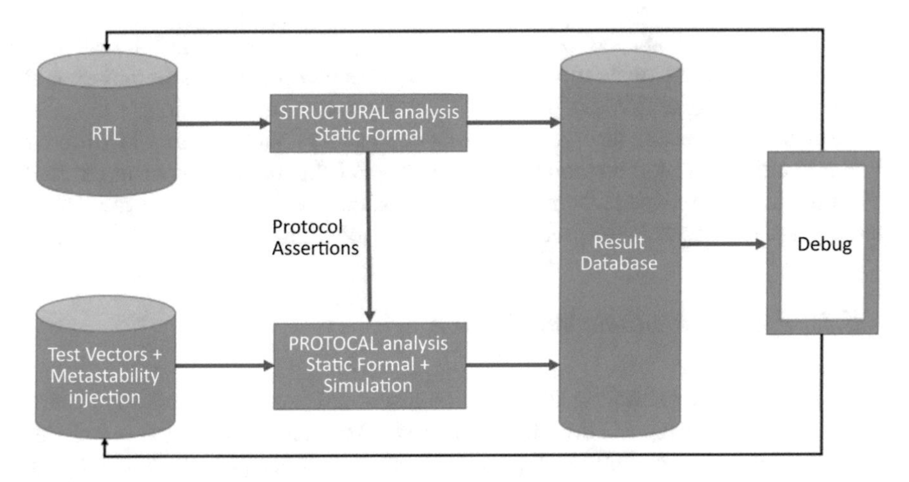

Fig. 8.9 Clock domain crossing—automated methodology

Let us see how we can combine structural analysis with protocol analysis to come up with an automated comprehensive methodology. The following is a generic diagram representing the automated process many EDA vendors now provide.

8.5.1 Automated CDC Verification

Figure 8.9 shows a proposed methodology. EDA vendors have implemented similar methodology (or are working toward).

8.5.2 Step 1: Structural Verification

Identify RTL blocks (not the entire SoC RTL) that have CDC signals at play. Feed such RTL blocks to the static formal structural analysis tool. This tool will identify CDC synchronization "structures" within your logic and analyze to see if they meet the requirements. For example, a single-bit CDC synchronization will work with a two-flop synchronizer. But for a multi-bit synchronizer, the two-flop solution won't work. You may need an asynchronous FIFO-based solution or a gray counter (where only 1 bit changes at a time). The tool will analyze such situations and provide a structural analysis report. The results are also stored in a UCDB style database for further debug analysis. This step should find issues with missing and incorrectly implemented synchronizers and potential re-convergence problems.

More important in this step is to automate derivation of SystemVerilog Assertions. For example, for a two-flop synchronizer, the input data should remain stable for at least 1.5x the receive clock. The structural analysis tool will (should) automatically write such assertions for the next stage of protocol verification. There are many such constraints/checks that need to be provided to the protocol analyzer. The structural analysis tool "knows" what type of structure has been designed and thereby should be able to create assertions for protocols that the structure needs to adhere to.

You need to evaluate the structural analysis results provided by an EDA vendor tool and either accept the recommendation or reject them and implement the best structure that you envision. Don't worry; the protocol analyzer will grab you if your structure does not meet synchronization protocol requirements.

8.5.3 Step 2: Protocol Verification

Once the structural analysis is complete, the assertions (either automatically created or manually) will be input to the protocol analyzer. The static formal method employed in the protocol analyzer will try all possible combinations of inputs (both in temporal and combinational domain) to the RTL block and see if any of the assertions FAIL. These assertions ensure that the CDC signal is stable when going from the TX to the RX domain; the multiple-bit CDC data is gray coded, or it is stable when it is sampled. The results will show failures which need to be analyzed to correct the synchronizer. Multiple iterations of this step will make sure that the logic will survive under all conditions of input and that the metastability has been addressed.

In addition to static formal, you may also want to simulate using the created assertions. For example, you feel comfortable with sweeping clocks to check for re-convergence logic. Or you want to deploy the so-called static + simulation hybrid methodology to check for the structural integrity against required protocol specification.

8.5.4 Step 3: Debug

Of course, debug is a big part of this strategy. The results from structural and protocol analysis are stored in an UCDB style database. The debug tool will associate the structure against the protocol and show the relationship. It will also help you debug failing assertions. EDA tools do support such debug capabilities.

Based on the debug results, you will either change the RTL or change the input test vectors and metastability injection strategy.

This loop will continue until there are no more assertions that fire and the metastability issues are completely resolved.

This is what I call a proposed methodology. You may discuss it with EDA vendors to see how close they come to it with their proposed solution.

8.6 CDC Verification at Gate Level

The next problem is CDC at gate level. Gate-level simulations are notorious in propagating an unknown "X," rapidly throughout the design. The two-flop synchronizer can cause the "X" propagation problem. See Fig. 8.10 to understand this issue.

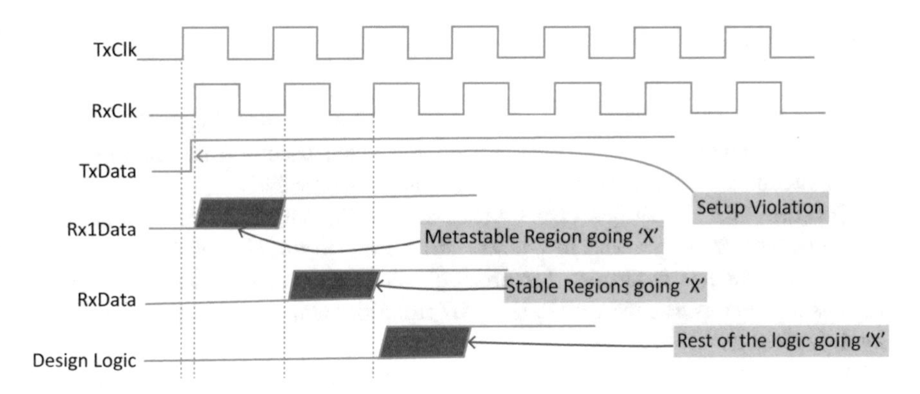

Fig. 8.10 Gate-level CDC

When doing gate-level simulations on a multi-clock design [(Clifford E. Cummings)], the ASIC library models of flip-flops are modeled with setup and hold time expressions to match the timing specifications of the actual flip-flops. ASIC libraries typically model flip-flops to drive X's (unknowns) on the flip-flop outputs when a timing violation occurs. When simulating gate-level synchronizers, setup and hold time violations might cause ASIC libraries to issue setup and hold time error messages, and the offending signals are frequently driven to an X value. These X values propagate to the rest of the design causing problems when trying to verify the functionality of the entire gate-level design (Fig. 8.10).

There are many techniques available to turn "OFF" such "X" propagation when doing CDC verification at gate level. If you are familiar with SystemVerilog and EDA simulators, you may be familiar with these techniques. But for the sake of completeness, here they are.

Change the flop setup and hold times to 0. This will obviously not give any timing violation and hence prevent "X" from being generated because of timing violation. BUT this will basically nullify the setup/hold of "all" the flops in your vendor-provided library. So this may not be a good strategy after all.

Turn OFF the timing check in the "specify" block of the flop cells. Many vendors provide a command line option "+no_notifier" to automatically turn OFF "X" generation due to a timing violation. I believe this is a preferred methodology, since you will indeed get a timing violation telling you that there is a synchronization issue but will not generate an "X."

8.7 EDA Vendors and CDC Tools Support

So what kind of industry tools are available to help a DV engineer tackle CDC verification? Synopsys SpyGlass CDC and Mentor's Questa CDC are two of the *many* tools available in the EDA market. I've described only Mentor's solution. Synopsys does not provide information on their SpyGlass CDC tool unless you register. So do not go there.

8.7.1 Mentor

Here's a brief description of Mentor's Questa CDC methodology (Fig. 8.11) [(MentorGraphics)]. The following description is taken directly from Mentor's marketing literature. Make sure that their claims are indeed valid!

Questa® CDC identifies errors using structural analysis to recognize clock domains, synchronizers, and low-power structures via the Unified Power Format (UPF). It generates assertions for protocol verification along with metastability models for re-convergence verification. All properties and design intent are inferred by the software.

The technology checks all potential CDC failures, statically verifying that all signals crossing asynchronous clock domain boundaries are guarded by CDC synchronizers. It then illustrates DUT issues found with familiar schematic and waveform displays. Additionally, in concert with Questa simulation, the CDC-FX app injects metastability into RTL functional simulation to verify if the DUT is tolerant of random delays caused by metastability.

Low-power intent awareness—Questa CDC accepts your UPF file without modification to ensure low-power circuitry does not introduce CDC-related issues. Specifically, Questa CDC considers all isolation and retention cells, power domains with dynamic voltage, and frequency scaling (DVFS) and verifies voltage domain crossing (VDC) paths.

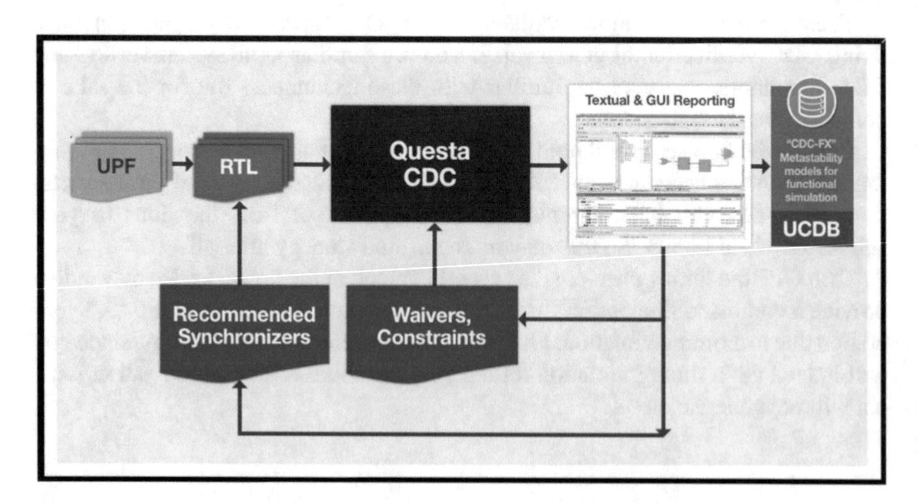

Fig. 8.11 Mentor Questa CDC methodology

Chapter 9
Low-Power Verification

Chapter Introduction
Low-power verification has become probably one of the most complex design verification problems to address under the design verification umbrella. Verification of low power is not simply restricted to checking for isolation cells, retention cells, and power domain ON/OFF conditions, but it also needs to check to see if the applied low-power techniques indeed improve the battery life without affecting performance!

9.1 Power Requirements: Current Industry Trend

Figure 9.1 describes a real dilemma that the industry faces. The number one issue in today's devices is the ever-shrinking form factor which means the room for large battery is obviously shrinking. This translates into designing devices that use as low power as possible. The requirements for power do not change, but the amount of logic onto VLSI chips keeps increasing. This means the power consumption trend keeps going up. The challenge is to narrow the gap between power requirements and power trend. Also, one can mention the ever-increasing need for green electronics, the success of mobile electronics operated from batteries with limited lifetime between recharges, and the emergence of applications requiring near zero-power energy (wireless sensor networks, implantable devices for health monitoring or smart cards).

Managing power consumption is even more challenging in the context of SoC (System on Chip) design. SoCs integrate various kinds of digital and analog blocks. Digital block may be either pure processor, microcontroller, or DSP, or it can be made of hardwired functions with a mixture of standard cells and SRAM memories. As a matter of fact, several dozens of CPUs, several hundreds of SRAM cuts, and few millions of standard cells may be integrated today in a single chip. For consistent power modeling, the external memory, battery, clock, and any other device on the platform that interfaces with the SoC have also to be considered.

A.B. Mehta, *ASIC/SoC Functional Design Verification*,
DOI 10.1007/978-3-319-59418-7_9

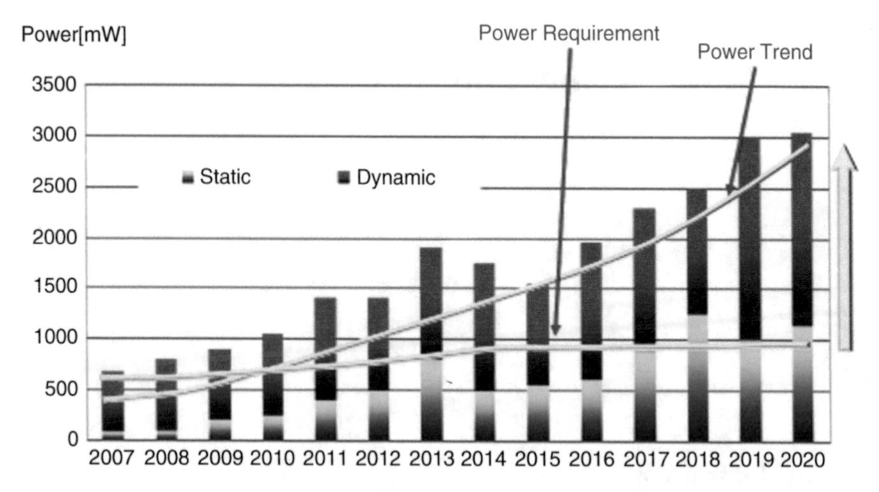

Fig. 9.1 Power requirements vs. power trend

Additional difficulty stems from the fact that the integrated IPs may come from multiple providers, using different methodologies, CAD tools, and description formats. When a SoC is built, some of the IPs are reused, and some of them are designed for the first time. For the former, accurate models can be obtained from existing data, whereas for the latter only rough estimates may be available at first. It is important to IP integrators to have tools able to cope with this heterogeneity.

There are two types of power that we need to worry about. These are the dynamic power (when transistors switch state from 1->0 or 0->1) and the static power (when the transistor is in quiescent state). With giga-large integration of transistors, a multitude of functions will be executing simultaneously to draw a large amount of dynamic power, and we need to architect a solution for it and verify that solution.

With process technologies below 14 nm, static power consumption has become a prominent and, in many cases, dominant design constraint. Due to the physics of the smaller process nodes, power is leaked from transistors even when the circuitry is quiescent (no toggling of nodes from 0 to 1 or vice versa). New design techniques have been developed to manage static power consumption. Power gating or power shutoff turns off power for a set of logic elements. Back–bias techniques are used to raise the voltage threshold at which a transistor can change its state. While back bias slows the performance of the transistor, it greatly reduces leakage. Logic design strategies to minimize static power and dynamic power are beyond the scope of this book.

We will focus on dynamic power *verification* in this book. What are high level strategies to lower dynamic power and more importantly how do you verify that those strategies do not affect SoC functionality? We will discuss power-management controllers, isolation cells that logically and/or electrically isolate a shutdown power domain from "powered-up" domains, level-shifters that translate signal voltages from one domain to another, and retention registers to facilitate fast transition from a power-off state to a power-on state for a domain.

9.2 Dynamic Low-Power Verification Challenges

How do you address the increasing demands of power due to increased integration of functions at smaller geometries? An entire discipline in academia and industry is vying to come up with technologies and methodologies to reduce dynamic power during functioning of a device.

Low-Power *Verification* Challenges

1. **Power vs. performance**

Requirement is to lower the power while maintaining or improving the performance. Techniques such as Dynamic Voltage and Frequency Scaling (DVFS) are being deployed to address this challenge. But how would you verify that the DVFS scheme (or any other technique for that matter) does not affect the functionality of the device? That's a verification challenge. You need to be able to verify functionality, power consumption, and performance improvement all together. Simply verifying that the functionality is correct is not sufficient. If the power consumption is high, correct functionality does not achieve the goal of low-power design.

2. **Dynamic power analysis and verification is dependent on use case**

In other words, dynamic power consumption is dependent on the activity level in the device. The higher the toggle frequency, the higher the dynamic power. One can indeed use scan-based vectors to induce such "activity" within the device, but experience has shown that such power analysis is too pessimistic and will require very pessimistic changes to device architecture.

What you need are dynamic simulation vectors that produce "real life" traffic for each subsystem (e.g., audio, video, graphics) to induce the highest possible "activity" (aka traffic) to get highest (worst case) power consumption numbers.

Power consumption is meaningful only when related to application use cases. For this, scenarios are built representing applications in terms of IP active/idle state sequencing and values of parameters considered for power computation (bandwidth, MIPS, etc.). Ideally, scenarios are described independently of implementation for several reasons:

- Same scenario may be run on several hardware platforms (low power or not, high end or low end…).
- Models are built by hardware designers or architects and scenario by software teams which may not be aware of hardware constraints.
- Separation of concerns principle is respected, and thus complexity is highly reduced.

To mimic real life use cases, there is also a need to run these subsystems *concurrently* to get real traffic scenarios during simulation. For example, in a cell phone, you could be talking to someone at the same time using GPS to navigate your travel. Real world devices handle heavy multitasking, and we must reflect such scenarios during simulation.

3. **Requires coordination with data from physical design. Iteration between physical design and RTL/verification**

For example, you need to coordinate power switching, gating, and isolation numbers from physical design with the power strategies you deployed during RTL design. If they are not in concert, the RTL (and above) architecture needs to change, and new verification effort needs to be undertaken thereof.

4. **DFT and low-power verification. What's the power consumption in scan mode?**

If the power consumption is way too high during maximum toggle scan patterns, you may need to change your DFT strategy to allow successful testing of the device on a tester. Verification of DFT architecture and RTL implementation and verification thus become ever more important.

5. **Dynamic lower power verification at ESL (TOM 2.0)**

Dynamic power analysis at RTL is proving to be too late in the design cycle. Changes in architecture and DFT strategies at RTL require significantly more effort than making such changes at higher level (transaction level) abstractions. The new methodologies are centered around building Transaction Level (TLM) Virtual Platform(s) upon which you can perform dynamic power analysis. We have devoted the entire Chap. 11 on ESL (Electronic System Level) TLM2.0 (Transaction Level Modeling—Standard 2.0) on ESL/TLM2.0 technology and methodology.

Analyzing power at ESL is one thing, but how would you verify that your defined architecture will indeed reduce dynamic power? How will you measure power at ESL? What kind of vectors (or software application) will you supply at ESL? How would you measure power vs. performance at ESL? Without such verification, you would not know if you need to make architectural changes to your design before committing to RTL.

9.3 UPF (Unified Power Format)

To address these challenges, among many possible solutions, the industry decided to come up with a Unified Power Format (UPF) that allows you to describe in a *standard* way the power intent of your design.

The idea behind coming up with a power format and simulation support thereof is to help verify functionality of the device with its power intent architecture *without* modifying RTL for power-reducing features such as power gating, power switch, isolation cells, retention cells, etc... UPF and the supporting simulator are designed such that the simulator uses UPF to "superimpose" power-related RTL code, based on UPF, without changing the RTL. If this does not make sense now, hold on. We'll soon dive into detail. The idea is also to have a common/unified format that RTL, gate, and even physical design can understand. Without such unified power format, the assumptions at RTL could be misinterpreted at gate level and then again at physical level. And finally, UPF is also an IEEE industry standard and not tied to any specific simulator or EDA vendor. (Amen to that!)

Here's a high-level summary of UPF.

- UPF = Unified Power Format = IEEE 1801–2015 Accelera Standard.
- UPF is intended:
  - For specifying power intent for an electronic design.
  - For use in verification of the structure and behavior of the design in the context of a given power-management architecture.
  - For driving implementation of that power-management architecture.
  - The main idea behind UPF is to keep the power specification separate from the functional specification and to use their ability to express power requirements and constraints to tools throughout the RTL-to-GDSII flow.
- UPF supports:
  - Incremental refinement of power intent specifications required for IP-based design flows
- Files written to this standard annotate an electric design with the power and power control intent of that design. Elements of that annotation include:
  - Power supplies: supply nets, supply sets, power states
  - Power control: power switches
  - Additional protection: level shifters and isolation
  - Memory retention during times of limited power: retention strategies and supply set power states

Why UPF?

- Support low-power design techniques through the entire design cycle using a single file format.
- Accurately define and capture lower power design intent, modes, and constraints.
- Support design implementation as well as functional verification.
- Support physical level design as well (e.g., floor plan and power grids).
- Serve as common constraint for all tools (synthesis, APR, timing, DFT).
- Allow RTL simulators (or even ESL for that matter) to superimpose UPF on design *without* actually modifying the power intent logic of the design.

9.3.1 UPF Evolution

The Unified Power Format (UPF) was developed to enable modeling of these new power-management techniques and to facilitate automation of design, verification, and implementation tools that must account for power-management aspects of a design. The initial version of UPF, developed by the Accelera Systems Initiative, focused primarily on modeling power distribution and its effects on the behavior of a system. In May 2007 that initial version was donated to the IEEE, and in March 2009 a new version, IEEE Std. 1801, was released. That update of UPF added many new features, including the concept of successive refinement, more abstract modeling of system power states, and more abstract modeling of supply networks.

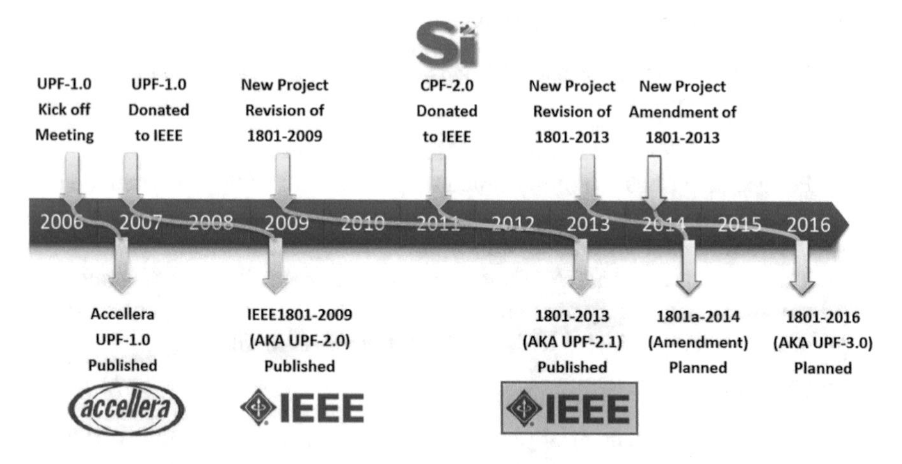

Fig. 9.2 UPF evolution (courtesy IEEE Standards Association)

IEEE Std. 1801–2016 includes enhanced concepts for modeling power states and transitions at all levels of aggregation; enhanced support for methodologies such as successive refinement and bottom-up implementation; and a detailed information model that serves as the basis for enhanced UPF functions and query functions. This current version also provides support for component power modeling for system-level (ESL/TLM2.0) power analysis in virtual prototyping applications.

Here's a simple evolution chart of UPF (courtesy IEEE). Refer to Fig. 9.2. Even though the effort started in 2007, it is only recently that the UPF methodology has taken off. All major design companies use UPF for their low-power design and verification.

The 1801 Full Revision PAR (2015/16) Project was approved at the June 2013 IEEE-SA board meeting. It will extend scope of "Power Intent" up toward system level and add power modeling.

9.4 UPF Methodology

To reiterate the discussion above, the Unified Power Format (UPF) provides the ability for electronic systems to be designed with power as a key consideration early in the process. UPF accomplishes this by allowing the specification of what was traditionally physical implementation-based power information early in the design process—at the register transfer level (RTL) or earlier. Figure 9.3 shows UPF supporting the entire design flow. UPF provides a consistent format to specify power-design information that may not be easily specifiable in a hardware description language (HDL) or when it is undesirable to directly specify the power semantics in an HDL, as doing so would tie the logic specification directly to a constrained power implementation. UPF specifies a set of HDL attributes and HDL packages to

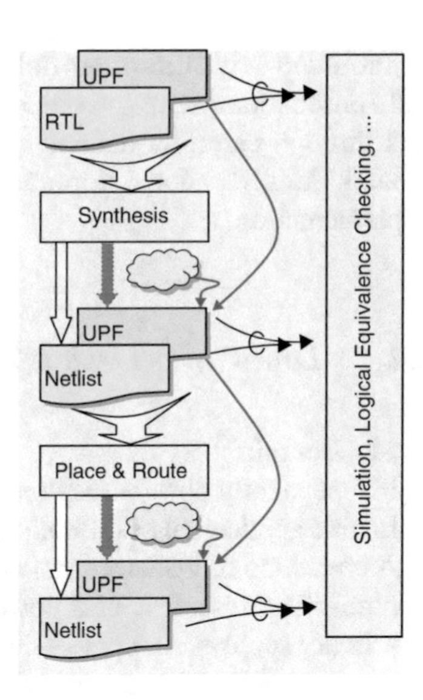

Fig. 9.3 UPF methodology

facilitate the expression of power intent in HDL when appropriate. UPF also defines consistent semantics across verification and implementation to check that what is implemented is the same as what has been verified.

As shown in Fig. 9.3 UPF methodology, UPF files are part of the design source and, when combined with the HDL, represent a complete design description: the HDL describing the logical intent and the UPF describing the power intent. Combined with the HDL, the UPF files are used to describe the intent of the designer. This collection of source files is the input to several tools, e.g., simulation tools, synthesis tools, and formal verification tools. UPF supports the successive refinement methodology where power intent information grows along the design flow to provide needed information for each design stage.

Simulation tools can read the HDL/UPF design input files and perform RTL power-aware simulation. At this stage, the UPF might only contain abstract models such as power domains and supply sets without the need to create the power and ground network and implementation details.

A user may further refine the UPF specification to add implementation-related information. This further-refined specification may then be processed by synthesis tools to produce a netlist and optionally update the UPF files accordingly.

In those cases, where design object names change, a UPF file with the new names is needed. A UPF-aware logical equivalence checker can read the full design and UPF files and perform the checks to ensure power-aware equivalence.

Place and route tools read both the netlist and the UPF files and produce a physical netlist, potentially including an output UPF file.

In the end, UPF is a concise, power intent specification capability. Power intent can be easily specified over many elements in the design. A UPF specification can be included with the other deliverables of intellectual property (IP) blocks and reused along with the other delivered IP. UPF supports various methodologies through carefully defined semantics, flexibility in specification, and, when needed, defined rational limitations that facilitate automation in verification and implementation.

9.4.1 Low-Power Design Terminology/Definitions

Power Domain

- Independently powered regions.
- Enable application of different power reduction techniques in each region.
- A collection of instances that are treated as a group for power-management purposes. The instances of a power domain typically, but do not always, share a primary supply set. A power domain can also have additional supplies, including retention and isolation supplies.

Composite Power Domain

A power domain consisting of subordinate power domains called subdomains. All subdomains in a composite domain share the same primary supply set.

Power Supply Network

- Abstract description of power distribution (ports, nets, sets, and switches)

Power State

- A subset of the functional states of an object that have the same characteristics with respect to power supply (for a supply set) or power consumption (for a power domain, composite domain, group, model, or instance)

Power State Table

- The legal combinations of states of each power domain. Table that specifies the legal combinations of supply states for a set of supply objects (supply ports, supply nets, and/or supply set functions)

State Retention

- To save essential data when power is off. Enhanced functionality associated with selected sequential elements or a memory such that memory values can be preserved during the power-down state of the primary supplies
- To enable quick resumption after power up

Isolation

- To ensure correct electrical/logical interaction between domains in different ON/OFF power states.
- Isolation cell is an instance that passes logic values during normal mode operation and clamps its output to some specified logic value when a control signal is asserted.

Level Shifting

- To ensure correct communication between different voltage levels.
- Level shifter cell is an instance that translates signal values from an input voltage swing to a different output voltage swing.

9.5 UPF: Detailed SoC Example

We will go through a complete SoC example and create a step-by-step UPF for it. The design has SoC as the top-level module which instantiates the Video Sub block, Video_SB, and the Audio Sub block, Audio_SB. We further show hierarchy under Video_SB.

9.5.1 Design/Logic Hierarchy Navigation

Let us first see how UPF allows us to specify the hierarchy of a design. Note that in this hierarchy, you may also include the testbench as your top-level module. That testbench module can go above the "SoC" module or can be hierarchical to the side of it (horizontal to it).

In this design, SoC is the top-level module that instantiates the Video and Audio subsystems. Under Video there is a logic hierarchy. Let us see how UPF allows us to define this hierarchy.

Note that all the names in Fig. 9.4 are *instance* names and not the module names.

Consider Video_SB as our DUT for this illustration. To set the design (DUT) top, the following command needs to be used:

```
set_design_top SoC/Video_SB
```

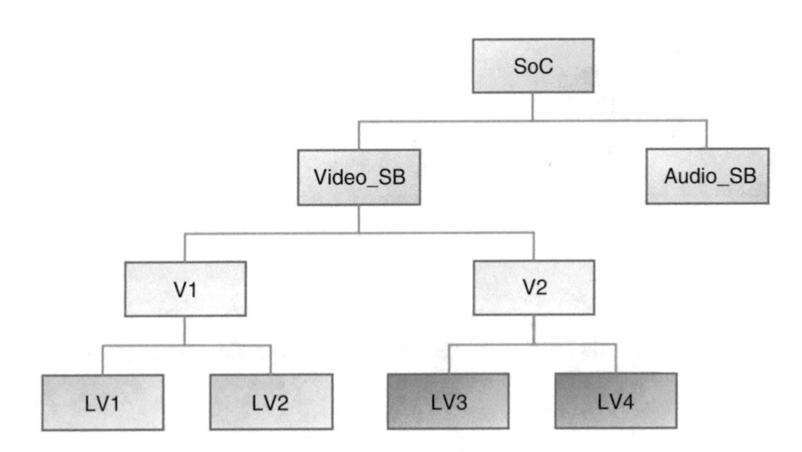

Fig. 9.4 UPF: design/logic hierarchy navigation

Video_SB now becomes the "current" scope for the subsequent commands. Now you can set a scope under this design_top and change it with the following commands.

To leave the scope to where it is, use the following command. Note that "." keeps the current scope unchanged.

```
set_scope .
```

To set the scope to "LV1":

```
set_scope V1/LV1
```

To change the scope one level up (i.e., the parent of the current scope, which is V1):

```
set_scope ..
```

So, now the scope is "V1."

To change the scope to 'LV2' and since the current scope is 'V1', we simply do the following:

```
set_scope LV2
```

9.5.2 Power Domain Creation

Now, let us see how we create power domains. In our example design (Fig. 9.5), we have identified the following power domains. All power domains are under the design_top Video_SB. The four domains are:

1. Video_SB/Video_PD
2. Video_SB/V1_PD
3. Video_SB/V2_PD
4. Video_SB/LV12_PD
5. Video_SB/LV34_PD

So, how does UPF allow us to identify these power domains?

First, set the scope to Video_SB.

```
set_scope Video_SB
```

Then create power domain Video_PD with the following UPF command. Note that *-include_scope* tells UPF to create Video_PD power domain in current scope.

```
create_power_domain Video_PD -include_scope
```

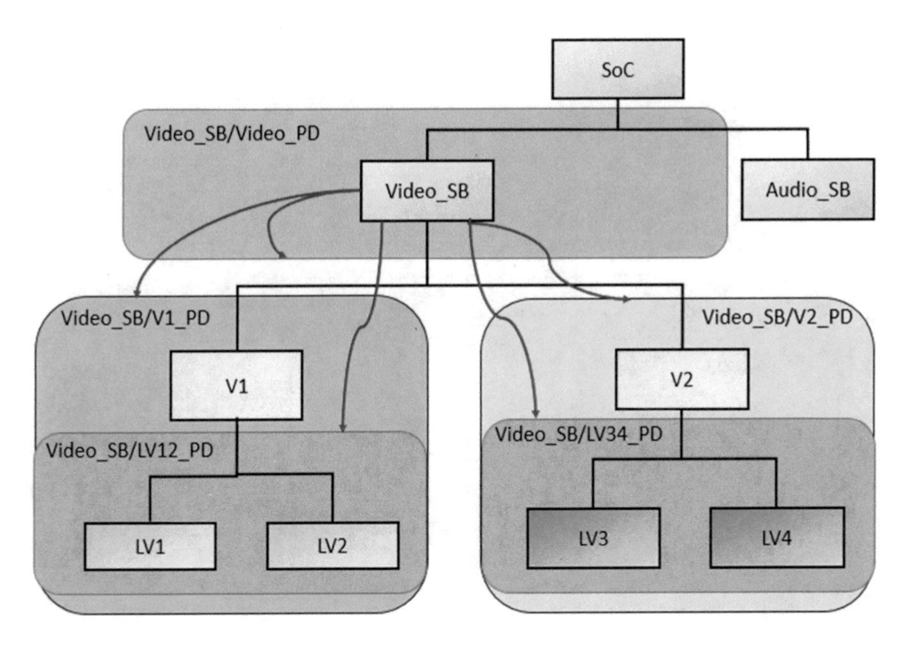

Fig. 9.5 UPF: power domain creation – 1

Now create power domain V1_PD. This command takes the instance V1 out of the hierarchy of Video_SB and creates its own power domain called V1_PD. In other words, in the UPF world, a more specific command takes over a more generic command. In this case, specifically calling "V1" takes precedence over the power domain created above ("Video_PD") which would have included the entire hierarchy.

```
create_power_domain V1_PD -elements {V1}
```

On similar thought process, the following creates power domain LV12_PD for instances LV1 and LV2.

```
create_power_domain LV12_PD -elements {V1/LV1 V1/LV2}
```

In similar fashion, you can create power domains for the hierarchy under "V2."

```
create_power_domain V2_PD -elements {V2}
create_power_domain LV34_PD -elements {V2/LV3 V2/LV4}
```

Here's another example of the same design where the power domains are created slightly different. This is to illustrate how UPF helps identifying power domains in different hierarchies.

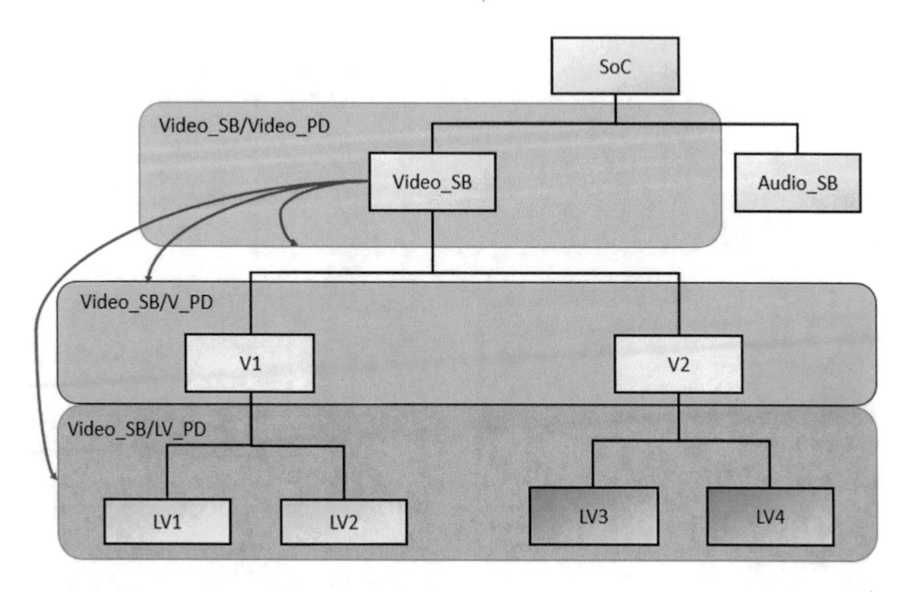

Fig. 9.6 UPF: power domain creation—2

The UPF commands to identify power domains in Fig. 9.6 are:

```
set_scope Video_SB
create_power_domain Video_PD -include_scope
create_power_domain V_PD -elements {V1 V2}
create_power_domain LV_PD -elements {V1/LV1 V1/LV2 V2/LV3 V2/LV4}
```

9.5.3 Supply Power to the Power Domains: Supply Network

Okay, now let's see how we supply power to these power domains (Fig. 9.7). Creating the supplying power requires the following four commands:

```
create_supply_port
create_supply_net
connect_supply_net
set_domain_supply_net
```

Let's look at each and see how they apply to the power domains shown in Fig. 9.7.

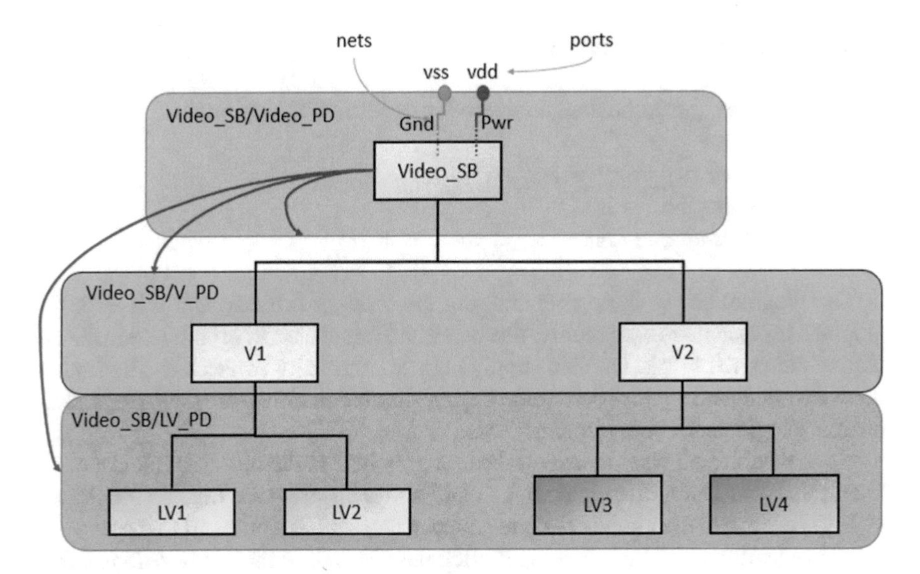

Fig. 9.7 UPF supply network

9.5.3.1 create_supply_port

As the name suggests, *create_supply_port* creates the power/voltage ports for the design. In this design, we have the power (VDD) and ground (VSS) ports. This is how they are declared in UPF.

```
create_supply_port VDD -direction in
create_supply_port VSS -direction in
```

9.5.3.2 create_supply_net

Now that we have created the supply ports, we need two nets to connect them to the design instances. So, first we create two nets. Here's how:

```
create_supply_net Pwr -domain Video_PD
create_supply_net Gnd -domain Video_PD
```

9.5.3.3 connect_supply_net

Now that we have created the supply ports, we need to connect those. Here's how:

```
connect_supply_net Pwr -ports (VDD)
connect_supply_net Gnd -ports (VSS)
```

9.5.3.4 set_domain_supply_net

We now must establish the domain name for these nets. Here's how:

```
set_domain_supply_net Video_PD \
-primary_power_net Pwr \
-primary_ground_net Gnd
```

The diagram shows these nets entering the Video_SB block. But this is only a graphics representation. In reality, these nets are connected to *all* module instances within Video_SB block. Domain supply nets are implicitly connected. The state of these Pwr and Gnd sources determine during simulation if they are ON/OFF and the resulting logic under both circumstances.

Now, what if you want to extend these supply nets to the other sub blocks of the hierarchy? You can either create new VDD and VSS ports for the sub block (e.g. V_PD) or "reuse" the already declared supply nets and extend them to the new sub block power domain. The following section describes the commands to do just that.

9.5.3.5 create_supply_net -reuse

Referring to Fig. 9.8:

```
create_supply_net Pwr -reuse -domain V_PD
create_supply_net Gnd -reuse -domain V_PD
```

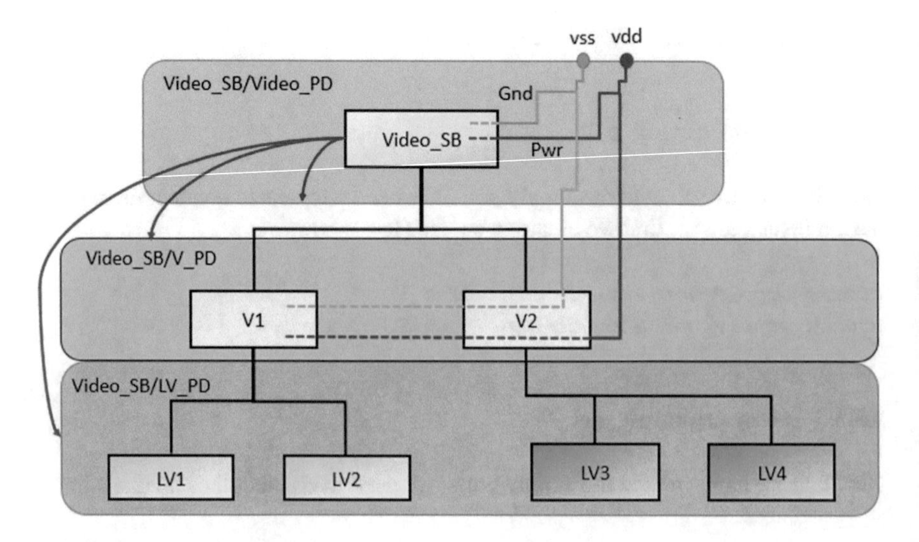

Fig. 9.8 UPF: supply network reuse

And again, as shown above we must establish the domain name for these nets. Here's how:

```
set_domain_supply_net V_PD \
-primary_power_net Pwr \
-primary_ground_net Gnd
```

9.5.4 Power Switch Creation

So, how do you turn power ON/OFF for a given power domain? You need a power switch for that embedded in the power domain that needs to be turned ON/OFF (Fig. 9.9).

First, we create a "switched" supply net called VDDsw. Here we plan to switch ON/OFF "vdd."

```
create_supply_net VDDsw -domain V_PD
```

Once that is done, the gist of power switch creation comes into picture. Here's how you do it. The explanation follows.

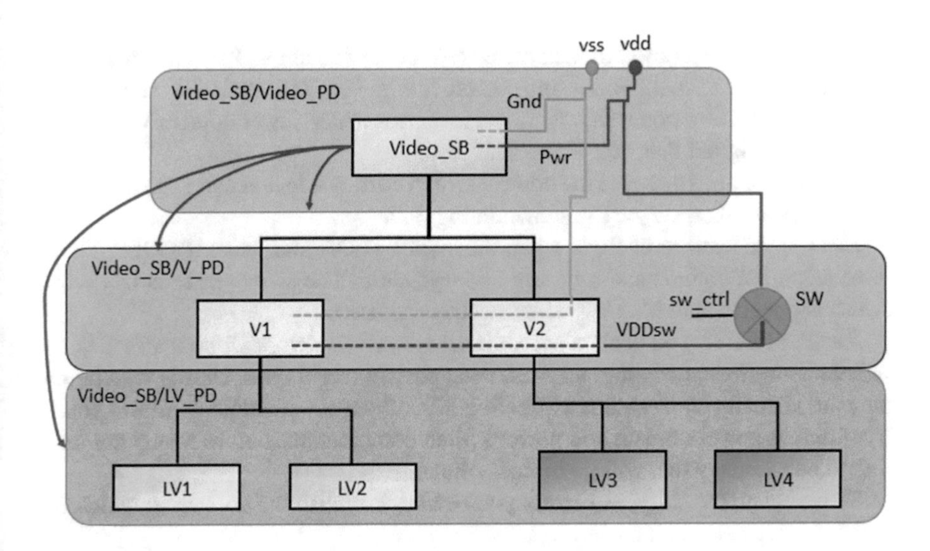

Fig. 9.9 UPF: power switch creation

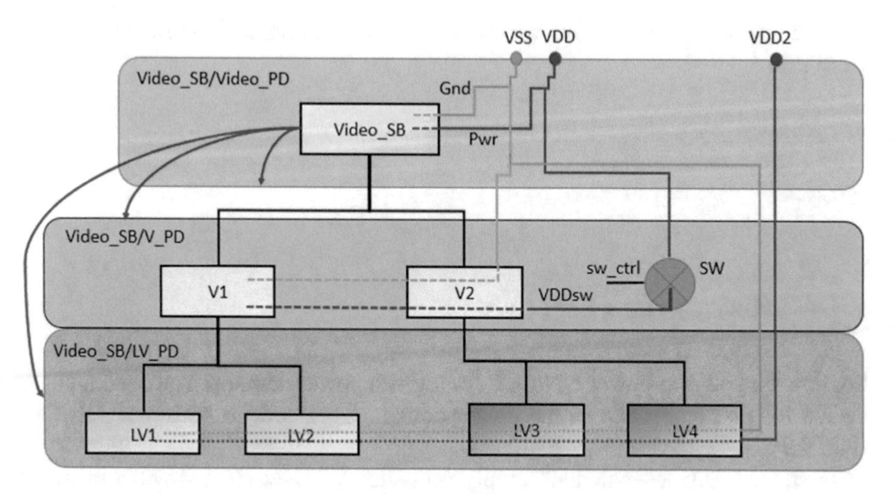

Fig. 9.10 UPF: supply port states

9.5.4.1 create_power_switch

```
create_power_switch SW -domain V_PD \
-input_supply_port {pwin Pwr} \
-output_supply_prt {pwout VDDsw } \
-control_port {swctrl sw_ctrl} \
-on_state {Pwon swctrl} \
-off_state {Pwoff !swctrl}
```

First, we create a power switch called SW in the domain V_PD. Then we assign input and output to this switch. As you note in Fig. 9.9, VDDsw is the output of the switch which is then connected to all the instances under power domain V_PD. The input is our original Pwr net.

Second, we need to create the control switch port. We do that using the command *-control_port* called sw_ctrl as shown in Fig. 9.9.

Finally, we need to declare, when the switch is ON and when it's OFF. That's done using the commands *-on_state* and *-off_state*. The switch state is ON when swctrl is positive, and it's OFF when swctrl is negative.

Rather intuitive, you declare such a power domain network without affecting (or rather physically connecting) all these nets, ports, and switches. During simulation, they are virtually connected not affecting RTL. Such a methodology allows you to try different power domain and power switch combinations before finally committing to one strategy that will go to final silicon.

To complete the story, as before, you need to establish the domain in which the power supply nets belong.

```
set_domain_supply_net V_PD \
-primary_power_net VDDsw \
-primary_ground_net Gnd
```

9.5.5 *Supply Port States*

Let us create one more supply port named VDD2 and connect it to the power domain LV_PD (Fig. 9.10). The first thing to note here is that we did not create a supply net (as we did for supply port VDD and ground port VSS in previous examples). A port name can be used where a net name is allowed in that you can directly connect the supply/ground ports to the instances of a given power domain.

9.5.5.1 add_port_state

Now that the supply for each power domain has been specified and connected to the instances in their respective domains, we need to specify the values these supply ports can take on. These "values" are referred to as port "states."

To define the state of a port, we use the following command:

```
add_port_state VDD -state {ON_10 1.0}
```

This command states that VDD is in full ON state with 1.0 volts. Since ON state is the only state defined for VDD, this supply is always ON.

For the internally switched voltage VDDsw (output of switch SW), we should specify both the ON and OFF states (since it is switched). Here's the command to do that.

```
add_port_state SW/VDDsw -state {ON_10 1.0} -state {OFF off}
```

Note that a positive (or zero) voltage value indicates an ON state while "OFF" states the OFF state of the port.

Similarly, for the VSS, the common ground port, we can specify that it's at 0 volts as follows:

```
add_port_state VSS -state {ON_00 0.0}
```

This states that VSS is full ON but at 0.0 volts.

Note also that the "labels" ON_10, ON_08, etc. are user defined and have no bearing on the state of the port.

9.5.6 *Power State Table*

In a typical SoC, there are tens of power domains, switched or unswitched. It is hard to manually keep track of all the different combinations of the state. You can always create a separate XL spreadsheet and constantly synchronize your UPF state definitions with that in the XL. But this is very error prone. Note that such a state table is useful in your state retention and isolation strategy. That will be described in the coming sections.

	Video_PD	V_PD	LV_PD	
State \ Supply	VDD	VDDsw	VDD2	VSS
Normal	ON_10	ON_10	ON_08	ON_00
Sleep	ON_10	OFF	ON_08	ON_00
Hibernate	ON_10	OFF	OFF	ON_00

Fig. 9.11 UPF: power state table

Here are the commands to create a Power State Table:

```
create_pst
add_pst_state
```

Let us assume with the table in Fig. 9.11 that we want to create and see how above commands help us do just that.

9.5.6.1 create_pst

create_pst defines the header of the table. It provides the supply port names, as follows:

```
create_pst PST1 -supplies {VDD, VDDsw, VDD2, VSS}
```

These values define the header (and columns) of the table.

9.5.6.2 add_pst_state

Now, let's add the Power State Table states for each of the state that the SoC will be in, namely, Normal, Sleep, and Hibernate. These states correspond to each of the supply port (net) specified in the header. This is how that is done:

```
add_pst_state Normal   -pst PST1 -state {ON_10, ON_10, ON_08, ON_00}
add_pst_state Sleep    -pst PST1 -state {ON_10, OFF, ON_08, ON_00}
add_pst_state Normal    -pst PST1 -state {ON_10, OFF, OFF, ON_00}
```

9.5.7 State Retention Strategies

State retention means that the "state" of the logic in a power domain is preserved before the power is shut down. When power is brought back up again, the state of the logic before power down is restored. This helps with very fast recovery of the block from power down to power-on state. Retention is also required if you simply cannot bring the block into the state that it was before it was powered down.

Let us continue with the state table of Fig. 9.11. This table tells us which power domains need state retention (if), isolation, and level shifting.

9.5.7.1 set_retention

The power domain V_PD is OFF in Sleep and Hibernate state (Fig. 9.12). Let us assume that we do need to retain the state of the V_PD block so that the V_PD block can restart quickly as soon as it comes out of Sleep or Hibernate state. That's done as follows:

```
set_retention V_PD_retention \
-domain V_PD \
-retention_power_net Pwr \
-retention_ground_net Gnd
```

This command specifies that all state elements of the V_PD power domain logic must be retained when it enters either the Sleep mode or the Hibernate mode. The retention logic should use the Pwr and Gnd nets in that block. *Note that you are powering the retention registers using the ALWAYS_ON VDD and VSS nets and not the switched power nets.* The switched power supply powers the rest of the logic in the V_PD power domain. Figure 9.12 shows the retention registers in blocks V1 and V2 in V_PD power domain. They are powered using VDD and VSS and will remain powered ON when V1 and V2 of the V_PD domain are shut OFF.

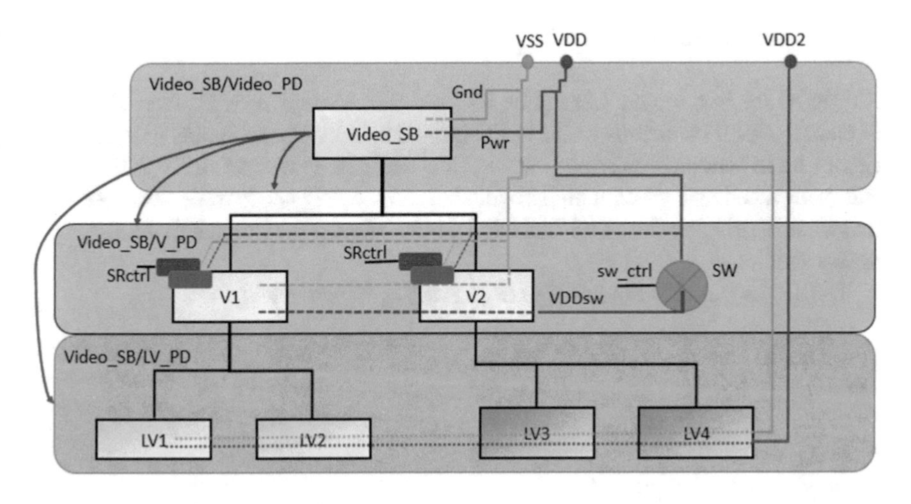

Fig. 9.12 UPF: state retention strategy

	Video_PD	V_PD	LV_PD	
State \ Supply	VDD	VDDsw	VDD2	VSS
Normal	ON_10	ON_10	ON_08	ON_00
Sleep	ON_10	OFF	ON_08	ON_00
Hibernate	ON_10	OFF	OFF	ON_00

Fig. 9.13 State table showing isolation requirements

9.5.7.2 set_retention_control

The set_retention_control command adds control to these retention registers, as shown in Fig. 9.12. The command for our design is as follows:

```
set_retention_control V_PD_retention \
-domain V_PD \
-save_signal {SRctrl posedge} \
-restore_signal {SRctrl negedge}
```

Note that we did not use retention registers in the LV_PD domain. This is just to point out that you don't have to have retention strategy for each power domain. Only if the state prior to shut down cannot be restored quickly (or reinitialized) that we need retention registers.

9.5.8 Isolation Strategies

9.5.8.1 set_isolation

As you notice in our state table (Fig. 9.13), V_PD domain is OFF when Video_PD domain in ON. That means when V_PD turns OFF, its outputs (inputs to Video_PD) cannot be in random (unknown) state. They must be isolated from V_PD outputs and maintained at a predictable known state. Video_PD needs to be designed such that when V_PD goes into OFF state and that its outputs are isolated that it functions as expected.

The set_isolation for our design (Fig. 9.14) is shown below.

```
set_isolation V_PD_isolation \
-domain V_PD \
-applies_to outputs \
-clamp_value 0 \
-isolation_power_net Pwr \
Isolation_ground_net Gnd
```

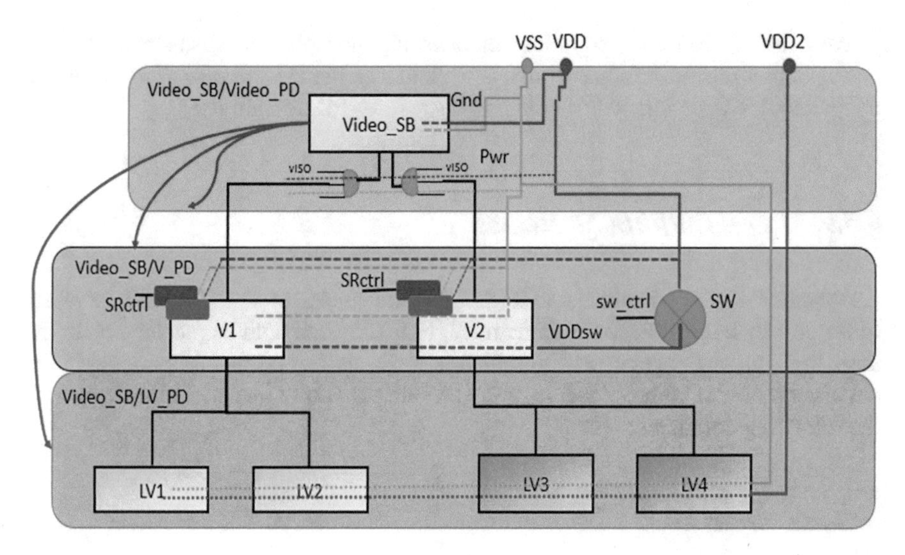

Fig. 9.14 UPF state isolation strategy

This command specifies that the power domain V_PD's outputs need to be isolated. *-domain V_PD* and *-applies_to outputs* facilitate this function. We also need to specify at what logic level the outputs of isolation cells will be "clamped" to, once the power is turned OFF. In our design, we have chosen the value logic 0. And finally, we specify the power and ground nets that will power the isolation cells when V_PD power is turned OFF.

9.5.8.2 set_isolation_control

set_isolation_control command specifies the control signals to be used to enable isolation cells when V_PD is turned OFF. We also need to specify the location where these cells will be inserted. Will they be inserted within the V_PD block or its parent block Video_PD? All this is accomplished by the following command:

```
set_isolation_control V_PD_isolation \
-domain V_PD \
-isolation_signal vISO \
-isolation_sense high \
-location parent
```

This command specifies that we want to apply isolation control to the domain V_PD. The control signal name is vISO, and the control will be enabled when vISO is in logic high state. Finally, we specify that the isolation cells should be placed in V_PD's parent block. That block in our design is Video_PD. Hence, as shown in Fig. 9.14, the isolation cells are placed in the Video_PD block and powered by Pwr and Gnd nets.

Note that *-location self* command dictates that the isolation cells be placed in the power domain itself. In other words, if we had specified *-location self* in the above command, the isolation cells would be places in V_PD domain itself.

9.5.9 Level Shifting Strategies

As we can see from our state table (Fig. 9.15), the voltage of V_PD domain is higher (different) than the voltage for domain LV_PD. They can both be powered up at the same time, but they operate at different voltage levels. For correct functioning of the circuit, we need a voltage level shifter. UPF allows you to insert level shifters with the following command:

```
set_level_shifter LV_PD_LS
-domain LV_PD \
-threshold 0.1 \
-applies_to both \
-rule both \
-location self
```

This command specifies that level shifters should be added to the inputs and outputs of the LV_PD domain, if the threshold between the two neighboring domains (i.e., LV_PD and V_PD) is at least 0.1 volts. "-applies_to both" indicates that the level shifters should be applied to both the inputs and the outputs. And "-rule both" specifies that the level shifters should be able to level shift both from High_to_low and Low_to_high. Finally, "-location self" indicates that the level shifters should be places in the LV_PD domain itself. Note that you do not need a control signal for level shifters to work. They simply sample the voltage and level shift them. Level shifters do require power (obviously), and UPF 1.0 explicitly supplies power to level shifters from the source and sink logic. But this limitation is removed from UPF 2.0, and you can explicitly provide Pwr and Gnd information for the level shifters.

The following Fig. 9.16 shows the inserted level shifters:

	Video_PD	V_PD	LV_PD	
State \ Supply	VDD	VDDsw	VDD2	VSS
Normal	ON_10	ON_10	ON_08	ON_00
Sleep	ON_10	OFF	ON_08	ON_00
Hibernate	ON_10	OFF	OFF	ON_00

Fig. 9.15 Level shifting strategy

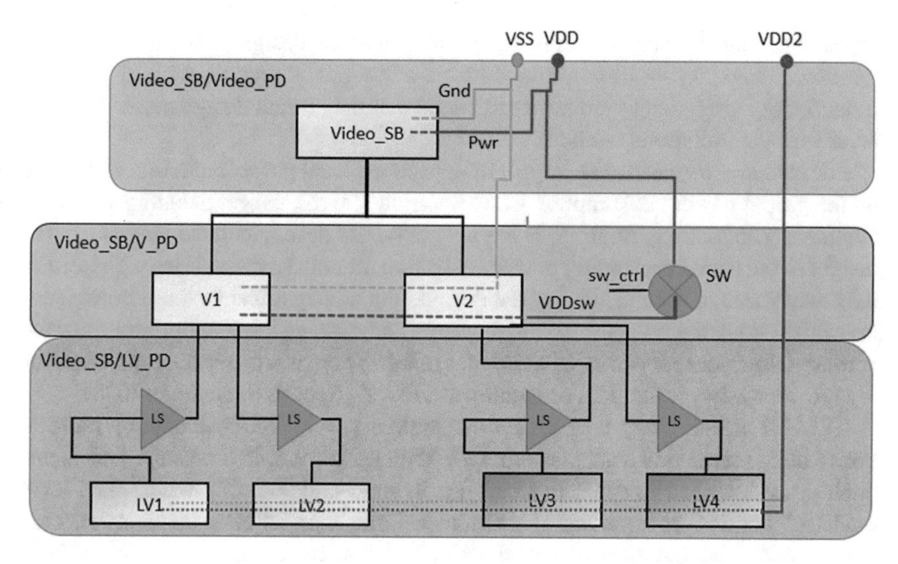

Fig. 9.16 UPF: level shifter strategy

9.6 Power Estimation at Architecture Level

One of the problems with power estimation at RTL is that it can be too late to make architectural changes to meet power requirements. Changes to internal memories, register files, interconnect topology, etc. are very time-consuming, error prone, and difficult overall. The burden on verification also increases since the testbench architecture may have to change as well and certainly you need to write new tests, modify response checkers (scoreboard), etc. In short, changing the RTL architecture and verifying it are a lot harder than making the same changes at Electronic System-Level architecture of an ASIC.

What about IP-based SoC designs? The typical system design now has known a paradigm shift with the reuse of Intellectual Properties (IP). An application can be now developed in a very short time with the association of existing MPSoC platforms. Although this design methodology enhances the designer efficiency and reduces the time to market, its weak point remains the consideration of the power consumption metric. Current system power estimation is obtained after design place and route or developing power models at the RTL level. At these levels, when the power estimation exceeds the power budget, the designer must backtrack on architecture and algorithm parameters. This operation is time-consuming, and the power estimation is not always obvious. Moreover, this estimation is not useful to design a new system or extend it for a complex embedded system. To improve the design flow effectiveness, it is necessary to adapt new approaches for considering the power metric in the design flow.

Reducing power consumption has undeniably become a popular subject in the EDA (Electronic Design Automation) field owing to its criticality. The issue is tricky because the goal of the existing flows is basically to implement some functionality, whereas power appears to be a nonfunctional property. It is a ubiquitous aspect that cannot be dealt with in a definite area of the design or at a given level in the flow. Hence trying to combine advanced low-power techniques and complex functionality implementation taking advantage of the current design methodologies faces serious limitations (Kaiser).

For instance, assessing the impact of applying a local power reduction technique on the global power consumption of a system in real life usage, meaning when the application software is running, is a tough task. The designers have indeed limited insight in the actual processing power repartition, in both space and time. Therefore, they may waste their time altering the design with nonsignificant overall power saving while taking the risk of introducing errors and missing the true opportunities for substantial power reduction. What is truly missing here is a consistent approach for low-power design at the ESL or architectural level (Report, September 2008).

TLM2.0 methodology (aka Electronic System Level, ESL methodology) allows you to build transaction level platform (aka Virtual Platform) in SystemC. This methodology is described in detail in Chap. 11. In short building a SystemC/TLM level model of the SoC runs orders of magnitude faster than its RTL counterpart. This allows quick trial and error of power estimation that can come close to 70–80% in ballpark of the RTL estimation. Transaction level traffic pattern recognition (e.g., IO traffic vs. memory traffic) is one of the ways to power estimate at TLM2.0 Virtual Platform level. There are other methods also available to reach the goal of power estimation and thereby power optimization at TLM2.0 level, such as Dynamic Voltage and Frequency Scaling (DVFS). These topics are beyond the scope of this book.

The graph in Fig. 9.17 shows the ability to optimize power at architectural level is almost 80%. This is because the Virtual Platform level architecture is much easier to modify and thereby improve power before committing to RTL.

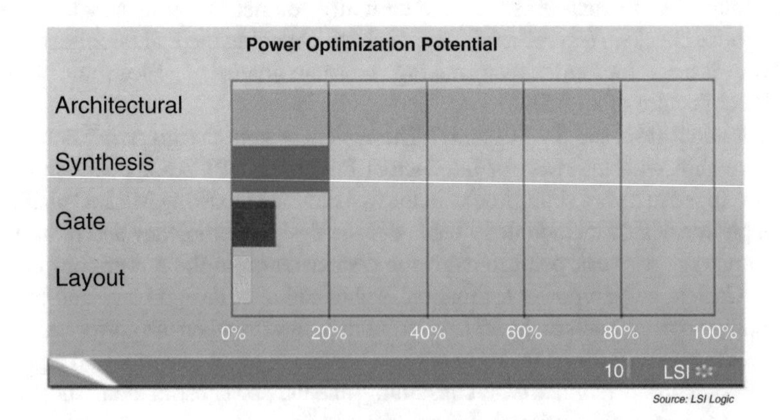

According to a study by LSI Logic, techniques available at the RTL synthesis phase have the ability to reduce power by 20 percent; those at the gate level offer a 10 percent reduction; while those at the layout level can reduce power by only 5 percent. Waiting until the RTL phase to begin optimizing for power is a wasted opportunity because power usage can be reduced by 80 percent at the ESL.

Souce Mentor / LSI

Fig. 9.17 Power estimation at architectural level

9.7 UPF Features Subset (IEEE 1801–2009)

Refer to Fig. 9.18.

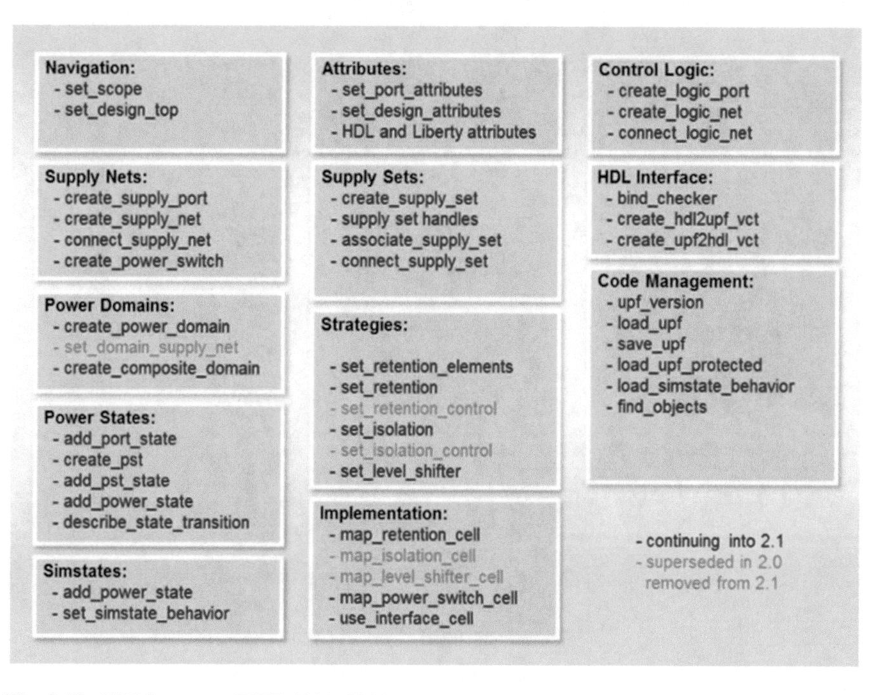

Fig. 9.18 UPF features—IEEE 1801–2009

Chapter 10
Static Verification (Formal-Based Technologies)

Chapter Introduction
Static verification is an umbrella term, and there are many different technologies that fall under it, for example, Logic Equivalence Check (LEC), Clock Domain Crossing (CDC) check, X-state verification, low-power structural checks, ESL ⇔ RTL equivalency, etc. This chapter will discuss all these topics and a lot more including state-space explosion problem, role of SystemVerilog Assertions, etc.

10.1 What Is Static Verification?

Static verification (compared to *dynamic* verification using simulation-based technologies) is the overarching term for a collection of techniques that use static analysis based on mathematical transformations to determine the correctness of hardware or software behavior in contrast to dynamic verification techniques such as simulation. As design size and simulation time have increased, verification teams have looked for ways to reduce the number of vectors needed to exercise the system to an acceptable degree of coverage.

SoC design complexity demands newer and better verification methods to accelerate verification and debug, as well as reduce overall schedule duration and improve schedule predictability. The verification of complex chips and systems is a highly challenging task. Techniques to reduce the verification schedule, improve predictability, and accelerate debug are highly desired by engineers and management alike.

Even the most carefully designed UVM testbench is inherently incomplete since constrained random methods can't hit every corner case. Unfortunately, this means that even after 100% functional coverage is achieved, there can still be showstopper bugs hiding in unimagined state spaces. Hence, formal verification plays a vital role in the verification of today's complex designs. Formal tools statically analyze a design's behavior with respect to a given set of properties (assertions), exhaustively

A.B. Mehta, *ASIC/SoC Functional Design Verification*,
DOI 10.1007/978-3-319-59418-7_10

exploring all possible input sequences in a breadth-first search manner to uncover design errors that would otherwise be missed.

One way to approach this problem is to move the bug identification earlier in the cycle as much as possible. When bugs are caught early, they are easier, faster, and cheaper to triage, debug, and fix. The problem is finding the tougher hard-to-stimulate bugs that result from seemingly impossible-to-think-of corner case scenarios. This is where formal methods come in, as they are less reliant upon the user to think of the potential scenarios that could trigger bugs. When combined with debugging tools and methods, the true power of formal verification is realized.

Static formal methods are techniques that can perform analysis on the design independent of or in conjunction with simulation and have the power to identify design problems that can otherwise be missed until very late in the project schedule, or even in the manufactured silicon, when changes are expensive and debug is highly challenging and time-consuming. When applied early in the design cycle, these methods can identify RTL issues such as functional correctness and completeness well before the simulation test environment is up and running.

The most fundamental difference between simulation and static formal is that the latter is implemented using mathematical techniques. This approach does not require a vector set and therefore *does not require a testbench* to be written to exercise the design. This saves time right away. However, the main advantage of this approach is that typically, a vector set is written to exercise only the behavior of the design in its expected mode of operation. In reality, the input to a block often deviates from the designer's initial expectations, and the design is then in an untested territory. There are of course practical reasons for this: it is hard to "expect the unexpected." SystemVerilog Assertions (properties) can test the design in all possible modes of operation and therefore can isolate bugs and undesired behavior that a designer might not have thought to test.

Additional applications of static verification technology can verify SoC connectivity correctness and completeness and help isolate differences between two disparate versions of the RTL design. Once the simulation environment is available, formal methods can complement simulation to add additional analysis for even better results, for example, for unreachable coverage goals. By employing formal techniques at the appropriate time in the design and verification process, bugs can be caught significantly earlier in the project, including hard-to-find bugs that typically elude verification until late in the project. The result is a higher quality design and overall schedule improvements as well as better predictability.

Of course, static tools are not that new to us, and we can compare this approach with something much more familiar, namely, "Static Timing Analysis." A timing simulation cannot possibly provide certainty we have hit the slowest set of vectors. If we don't pick the worst-case set of vectors, we won't get a worst-case result. It's therefore going to be impossible on a large design to learn much with timing simulation. The static timing tool will automatically isolate any paths that are too slow. So, wouldn't it be great if we could do the same thing for the design's functionality?

That's how the static formal technology was born.

10.2 Static Verification Umbrella

Let us first take a quick snapshot of the umbrella of static verification technologies.

1. Static formal verification (i.e., see that a piece of logic does not fail under *any* given input condition, and do this statically without the need for input simulation vectors).

   Note: *This type of verification is also known as model checking, static functional verification, or property checking. We will use static formal as the preferred terminology in this book.*
2. Static formal plus simulation *hybrid* verification.
3. Logic Equivalence Check (LEC) (discussed in Sect. 5). This is where you formally (i.e., without vectors) check the following for equivalency:

   a. RTL ⇔ RTL (critical path optimization, for example)
   b. RTL ⇔ gate (post-synthesis)
   c. Gate ⇔ gate (clock tree insertion, delay buffers for hold, etc.)
   d. Layout vs. schematic (LVS)

4. Clock domain crossing verification (Chap. 8).
5. RTL lint.
6. Structural checks.
7. Low-power structural checks.
8. X-state verification.
9. Connectivity verification.

There are a couple of big stumbling blocks with dynamic simulation that we have been using for ages. As the design complexity and size grow, these simulation techniques are dragging design verification to its knees.

1. Simulation runs *very* slow (100Hz–1Khz, if you have the fastest server, no other processes running, truck load of dynamic and static memories, and you are *very* lucky).
2. Even with the coverage tools available, one cannot guarantee that you have fully exhausted all verification possibilities for a given piece of logic, for example, something as simple as an asynchronous FIFO or as complex as a second-level cache coherency algorithm.

   a. Coverage is only as good as your model of the coverage. If you forgot to put all coverpoints, your coverage will be 100% before it really is.

3. Acceleration and emulation do speed up simulation by orders of magnitude, but they require test vectors. In other words, if you didn't write enough tests for all possible corner cases, what good will it do that you simulate your vectors fast? And what about compile times? You run a simulation in seconds and then wait for hours for compile to complete for the next round of simulation.

10.3 Static Formal Verification (Aka Model Checking Aka Static Functional Verification)

What we need is a way to "automatically" verify your logic under test with formal proof. Simply put, we need a tool that applies *all* possible combinations of input both in combinatorial as well as temporal domains. In other words, given a model of a system, exhaustively and automatically check whether this model meets the given specification.

So, what does static formal flow look like? Referring to Fig. 10.1:

1. Write a property in SystemVerilog Assertion (SVA) language (or PSL—Property Specification Language or OVL for that matter).
2. Supply assumptions (SVA "assume" statement) to the static formal tool.
3. The static formal tool mathematically derives a model for your RTL logic under test.
4. It applies all possible "stimuli" in combinational and sequential domain. It verifies that the property does not *fail* under any circumstance. It exercises all possible "logic cones" of a given logic block and proves that the assertion(s) are not violated.
5. If the property/assertion *fails*, it will give the exact scenario under which the property fails. You can then simulate this scenario and debug the error at hand.

This eliminates the need for a testbench and vectors. It makes sure that the logic never fails under *any* circumstance.

However, currently static formal is limited to small design sizes because of the so-called logic cone state-space explosion. When the static formal algorithm evaluates the logic under test and creates a model (let us say temporal 100 clocks and combinational 100,000 gates), its ability to explore all possible corners of logic hits the so-called state-space explosion problem. The number of automatically created input stimuli combinations reaches a prohibitive limit and would take forever to prove that the SVA properties for a given logic block never fail. In addition, as shown in Fig. 10.2, the number of "logic cones" to cover all states in the temporal space will explode, and there isn't enough memory in the universe to cover all states. So, currently (as of the publication of this book), you need to subdivide your logic into smaller chunks of logic, write SVA properties for those blocks, and run static formal tool against those blocks.

10.3.1 Critical Logic Blocks for Static Formal

For static formal to be effective in your verification methodology, you should consider targeting the following critical areas first. That will keep the logic block small and help verification tremendously because these areas are trouble porn and hard to reach with simulation alone.

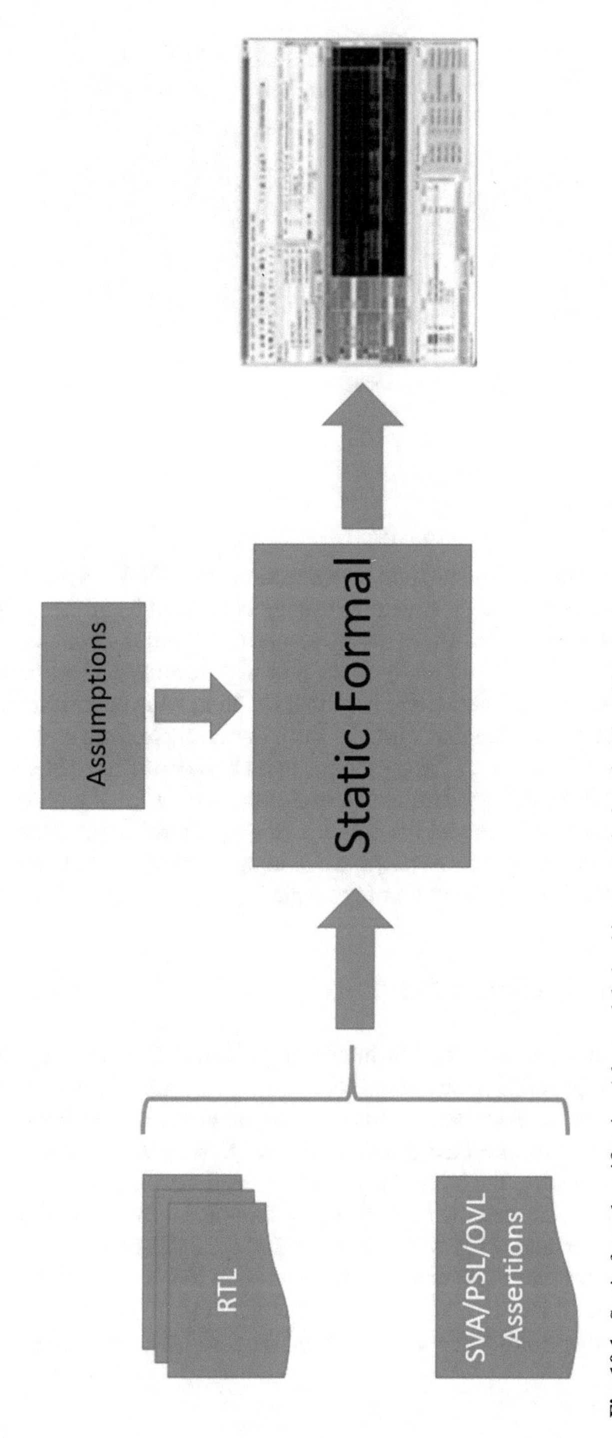

Fig. 10.1 Static formal verification (aka model checking or static functional verification)

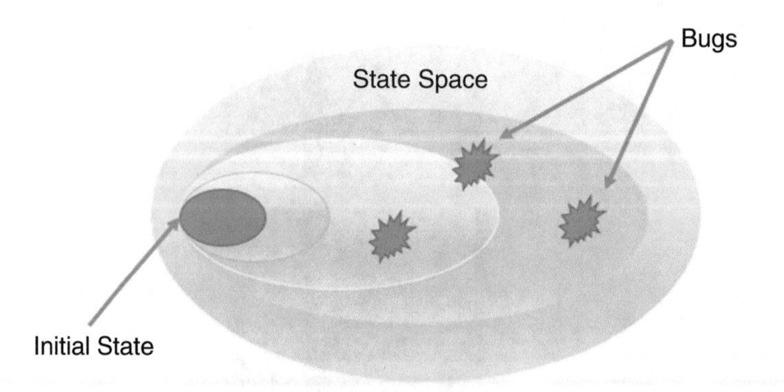

Fig. 10.2 State-space of static formal verification

10.3.1.1 Control Logic

SoC interconnect such as point-to-point connection bus or NoC (Network on Chip), buffers, and memories is logic structures usually controlled by arbiters, FIFOs, and other complex control logic. These design elements are better understood at micro-architecture level. But DV normally focuses at high-level specifications. Directed tests go for end-to-end scenarios and hope that the underlying control structures will be verified as part of it. But that's not the case. For example, with end-to-end tests, you may verify only a few of the conditions of an asynchronous FIFO. It would be much better to throw the asynchronous FIFO at the static formal tool and vet it out completely. Hence, to cover all the corner cases of control logic, carve out small pieces of control logic, write assertions and assumptions for these blocks, and apply static formal to test out each corner of the logic.

10.3.1.2 Inter-module Interfaces

There are thousands of inter-module interfaces in each SoC. Each block "assumes" certain interface protocol from the next block. Many a time, these assumptions are either wrong or never exercised. Again, end-to-end tests may or may not cover all cases of inter-module interface protocol. Hence, SystemVerilog Assertions must be written for checking the inter-module interface protocol. Once such assertions are written, small subsystems comprising of few of the low-level blocks can be static formally verified. This will guarantee that the inter-module interfaces survive under all possible input conditions. Once such low-level protocols are verified, your top level end-to-end tests will pass much more comfortably. You will have much better confidence in inter-module interface correctness with this approach.

10.3.1.3 Finite-State Machines

Code coverage will tell you if the *states* of a state machine are verified. But it will not tell you if all the state *transitions* of the state machine have been verified and that they are the correct state transitions. SystemVerilog Assertions must be written to check for state transitions, stuck-at states, live locks, or dead locks between two state machines, etc. It will be very time-consuming for directed (or random constrained for that matter) tests to cover all the state transitions and their corner cases. That's where static formal proves its usefulness. Take small subsystems of control logic or SoC peripheral interfaces where state machines play a critical role, and submit these smaller blocks to static formal. That way you have verified all possible state transitions and that those are valid state transitions. Please refer to Chap. 6 to understand in detail how to write such assertions.

10.3.1.4 Data Integrity

Devices such as bus bridges, DMA controllers, routers, and schedulers transfer data packets from one interface to another. One goal of verification is to ensure the integrity of the data during these transfers. Data packets must transfer correctly—even if they are reordered, demultiplexed and multiplexed, or segmented and reassembled during the process. Unfortunately, in a system-level simulation environment, data integrity mistakes are not readily observable. Usually problems are not evident until corrupted data is used. With assertions, the integrity of data along the entire data transfer path can be checked. A lost or corrupted data packet is detected immediately. Static formal is the perfect tool to deploy for such data paths.

10.3.2 SystemVerilog Assertions and Assumptions for Static Formal and Simulation

Figure 10.3 shows how SystemVerilog Assertions and assumptions play a role in static formal and simulation. SV "assume" statement, even though valid for simulation, is mainly invented for static formal. As we discussed before, to prevent the so-called state-space explosion problem with static formal tools, you need to "assume" certain inputs to be in certain logic state before you submit the block to static formal. Hence, you notice that the SV assumptions are being fed to the static formal part of Fig. 10.3. SV Assertions are useful for both static formal as well as simulation. This is one of the key advantages of SVA in that you write it once and use it for both the static formal and the simulation methodology components.

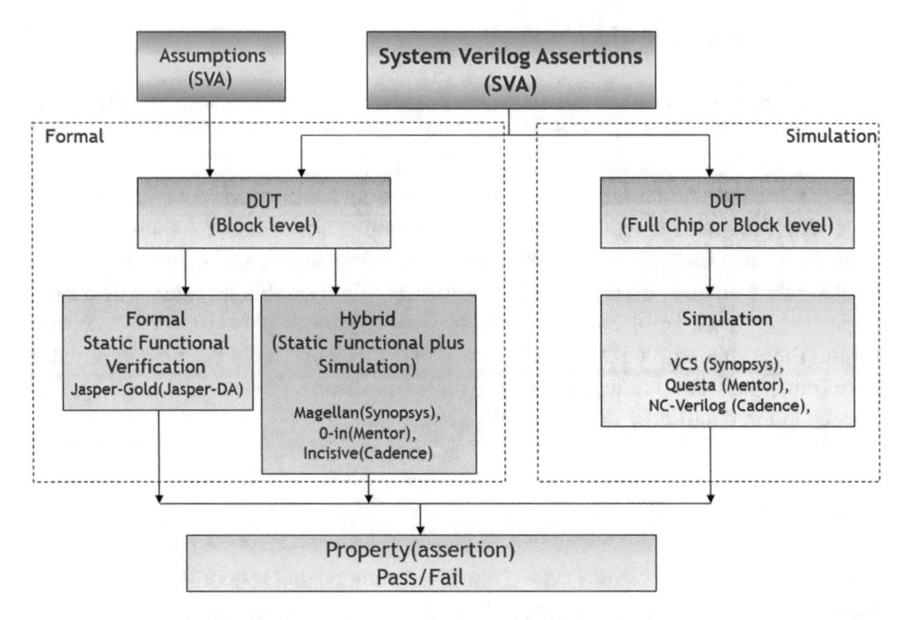

Fig. 10.3 Static formal, static formal + simulation hybrid, and simulation-only-based methodologies

10.3.3 SystemVerilog "Assume" and Static Formal Verification

This is an interesting operator. As shown in Fig. 10.4, "assume" specifies the property as an assumption for the environment. They may be used by simulators to constrain the random generation of free checker variable values or by formal tools to constrain the formal computation. The most useful environment for "assume" is that of static formal verification. As we have been discussing, static formal is a method whereby the formal algorithm exercises all possible combinational and sequential possibilities of inputs to exercise all possible "logic cones" of a given logic block and checks to see that the assertion holds. During such verification if you do not specify any constraints (i.e., for a 5 input (a, b, c, d, e) block and 100 clock temporal range, if you don't specify any constraints such as "assume" $a = 0$ and $b = 1$), then the static formal will try to explore all possible combinations of the 5 input both in combinatorial and temporal domain. Without any constraints provided via "assume," the static formal tool may experience something called "state space explosion" problem. As the description suggests, the tool may give up if too many inputs are unconstrained. This is where the "assume" statement comes into the picture.

'assume' is useful mainly for 'static formal' and 'constrained random' dynamic simulation where you need to constraint the environment using the conditions specified in a property.

- *'assume' statement allows properties to be considered as assumptions.*
- *When a property is 'assume'd the tools constrain the env. so that the property holds.*
- *For 'formal' analysis, there is no obligation to verify that the assumed property holds. The statement is simply assumed true and the scope of formal is constrained according to the assumed property.*
- *For simulation, the 'assume'd property must be checked (as in 'assert') and reported if it fails to hold.*

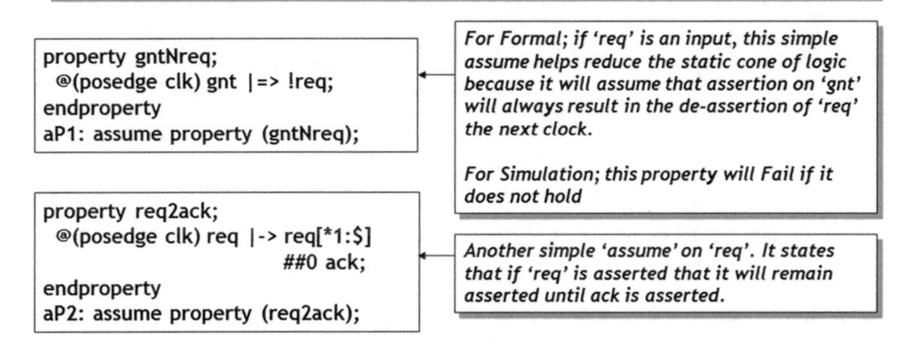

Fig. 10.4 SystemVerilog "assume" for static formal

10.3.4 Static Formal vs. Simulation

Having said all this, you may be thinking, "Is the author suggesting that we completely change our verification strategy in favor of something based on a static approach?" Well we know that's neither practical nor plausible. As mentioned above, the static formal only works on relatively smaller blocks of logic because of the state-space explosion problem. A static formal methodology is of most benefit at block level, where bugs and undesired behaviors can be eliminated at this early stage. The clean blocks would then be integrated together, and simulations would still need to be run on the entire system as always. The advantage of using this approach is that the block-level bugs have already been addressed and will not have to be fixed during system simulation time. This will ultimately make the use of simulation resources far more efficient.

Figure 10.5 (Andrew Jones and Jeremy Sonander) shows a typical advantage scenario of static formal vs. simulation. Static formal verifies block-level logic much more exhaustively since it applies all possible combinations of inputs to the

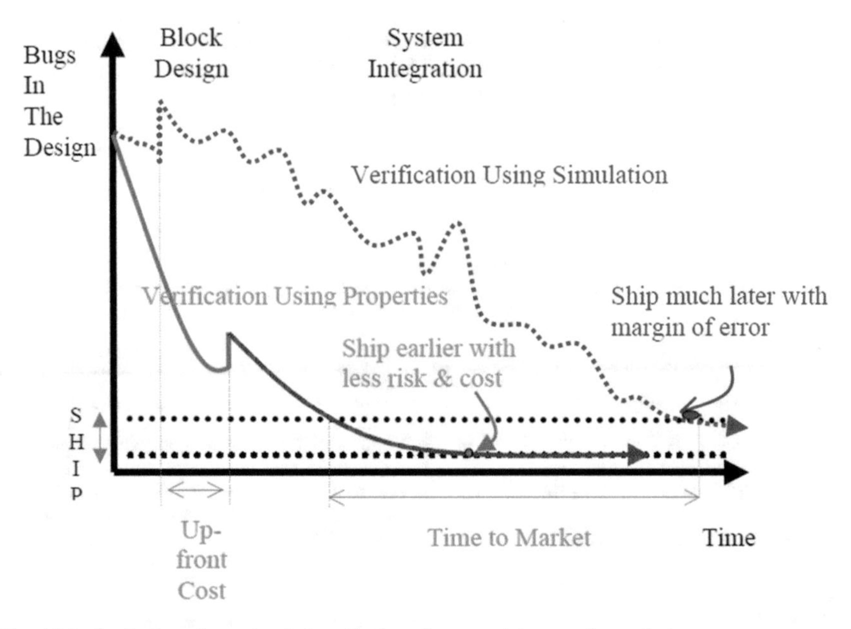

Fig. 10.5 Static formal vs. simulation (Andrew Jones and Jeremy Sonander)

block and proves that the assertions/properties applied to the block never fail. Static formal also runs thousands of orders of magnitude faster than simulation. End result is that you ship your SoC earlier with high level of confidence, less risk, and cost. Simulation-based verification will not only take longer, but you may not even be sure of having found all the corner case bugs.

10.4 Static Formal + Simulation Hybrid Verification Methodology

So, what do we do about this state-space explosion problem that limits the size of the logic block that can be verified using static formal methods/tools? That's where static + simulation hybrid technology/tools come into picture.

To reduce the logic cone that needs to be formally evaluated, the static + simulation technology strives to reduce the number of combinational and temporal domain "vectors" for a given logic block. A simulator that deploys this technology will simulate the logic block to reach a certain *known* state and thereby constrain (i.e., "assume") the inputs to be in a certain range. The static formal then takes over with those assumed input logic states and formally verifies all "remaining" combinations of inputs to the logic block to see if any of the SVA properties of the block fail.

Figure 10.6 shows that simulation is run on logic leading to the logic under formal. Simulation results in known states on the input of the logic cone of the state

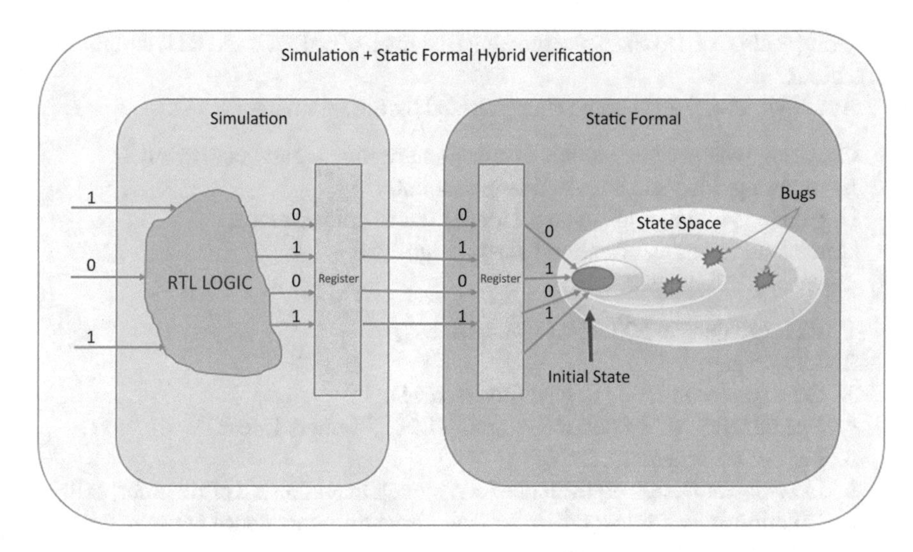

Fig. 10.6 Static formal + simulation hybrid verification methodology

space that formal needs to verify. These states are considered the initial states. These in turn are considered "assumptions" by the formal tool. Once such initial state is defined, the formal tool does not have to try "all" possible combinations of the inputs. This reduces the number of vectors it needs to generate. That in turn reduces the time to formally analyze the formal model and thereby the memory requirements.

Hybrid has become a very popular technology in recent years, and all major EDA vendors support it. With this technology, you can formally verify much larger blocks of logic (compared to static formal only).

10.5 Logic Equivalence Check (LEC)

LEC uses formal methods to prove that two versions of a design are, or are not, functionally equivalent. Some forms of formal verification are already widespread in design. Equivalence checking has been used for over a decade to check that RTL and gate-level descriptions of a design represent the same design. Equivalence checking was introduced in response to the problem of larger designs exceeding the effective capacity of gate-level simulation tools and quickly took over from hardware–acceleration solutions as well as software-only gate-level simulators. For users, the equivalence checking technology is relatively easy to use in the way it has been packaged by vendors, in tools such as Formality from Synopsys.

Equivalence checking has moved beyond SoC RTL design, migrating into FPGA design because of the use of very large devices and the time it takes to compare simulation with hardware given the limited internal visibility that a programmed FPGA offers. Through tools such as SLEC from Calypto Design Systems, equivalence checking is also used to check the functional equivalence of ESL and RTL descriptions of a block.

At a high level, LEC can be characterized by the following points:

- Checking whether two models of a design are functionally equivalent.
- Involves a golden and a revised target model.
- Objective is to find bugs in target model wrt to golden model.
- Crucial step in transformation-based design flow.
- Types of LEC checks:

  1. RTL (golden) to RTL (ECO—bug fixed)
  2. RTL to gate
  3. Gate (pre-scan-DFT) to gate (post-scan-DFT)
  4. Specification (C model or SystemC TLM ESL model) to RTL
  5. Layout vs. schematic (LVS)
  6. Low-power design equivalence (e.g., check that addition of retention cells or isolation cells, for low power will not affect the equivalence between the non-low-power netlist and low-power netlist)

- Mainstream in today's design flows.
- Current LEC tools can handle very complex designs.
- LEC is orders of magnitude faster than simulation.
- Support for multiple design languages such as SystemVerilog, VHDL, etc.

10.5.1 LEC Technology

LEC is a static technology, in that, the equivalence check between two representations of a design is not simulated. They are checked using formal (static formal) technology. LEC ignores timing and does only Boolean equivalence. Just as in static formal, LEC employs formal, mathematical techniques. There are four basic stages that LEC goes through to prove the equivalency between two forms of design.

1. Read
2. Match
3. Verify
4. Debug

10.5.1.1 Read

Logic Cones

During the Read stage, both versions of the design are read into the LEC tool. It then segments design into manageable sections called logic cones. As shown in Fig. 10.7, logic cones are groups of logic bordered by registers, ports, or black boxes (BB). The output border of a logic cone is referred to as the compare point.

Fig. 10.7 Logic cone

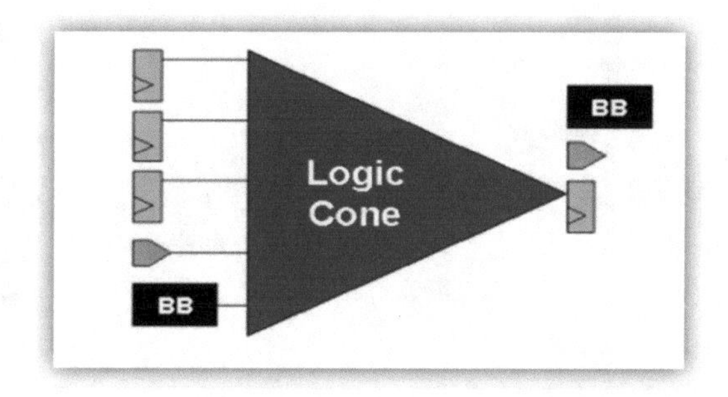

Black Boxes

A note on declaring black boxes as input to LEC tool is that during synthesis, there are several analog IP, memory blocks, pads, etc. which are not meant to be synthesized. Their model (library/lef) needs to be picked during the synthesis. It is essential to use the exhaustive list of black box modules in LEC setup as these are the modules which don't require internal verification, but their interface must be exhaustively verified to confirm their interaction with the rest of the design.

If any module is missed from this list of all the black boxes in the design and if there is any connection between these modules with other part of the SoC, then LEC tool will not check for such connections, which sometimes misses the genuine non-equivalence like broken connections between these modules.

10.5.1.2 Match and Verify

Compare Points

Compare points are those where LEC will perform comparison between RTL and gate netlist (for that matter any two forms of netlists, for example, gate to gate or RTL to RTL). The compare points are:

1. Primary outputs
2. Internal registers
3. Inputs to black boxes
4. Nets driven by multiple drivers

Compare points and logic cones go hand in hand. As mentioned before, the small segments called logic cones of a design must be surrounded by compare points. LEC will check equivalency at these compare points. After breaking a design into logic cones, LEC attempts to match (or map) between two different netlists. This is called the matching process. Both name-based (i.e., nonfunction) and function-based matching methods are deployed by a LEC tool. Name based means that synthesis will (during optimization) change the hierarchy or names of IO of registers or

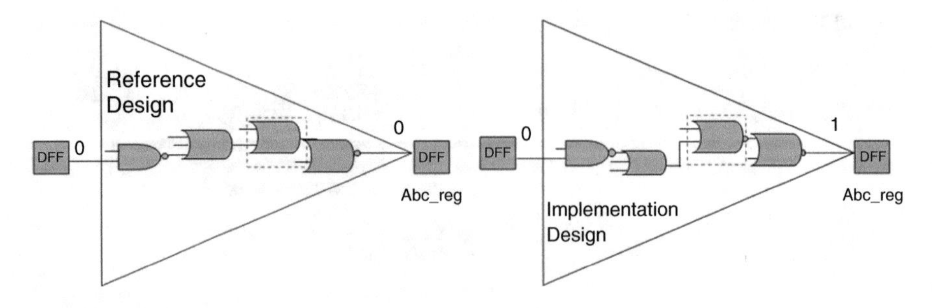

Fig. 10.8 Logic cone: pass-and-fail scenario

net names. LEC tool should be able to check, at compare points, the functional equivalence of RTL with gates even with such net name changes.

After "match" comes verification of logic cones. This is mainly a theoretical/algorithmic subject beyond the scope of this book. But LEC tools may use algorithms such as BDD, isomorphism, ATPG, etc. to verify the logic equivalence between two logic cones.

10.5.1.3 Debug

During debug phase, LEC will generate logic cones of the logic that fails equivalence check as shown in Fig. 10.8 (Synopsys).

This figure shows a very small representative logic cone. In reality, logic cones are very big (e.g., datapath logic). For this, LEC generates functional vectors to pinpoint the bug caught during LEC.

Let us now dive into each type of LEC technology and check what is available today for SoC design verification.

10.5.2 RTL to RTL Verification

There are many reasons you need to do RTL to RTL equivalence check. For example:

1. Low-power optimizations (gated clocks, power domains, etc.)
2. RTL "data change" (e.g., addition of new pipe stages)
3. Critical path optimization
4. "C" to RTL synthesized RTL

Under all these cases, LEC will check that once out of Reset, the two RTL versions are functionally equivalent.

10.5.3 RTL to Gate Verification

RTL to gate essentially entails equivalence check between golden RTL and the post-synthesis netlist. As mentioned before, the check is static and does not guarantee functional accuracy of either model. RTL must be golden, thoroughly verified. LEC simply makes sure that the RTL and the post-synthesis netlist are functionally equivalent.

RTL to gate equivalence checks for the following, at a minimum:

1. Low-power structural changes

   Low power requires the addition of retention cells, isolation cells, level shifters, etc. to power up and power down different SoC power domains. Such elements may not exist in RTL. LEC tool must make sure that RTL without low-power logic and netlist with low-power logic are functionally equivalent.
2. Synthesis optimization

   During synthesis, you may have turned on optimization knobs such as retiming, register merging, register inversion, etc. This will make it difficult for the LEC tool to do equivalence check with the post-synthesis netlist. RTL won't be optimized, but gates will be. The LEC tool should support such optimized post-synthesis netlists.
3. ECO changes
4. Complex datapaths
5. Phase inversion

10.5.4 Gate to Gate Verification

Gate to gate equivalence checks are needed because of the following transformations applied to a gate-level netlist:

- Buffer insertion for retiming
- P&R buffers (hold time violation) insertions
- Test logic insertion
- Clock trees insertion
- Scan chains insertion

10.5.5 ESL (C/ C++/ SystemC model) to RTL (Sequential Equivalence Checking—SEC)

C, C++, and SystemC (Electronic System Level, ESL, model) to RTL synthesis is done through tools that fall under High-Level Synthesis (HLS) domain. Sequential equivalence checking is a key enabler for the move to system-level design by allowing RTL models to be checked for equivalence with golden system-level models. It also enables sequential RTL changes for exploring alternate microarchitectures and is required for the successful deployment of behavioral synthesis tools for generating RTL from system-level models.

Sequential equivalence checking (SEC) is a formal technique that checks two designs for equivalence even when there is no one-to-one correspondence between the two designs' state elements (Anmol Mathur). In contrast, traditional combinational equivalence checkers need a one-to-one correspondence between the flip-flops and latches in the two designs. ESL models can be untimed C/C++ functions and have very little internal state. RTL models, on the other hand, implement the full microarchitecture with the computation scheduled over multiple cycles. Accordingly, significant state differences exist between the ESL and RTL models, and ESL to RTL equivalence checking clearly needs SEC. Researchers have investigated SEC techniques and commercial SEC tools that are now available, such as the one from Calypto Design Systems.

SEC is a key technology needed to keep ESL and RTL models consistent and to quickly weed out any RTL or ESL bugs without the need to write testbenches at the block level. As design teams deploy HLS-based flows, SEC fills several critical verification needs. SEC has been deployed in both HLS-based flows and flows in which RTL is manually created. SEC technology must continue to evolve to ensure that it can handle larger block sizes and that it can check designs with larger latency and throughput differences for equivalence.

The use of equivalence checking to verify RTL functional correctness has two key advantages. The first advantage is the complete verification of the RTL model with respect to the ESL. Unlike simulation-based or assertion-based approaches, in which functional coverage of the RTL model is an issue, SEC checks that all the RTL behaviors are consistent with those in the ESL. This results in a very high coverage of the RTL behaviors. It should be noted that ESL to RTL equivalence checking verifies only the RTL behaviors that are also present in the ESL. Thus, if the ESL has a memory implemented as a simple array while the RTL model implemented a hierarchical memory with a cache, equivalence checking will not verify whether the cache is working as intended as long as the overall memory system works as expected.

The second advantage of equivalence checking is simplified debugging. In case of a functional difference between the ESL and RTL, SEC produces the shortest possible counterexample that shows the difference. This contrasts with traditional simulation-based approaches, which may find the difference but only after millions of cycles of simulation. The conciseness of the counterexample makes the process of debugging and localizing the error much more efficient.

Here's a methodology perspective for SEC to work (Anmol Mathur):

Consistent Design Partitioning

SEC is a block-level verification tool due to capacity limitations of formal technology. To effectively use SEC, it is crucial that the ESL and RTL model be consistently partitioned into subfunctions and submodules. Clean and consistent design partitioning provides an opportunity to use sequential equivalence checking at the level of individual ESL/RTL blocks.

Creation of ESLs with Hardware Intent
To perform ESL to RTL equivalence checking using a sequential equivalence checker, the ESL must be written to let the tool infer a hardware-like model statically from the source. This requires that the team creates the ESL to follow certain coding guidelines that allow the ESL's static analysis. The use of statically sized data structures instead of dynamically allocated memory, explicit use of memories to reuse the same storage for multiple arrays instead of pointer aliasing, and statically bounding loops are some examples of constructs that make ESLs more amenable to SEC and HLS tools.

Orthogonalization of Communication and Computation
Clear separation between the computational and communication aspects of the ESL allows easier refinement of the communication protocol, if needed, to make the interface timing more closely aligned with that of an RTL model.

Key challenges in ESL to RTL (SEC) usage. (Anmol Mathur)

The main challenges in the usage of SEC-based flows stem from the fact that SEC has capacity limitations. SEC's complexity is a function of the following factors:

- Size of the ESL/RTL blocks being compared.
- Latency and throughput difference between the ESL and RTL. The greater the sequential differences between the ESL and RTL, the larger the sequential depth to which SEC needs to explore the ESL–RTL state machines, and, hence, the larger the run time and memory usage is in SEC.
- Difference in levels of arithmetic abstraction. If both the ESL and RTL represent their computations at the operator level, the complexity of using SEC is lower than the cases where, in the RTL model, the arithmetic operators have been decomposed into bit-level constructs.
- Amount of correspondence between internal states of ESL and RTL model. If the SEC tool can detect internal states and signals in the ESL and RTL model that are identical, then the verification problem can be decomposed and simplified.

Figures 10.9 and 10.10 ESL to RTL Equivalence Flow show a very simple C model and an RTL model. RTL model could be derived using High-Level Synthesis (or manually created). But we need to make sure that these two models are equivalent. Just as with other equivalence checks, the timing and internal structure between system-level model and RTL can differ significantly, but the outputs at compare points must be the same. Such a C/C++/SystemC to RTL equivalence checker will check for:

1. Combinational equivalence
2. Cycle-accurate equivalence
3. Pipelined equivalence
4. Stream-based equivalence
5. Transaction equivalence

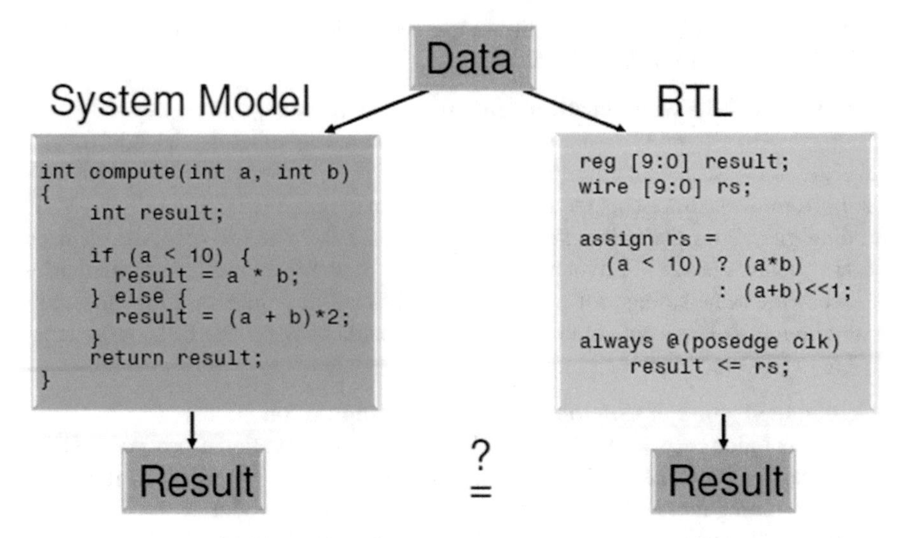

Fig. 10.9 ESL to RTL equivalence example

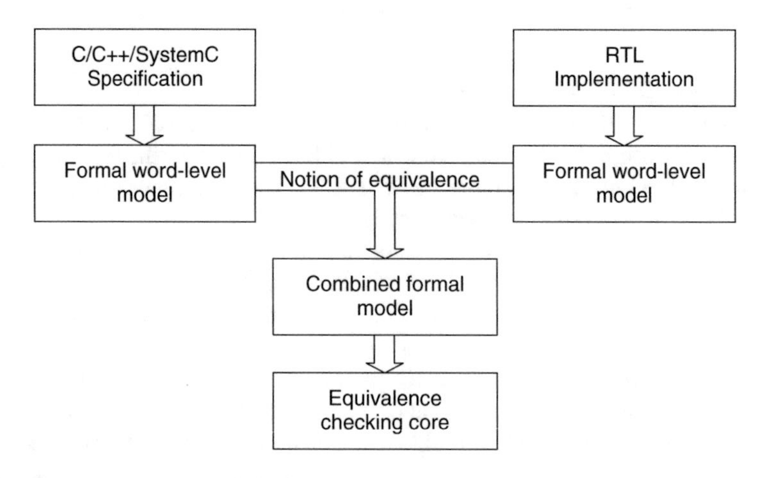

Fig. 10.10 ESL to RTL equivalence flow

The first step in any system-level equivalence checking system is the construction of a formal model from the C++/SystemC description as well as from the RTL. Preserving word-level information in this step enables the use of powerful word-level solvers in the equivalence checker. For the RTL, constructing a formal model is usually straightforward as most existing RTL front ends already produce a synthesized netlist which preserves word-level information. For the C++/SystemC description, however, the restriction to a synthesizable subset of the language as required by many C++ synthesis tools is far too rigid for a verification tool. New HLS algorithms are being developed to get around the ESL subset limitations.

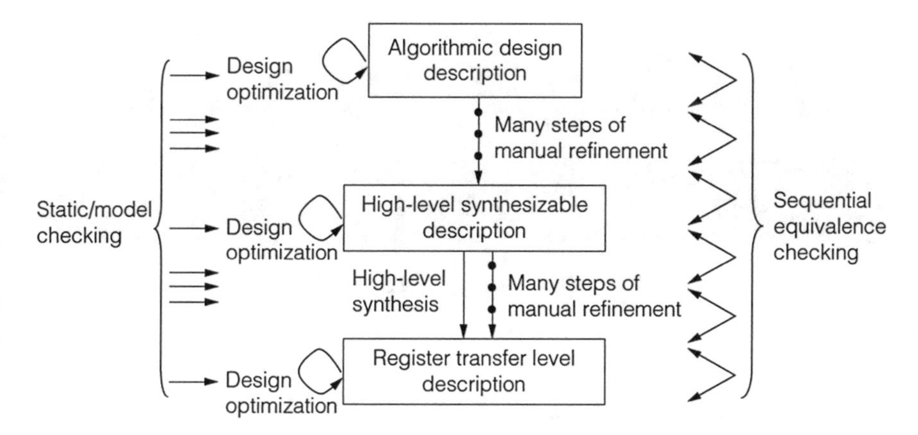

Fig. 10.11 Sequential equivalence checking (Anmol Mathur)

Figure 10.11 shows a high-level design flow for sequential equivalence checking (Anmol Mathur).

The prominent tool in this category is SLEC from Calypto (now Mentor Graphics). SLEC stands for Sequential Logic Equivalence Checker. SLEC compares a synthesizable subset of C, C++, and SystemC with a synthesizable subset of Verilog, SystemVerilog, and VHDL RTL. SLEC lets you compare C/C++/SystemC to C/C++/SystemC, C/C++/SystemC to RTL, or RTL to RTL.

A tool called HECTOR (Synopsys) is another widely used tool for system level (C/C++/SystemC) to RTL equivalence checking. HECTOR is designed to help find bugs and reduce RTL bring-up time, verify algorithmic consistency through design changes without running simulation, and increase functional sign-off certainty by verifying transaction equivalence between high-level models and RTL implementations. Its patented compilers improve capacity by generating word-level RTL and formal C/SystemC models. The powerful formal algorithms it provides find proofs in minutes or hours, not days, and its unique multiple leaf-level solvers solve complex logic.

HECTOR uses hierarchical equivalence, automatic design partitioning, efficient and patented memory models, and multiprocessor support to rapidly converge on proofs. HECTOR also supports multiple languages: Verilog, VHDL, SV, C, C++, and SystemC. HECTOR is well suited to algorithm-intensive blocks including video processing, wireless media, cameras, encryption/decryption logic, transformations, GPUs, and floating point. HECTOR can be targeted at datapath-dominated designs and arithmetic blocks including floating point and, uniquely, bit-serial division. HECTOR is also unique in the formal space in that it can do consistency checks on complete blocks including control logic.

Figure 10.12 shows the HECTOR product overview (Synopsys website).

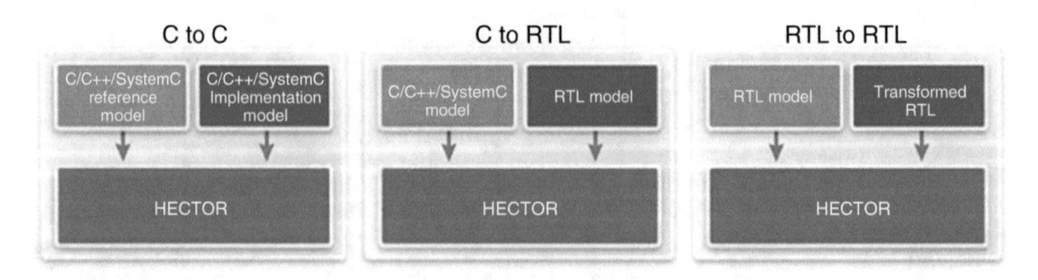

Fig. 10.12 Synopsys HECTOR: ESL to RTL equivalence product

10.5.6 Layout vs. Schematic (LVS) Physical Verification

LVS determines whether an IC layout corresponds to the original schematic. An LVS tool enables accurate circuit verification because it can measure actual device geometries across a full chip for a complete accounting of physical parameters. The measured device parameters supply the information for back annotation to the source schematic and comprehensive data for running simulations.

In the nanometer era, die areas are getting larger as the designs are getting more and more complex. To ensure the correctness of the implemented design, bigger layout databases need to be checked during the physical verification stage in the same ambitious project time frames as before. Any failure identified after the design is manufactured will result in expensive mask changes and delays in getting the System on Chip (SoC) to market. Physical verification is performed to check whether the design layout is equivalent to its schematic and checks the layout against process manufacturing guidelines provided by the semiconductor fabrication labs to ensure it can be manufactured correctly.

Physical verification includes:

- Design rule check (DRC): It verifies whether the designed layout can be manufactured by the fabrication lab with a good yield.
- Layout versus schematic (LVS): It is a method of verifying that the layout of the design is functionally equivalent to the schematic of the design.

It is important to note that DRC does not ensure the intended functionality of layout. DRC is only limited to checking whether the given layout conforms to design rules provided by the silicon foundry to ensure the faultless fabrication but without warranting whether the circuit will behave in a way that it was intended to. The idea of LVS originated from this very requirement. DRC is out of the scope of this book. We will focus on LVS instead in this section.

An EDA tool performs LVS by taking a set of instructional code input, commonly known as LVS rule deck, in the following two steps: extraction and comparison. The LVS rule deck guides the verification tool by providing the instructions and

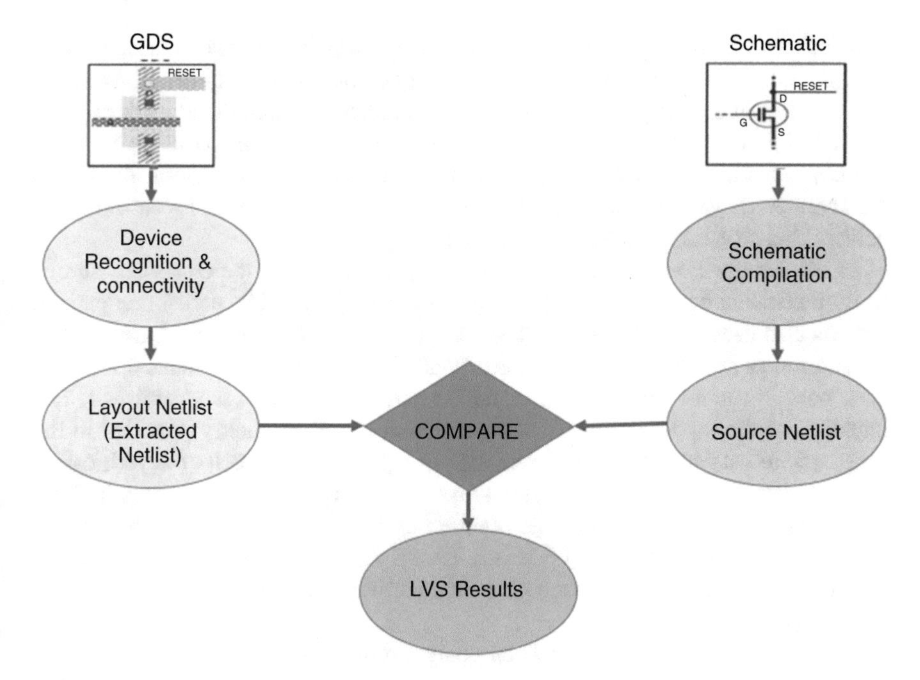

Fig. 10.13 LVS design methodology flow

identifying files which are needed for LVS. Design inputs needed for running LVS are as follows:

- Graphical database system (GDS) layout database of the design
- Schematic netlist of the design
- Cell definition file including intellectual property files and standard cells
- Pad reference file

An LVS rule deck is a set of code, which is written in Standard Verification Rule Format (SVRF) or TCL Verification Format (TVF), which guides the verification tool to extract the devices and connectivity of the integrated circuit (Rishabh Agarwal). The LVS rule deck contains the layer definitions for the identification of layers used in the layout file and matches description of a layer to the location of the layer in the GDS file. This helps in the recognition of the electrically connected regions in the layout, namely, the nets. Nets are recognized from the layout shapes through analysis between layout shapes in layers. LVS rule deck also contains device structure definitions.

Referring to Fig. 10.13, the verification tool takes the GDS file as input and breaks it down into basic design devices like transistors, diodes, capacitors, resistors, etc. These devices are identified in the GDS file by recognition of the layers and shapes that make up the circuit or by the cell definition of the devices/circuits provided in the cell definition file of the intellectual property blocks or in the LVS rule deck itself. It also extracts the connectivity information between these devices from the GDS file. The next step in connectivity extraction is uniquification of nets. Each electrical net is given a unique node number for identification during the extraction process. Net names can also be named based upon the presence of layout text objects or text statements in the control file. This device information along with their connectivity is written into a layout netlist file, generally called layout extracted netlist. This process is known as extraction.

In the comparison phase, the verification tool compares the electrical circuits from the schematic netlist and the layout extracted netlist. The netlist comparison process also uses the LVS rule deck. After the successful comparison between layout and source netlist, a one-to-one correspondence between the elements (instances, nets, ports, instance pins) of source netlist and layout netlist is established. The intention of the layout designer is to implement the functionality provided in the schematic into a geometrical representation of layout. Therefore, for the verification process to complete without error, both layout and source netlist must match. If the two netlists differ, discrepancies are reported in the form of an LVS result database which can be used to debug LVS issues. Result database would contain the list of incorrect elements and the reason of mismatch like incorrect nets, incorrect ports, and incorrect instances.

Here are some rules to follow when going through LVS flow:

1. Always verify the operation of a circuit via simulations at the schematic level before attempting to layout the cell. LVS only verifies the schematic and layout match, so if the schematic does not work, the layout will not either. If the schematic does not function properly, there is no reason to spend time debugging the LVS.
2. Always design in a hierarchical fashion, building smaller (lower level) cells before constructing larger circuit blocks from the lower-level cells. Performance improvements in LVS tools are achieved through hierarchical processing, that is, processing a repeated block only once, and hardware scaling, or the ability to divide the LVS job across many CPUs.
3. Always pass LVS on lower-level cells before attempting to check LVS on a higher-level cell. If the lower-level cells do not pass LVS, it is much easier to debug them on their own than after you have added the cell to a higher-level circuit.
4. Always recheck LVS on a cell if you make any changes to the schematic or layout.
5. If you modify a layout to correct a problem found in an LVS check, always re-extract the layout, and save it before running the LVS checker again.

Here are a few LVS tools from the EDA industry:

Mentor Graphics Calibre nmDRC and Calibre nmLVS

Cadence Design Systems Physical verification system (PVS). PVS integrates with Cadence Virtuoso® custom/analog, Cadence Innovus™ digital design, and mixed-signal flows. This provides for an end-to-end design and sign-off physical verification solution integrated with all Cadence tools.

Synopsys IC Validator (physical verification with IC Validator in the Synopsys Galaxy™ Design Platform) provides technology-leading, production-proven sign-off solutions for design rule checking (DRC), connectivity verification layout vs. schematic (LVS), metal fill insertion, and design for manufacturability (DFM) enhancements.

10.5.7 RTL Lint

This technology has been in use for over 20 years and is well understood. RTL Lint checks that the design code adheres to guidelines. It checks HDL code for synthesizability, simulatability, testability, reusability, and RTL/gate sign-off. Besides helping to enforce some known-good naming schemes, the checks are designed to explore design and coding deficiencies that impact simulation, synthesis, test, and performance.

RTL Lint checks for:

- Unsynthesizable constructs
- Unintentional latches
- Unused declarations
- Multiply driven and undriven signals
- Race conditions
- Incorrect usage of blocking and non-blocking assignments
- Incomplete assignments in subroutines
- Case statement style issues
- Set and Reset conflicts
- Out-of-range indexing

RTL Lint checks can run very quickly, because they do a shallow analysis of the HDL code itself rather than trying to understand the design represented by that code. Other static verification techniques look beyond the HDL representation to analyze the design and verify its characteristics.

Lint can be a highly effective tool when used in pre-simulation. It can catch bugs without requiring specific test vectors and so reduce the number of simulation cycles needed to achieve coverage of a logic block. A further strength of lint tools is that the rule decks they have assembled contain decades of experience and knowledge. The sheer number of error checks, however, can make parsing the error reports time-consuming and difficult.

It is ultimately the responsibility of the user to review the report generated by the lint tool and then decide which of the potential bugs can be waived and which need to be fixed. Because lint tools contain so many accumulated rules, designers continue to complain that they generate too many false positives. In this scenario, an obvious concern is that much of the simulation time saved may still be eaten up during analysis of the lint tool's output.

There are then two further issues that designers raise.

First, as lint is based on accumulated knowledge, it is sometimes the case that a significant number of the checks are duplicates while others have become obsolete or unnecessary.

Second, designers say that while lint tools are excellent for checking compliance with coding best practices, they can lack the finesse to accommodate the subtle differences present in all in-house coding styles (sometimes also the differences between coding styles in different divisions of the same company). This comparatively long-standing complaint has gained greater force of late, as SoC designs have been increasingly dependent on IP supplied by third parties—which, again, use differing coding styles.

As a result, designers say that they often have had to spend too much time pre-configuring lint tools to exclude or overcome these last two issues. However, companies within the EDA industry are responding to these criticisms.

At the most basic level, tool vendors have placed lint rules under close review to deliver the most compact decks that they can. They have also taken advantage of increasing computational power to reach a point where lint tools can analyze designs of, say, 300 million or more gates in a matter of minutes.

User interfaces have then been simplified so that it is much easier for designers to tweak a lint tool per actual requirements.

Companies such as Real Intent and Synopsys have decided to make the reports easier to use with hierarchical reporting and integration.

Real Intent's Ascent Lint addresses designers' fear of being overwhelmed by the number of lint flags raised by prioritizing potential bugs in its reports "so that fixes will produce the greatest improvement in the quality of the HDL." It also has debug hooks into the Synopsys Verdi platform that cross-probe the RTL to more closely identify where the lint flags are located. These themes form part of a "smart reporting" concept that Real Intent is introducing across all its products.

Synopsys incorporates lint within its SpyGlass platform, providing a methodology together with the lint rule sets. This, the company says, "provides an infrastructure for rule selection and methodology customization aligned with design milestones." Synopsys' approach is based on the idea that different rule sets apply during different phases of design. For example, it makes sense to run a check on synthesizable constructs before converting RTL to gates, but different rules would be prioritized to perform state-machine checks, for example, before simulation.

The result of these changes is that lint has become not just an aid to streamline verification and block-level RTL creation but part of the drive toward what is variously called "RTL sign-off" or "SoC sign-off."

10.6 Structural Checks

Structural checking tools perform a pseudo-synthesis of the design that identifies combinational and sequential elements and recognizes finite-state-machine structures. Such tools reduce the design to a generic netlist consisting of registers, latches, logic, and RAM, and most structural checks are performed on this netlist representation. Hence, structural checks can recognize many deficient or incorrect coding styles for synthesis and, at the same time, identify any potential simulation versus synthesis mismatch issues. As these checks are applied to a generic structural netlist representation of the design and not just to the syntax of the design description, they can more easily explore connectivity issues, fan-in and fan-out (driver-reader) relationships within a design, and FSM state/transition issues. However, it is important to realize that if part of the design is not synthesizable, that part will be treated as a black box and some structural checks will not be performed.

Some of the common structural checks include:

- Combinational loops
- Full and parallel case issues
- Clock gating and usage issues
- Bus conflicts and floating bus
- Dead code and unreachable blocks or states
- Unused input and undriven output ports
- Unresettable registers
- Arithmetic overflow

10.7 Low Power Structural Checks

A common problem that exists with power-aware verification of designs with active power management involves the accurate placing of level shifters and isolation cells. Often these are done manually or through scripts and are inserted before or after synthesis. These techniques are error prone and cause unique verification problems that are difficult to rectify and typically require costly respins. The Unified Power Format (UPF) low-power specification standard [("IEEE Standard for Design and Verification of Low Power Integrated Circuits)] allows designers to explicitly specify the insertion of isolation cells and level shifters at the RTL, both for verification and for synthesis [(Freddy Bembaron)].

Given a UPF description of the power intent, structural checks can be used effectively to verify that isolation and level shifting cells have been inserted where needed or will be when the design is synthesized along with its UPF specification. The UPF file typically contains a specification of the power domains that will exist in the design, the boundaries of which may require isolation or level shifting cells. The UPF file also typically defines the power states of the system, each of which is defined in terms of the power states of each power domain in the system. For any two adjoining power domains, their respective states in any given system power state determine whether isolation or level shifting will be required in that system power state.

For example, if an output of power domain A is an input to power domain B and A is powered down in some system state in which B is powered up, then isolation is

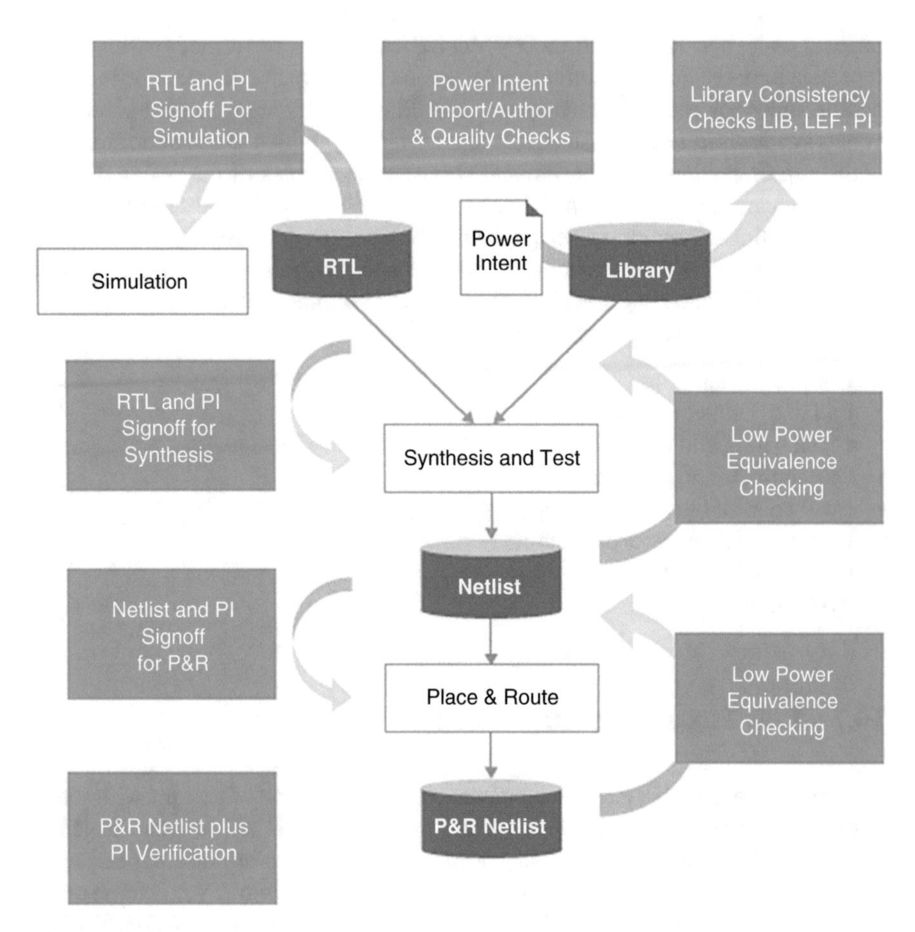

Fig. 10.14 Cadence Conformal Low Power XL© (Cadence) Equivalency Check methodology

required on that input to B. Structural checks can easily identify missing, unnecessary, or redundant isolation cells based on the power state table, power domain definitions, and design connectivity. Similarly, structural checks can identify missing or unnecessary level shifters at such interfaces, as well as verify that the direction of level shifting is correct, based on respective voltage levels defined in the power state table.

Structural checks provide more accurate results than RTL lint checks. This is because rather than just looking at the RTL code itself, such checks consider the structure of the design that will result from synthesis of the RTL code. However, structural checks alone are not sufficient; it is also important to consider the behavior of the design represented by the RTL.

Here's a brief description of Cadences Conformal Low Power XL© (Cadence) Equivalence Check methodology/product Fig. 10.14.

Low-power verification (equivalence) is amplified by the fact that most of the low-power function is introduced into the gate netlist during synthesis and physical implementation. Most simulation-based verification takes place at the RTL. Full-chip, gate-level simulation is neither a practical nor scalable methodology for verifying the logic function of today's designs due to their size and complexity.

During development, a low-power design undergoes numerous iterations prior to final layout, and each step in this process has the potential to introduce logical bugs. Conformal Low Power checks the functional equivalence of different versions of a low-power design at these various stages and enables you to identify and correct errors as soon as they are introduced. For example, it validates post-synthesis netlist and instantiated power intent back against the verified golden RTL and its associated power intent. It supports advanced dynamic and static power synthesis optimizations such as clock gating and signal gating, multi-Vt libraries, and de-cloning and re-cloning of gated clocks during clock tree synthesis and optimization.

Conformal Low Power supports the Common Power Format (CPF) specification language (UPF was not supported as of the writing of this book). It uses CPF for guidance to independently model how implementation inserts and connects low-power cells—level shifters, isolation, and state retention registers—into an RTL design, thus enabling true low-power equivalence checking from RTL to the gate level. Conformal Low Power can also model level shifters and isolation cells as domain anchor points during equivalence checking to detect whether logic gates have erroneously crossed domain boundaries from one version of the netlist to another. Conformal Low Power supports other power intent standards as well.

Conformal Low Power XL © (Cadence) reports the following:

- Power- and ground-domain-assignment-related problems and floating connections
- Level shifters: missing, redundant, wrong domain location, or wrong connectivity
- Isolation cells: missing, redundant, wrong gate type, wrong location, and wrong isolation enable polarity
- Control signals that are not powered appropriately
- Incorrect power and ground connectivity, including shorts and opens
- Instances with undefined power domains or mixed power domains
- Missing, redundant, and incorrect power connection and wrong level shifter types
- Missing, redundant, and incorrect isolation cell power connectivity
- Power control signals to power switches, isolation cells, and state retention registers that are not powered
- Incorrect power connection to state retention registers

10.8 X-State Verification

Some other uses of formal checks are related to initialization, x-generation, and x-propagation [(Turpin)]. The goal of such checks is to eliminate pessimistic x-propagation as seen in simulation and to make sure an unknown or x-state is not generated or consumed unintentionally in the design. When an unknown state or an uninitialized state is sampled, the resultant value is unpredictable. Hence, it is also important to ensure that registers are initialized before they are used. Connecting a global reset to all the registers is ideal. However, due to routing congestion, this may not be always possible. Instead, partial reset may be used, in which case we need to verify that the whole block eventually reaches a predictable reset state and that unpredictable register values do not propagate before they are overwritten with predictable values.

The common formal checks related to X-state verification are:

- Reachable x-assignment
- Conflicting drivers
- Unguarded x-propagation
- Uninitialized registers
- Use of uninitialized values

Formal checks are the most precise form of generic or automatic static checking available, because they consider the functionality of the design in addition to its structure. Thus, formal checks produce few false failures, whereas RTL lint checks and even structural checks may produce many false failures. However, generic formal checks are limited to detecting relatively simple, common errors that can occur in most any design.

10.9 Connectivity Verification

Connectivity checking involves the validation of the internal wiring of a device (Erich Marschner). It verifies the connections among blocks of logic in a design are correct. Checking the connectivity with dynamic simulation using a directed or constrained random approach will certainly find some of the connectivity bugs. However, designs can contain tens of thousands of wires controlled by configurations, all of which potentially need to be checked for correctness. Even small SoCs can have tens of thousands of static and dynamic interconnections due to BIST, low-power isolation circuitry, and multiplexing of I/O's layered on top of the baseline point-to-point IP interconnections. Even worse, add in the constant stream of bug fixes and ECOs throughout a project's lifecycle and connectivity verification has become a high-risk, high-cost testbench creation and debug project requiring weeks of man–hours and simulations all on its own.

Formal methods offer a solution that is quick and exhaustive and allow for efficient debug. With the connectivity specification captured in a table or spreadsheet, assertions can be generated automatically. Formal methods process all the assertions at the same time. Any assertion failure will pinpoint an issue with a particular connection.

Chapter 11
ESL (Electronic System Level) Verification Methodology

Chapter Introduction
Electronic System Level refers to simulating a design at abstractions higher than RTL (register transfer level). Higher level means at transaction level where the low-level implementation detail is not of consequence, only the raw functionality and hardware-based concurrency.

This chapter will discuss OSCI TLM2.0 standard definition, virtual platform examples, and how to use a virtual platform for design verification, among other topics.

11.1 ESL (Electronic System Level)

Electronic System Level is now an established approach at most of the world's leading system-on-a-chip (SoC) design companies and is being used increasingly in system design. From its genesis as an algorithm modeling methodology with "no links to implementation," ESL is evolving into a set of complementary methodologies that enable embedded system design, verification, and debugging through to the hardware and software implementation of custom SoC, system-on-FPGA, system-on-board, and entire multi-board systems.

ESL concepts and terminology have been around since 2001. But it's only recently (onward) that the methodology has taken a standard shape and has become widely adopted. There has been a significant maturing and development in many areas of ESL flow.

The most significant development that changed the ESL landscape is the emergence of OSCI TLM 2.0 specification, which creates a standardized way to connect models described at the loosely timed or approximately timed (or untimed for that matter) transaction level. The ramifications of its introduction have been huge. Instead of every vendor of system-level virtual platforms having their own proprietary languages, models, methodology, and tools, every major user developer is now

A.B. Mehta, *ASIC/SoC Functional Design Verification*,
DOI 10.1007/978-3-319-59418-7_11

beginning to standardize on the use of TLM 2.0 as the way in which to interconnect models. Models developed for one system will be able to work on another, meaning that the problem of model availability and true interoperability are now being solved.

Transaction-level modeling (TLM) is a high-level approach to modeling digital systems where details of communication among modules are separated from the details of the implementation of functional units or of the communication architecture. Communication mechanisms such as buses or FIFOs are modeled as channels and are presented to modules using SystemC interface classes. Transaction requests take place by calling interface functions of these channel models, which encapsulate low-level details of the information exchange. At the transaction level, the emphasis is more on the functionality of the data transfers—what data are transferred to and from what locations—and less on their actual implementation, that is, on the actual protocol used for data transfer. This approach makes it easier for the system-level designer to experiment, for example, with different bus architectures (all supporting a common abstract interface) without having to recode models that interact with any of the buses, provided these models interact with the bus through the common interface.

So, ESL design means modeling at transaction level in SystemC-TLM 2.0, C/C++, MATLAB, etc. The transaction-level models (TLM) allow designers to:

- Develop software and explore hardware platform architecture alternatives *before* committing to RTL, for example, firmware, OS, drivers, and applications code.
- Break down SoC complexity into manageable TLM level modular functionality.
- Start DV test development *before* SoC RTL is ready. ESL platform allows you to verify your tests and remove wrinkles within, so that the tests are ready for deployment when RTL is ready.
- Simulate orders of magnitude faster than RTL (e.g., boot small Linux kernel on a very basic SoC in under 10 wall clock seconds). More on this coming up in later sections.
- Improve power and performance estimation at system level *before* committing to RTL.

The author highly recommends the book written by Brian Bailey to understand in detail the ESL models, how they are developed, and how they are used and applications of TLM models. This book will focus on the advantages of ESL toward functional verification.

11.1.1 How Does ESL Help with Verification?

Let us now see where does ESL fit in the design verification paradigm. Virtual platforms created with TLM 2.0 models are extremely fast, and they help with software development in parallel to hardware development. But there is also the great benefit of being able to create tests for the SoC *before* the RTL is ready.

A virtual platform functionally models the entire architectural state of the SoC. In other words, you can create functional tests on the virtual platform which will eventually be run on RTL. You can create UVM agent to drive the ESL virtual platform (i.e., the virtual platform acts as the DUT) with transaction-level tests and compare the response from the virtual platform with expected response in a scoreboard. The virtual platform is architecturally 100% compatible to your SoC-DUT, and hence such a methodology will allow you to create tests in the absence of RTL, a great advantage of ESL to design verification.

At a high level, here are the advantages of ESL for design verification:

- Faster development:
  - ESL transaction-level abstraction reduces testbench development time.
- Faster simulation:
  - ESL testbench + ESL design (i.e., Virtual Platform) simulates orders of magnitude faster than RTL.
- Faster debug:
  - Debug at ESL/TLM is lot easier than debug at cycle accurate RTL.
- Earlier verification:
  - Find functional/architectural bugs *before* RTL is ready for verification.
- Faster time to production because of faster Develop->Simulate->Debug->Cover Loop

11.1.2 ESL Virtual Platform Use Cases

Fig. 11.1 shows the multi-useful virtual platform. It allows you to develop APPS, OS, firmware, drivers, etc. in parallel to hardware development. Note that if you deploy emulation or acceleration methodology for software development, you still need to wait for RTL to be very stable; else the emulated model will be buggy, and software development will suffer.

Virtual platform also allows for power and performance analysis and design before committing to a final power architecture of the SoC. There are many techniques available to measure power and performance on a virtual platform.

For example, a DVFS (Dynamic Voltage and Frequency Scaling) methodology measures the power during IO bound transactions vs. CPU transactions vs. memory transactions. Whenever an application is IO bound, the CPU frequency (and hence voltage) is lowered for the CPU as well as blocks which are not involved in IO transfer. Similarly, memory bound and CPU bound transactions dictate different voltage and frequency strategy. With virtual platform speeds reaching hundreds of MIPS, allowing you to run applications in real time, you can experiment your power domain strategy with real-life applications. Power and performance go hand in

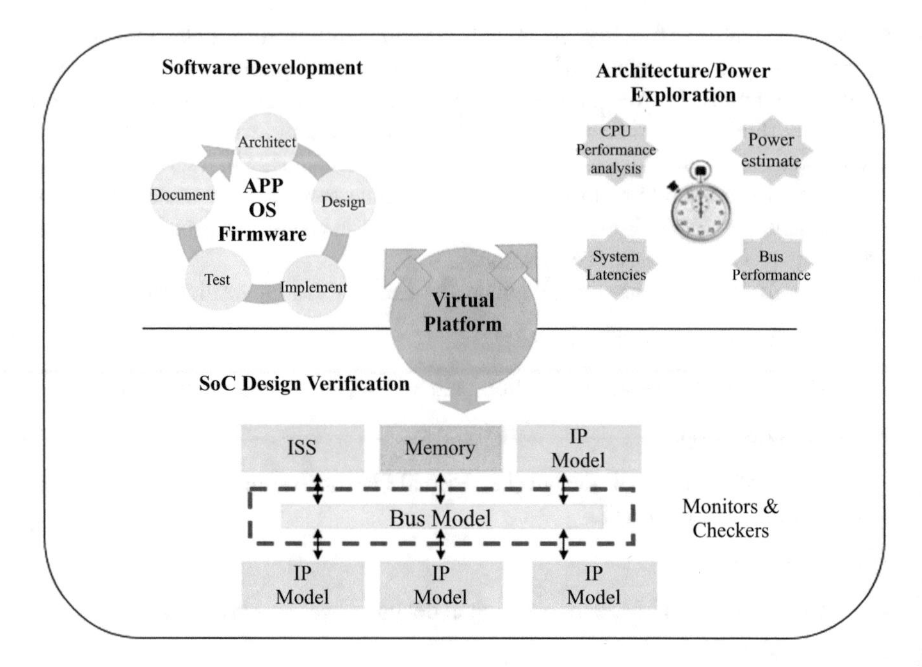

Fig. 11.1 Virtual platform: multiple use cases

hand. So, you design the power strategy along with performance verification. Without knowing the effect of power strategy on device performance, the strategy will be weak at best. If the performance requirements are not too stringent (as in IO case), you lower the power. Else you increase the power, thus, reducing the average power consumption.

And as we mentioned above, the virtual platform is a great boon to design verification. As shown in Fig. 11.1, a virtual platform models the SoC in its entirety at transaction level. The verification team starts developing tests on this virtual platform SoC way before the SoC-RTL is ready. When SoC-RTL is ready, the tests are ready as well (and the tests themselves have been verified for correctness). This is the best methodology for true parallel development of design and testbench development. More on this subject is discussed in upcoming sections.

11.2 OSCI TLM 2.0 Standard for ESL

As mentioned above, OSCI TLM 2.0 standard established modular transaction-level modeling (TLM) in SystemC. By utilizing the communications mechanism introduced in SystemC TLM 2.0, using ports, interfaces, and channels, users could implement different transaction-level interfaces. Without such a standard, a model written for the *same* functionality (e.g., a memory subsystem) by two different groups will most likely be non-interoperable. For example, one model may order the function call parameters as {address, data}, while the other may order them as {data, address}. When you want to use a model for verification, you need to know exactly how the parameters are ordered. Else your functional calls will fail. In other words, the testbench is now dependent on the model (whoever created it). Even though the users are implementing the exact same transaction semantics for the exact same memory subsystem, their method prototypes would be incompatible, and thus their models would not be interoperable (Brian Bailey).

TLM2.0 fixes this issue by standardizing the interface method calls. It defines a set of transport calls, which must be used to be compliant. The goal is to enable interoperability between high-level component models, which can then be plugged into any TLM2.0 compliant system model. Currently, engineers create ad hoc adaptors and wrappers for model integration. In addition to the standardized interfaces, TLM2.0 also defines a set of modeling styles, a generic payload as well as over a hundred rules for the expected behavior of TLM2.0 compliant models.

The TLM-2.0 classes are layered on top of the SystemC class library as shown in Fig. 11.2. For maximum interoperability and particularly for memory-mapped bus modeling, it is recommended that the TLM-2.0 core interfaces, sockets, generic payload, and base protocol be used together in concert. These classes are known collectively as the interoperability layer. The full scope of TLM2.0 discussion is beyond the scope of this book. Please refer to OSCI TLM-2.0 Language Reference Manual (TLM2.0)

To complete the story on TLM2.0, Fig. 11.3 shows the use cases, coding styles, and mechanisms applied by TLM2.0 in building and using virtual platforms.

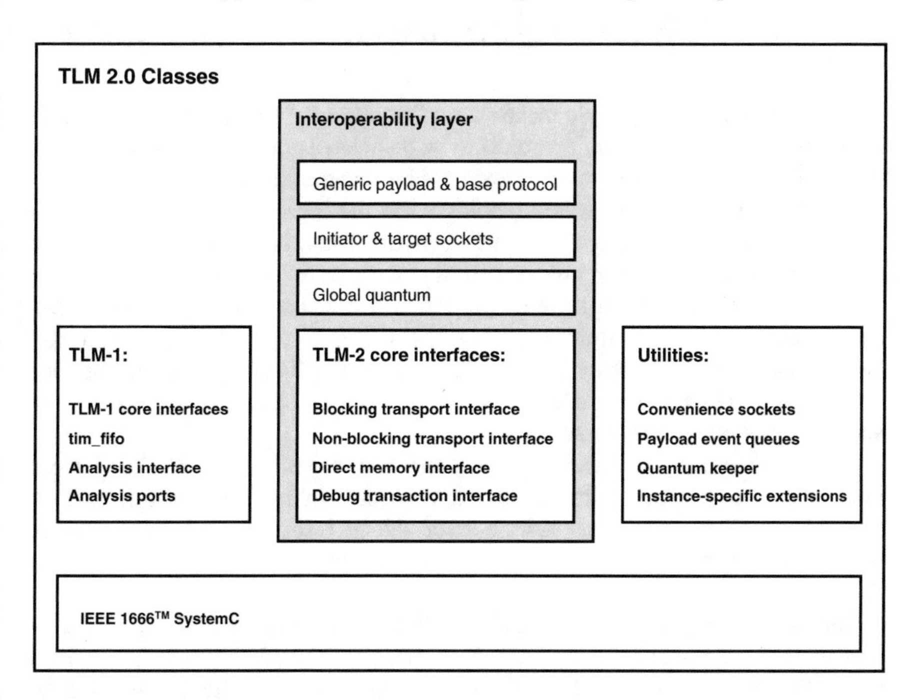

Fig. 11.2 TLM2.0 interoperability layer (OSCI TLM-2.0 LRM)

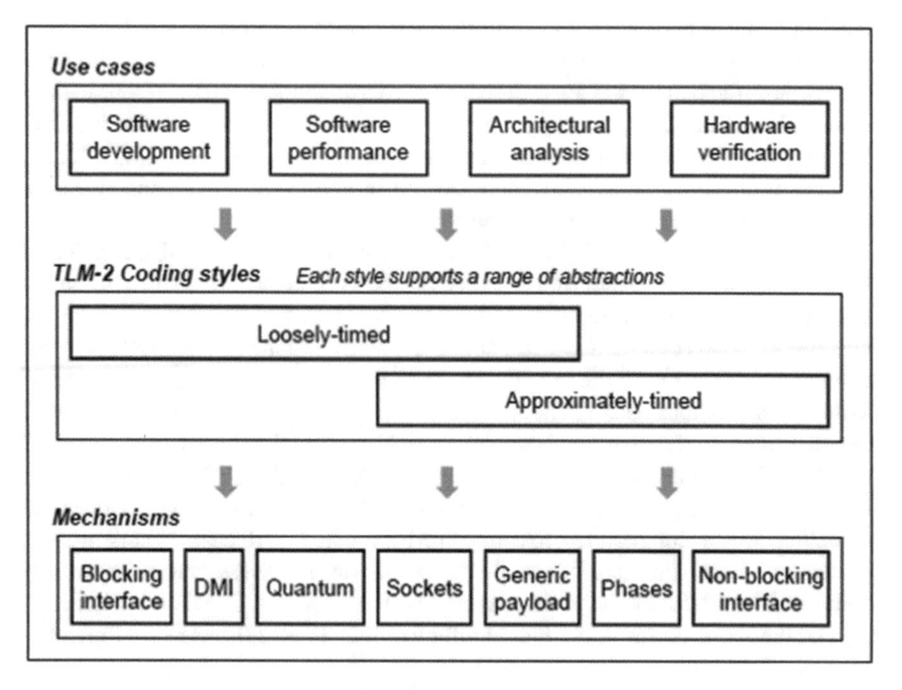

Fig. 11.3 TLM2.0 use cases and coding styles (OSCI TLM2.0 LRM)

11.2.1 Loosely Timed (LT) TLM 2.0 Transaction-Level Modeling

The loosely timed coding style makes use of the *blocking transport interface*. This interface allows only two timing points to be associated with each transaction, corresponding to the call to and return from the *blocking transport function*. In the case of the base protocol, the first timing point marks the beginning of the request, and the second marks the beginning of the response. These two, timing points could occur at the same simulation time or at different times.

The loosely timed coding style is appropriate for the use case of software development using a virtual platform model of an SoC, where the software content may include one or more operating systems. The loosely timed coding style supports the modeling of timers and interrupts sufficient to boot an operating system and run arbitrary code on the target machine.

LT is useful when cycle accuracy is not required, rather functional validation per SoC architectural specification is required. Some "timing" information may be available in LT models. But arbitration of shared resources and the impact of resource conflicts and contention on the system performance is not modeled and thus not considered.

Referring to Fig. 11.4, the LT model defines two timing points, the beginning time and the end time of a transaction. As is evident, LT modeling style is simple in that if it is used to model hardware where concurrent accesses to hardware resources play a significant part in the "functionality" of the working of an SoC, then the LT model will not be accurate.

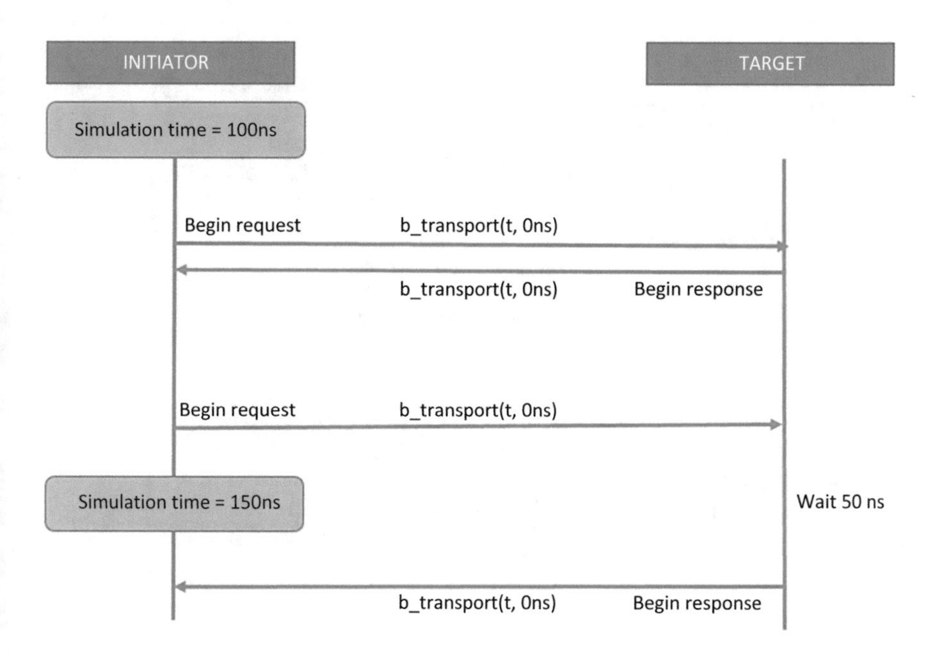

Fig. 11.4 LT model with blocking transport call

The blocking transport method may return immediately (i.e., in the current SystemC evaluation phase) or may yield control to the scheduler and only return to the initiator at a later point in simulation time.

The TLM-2.0 blocking transport interface is intended to support the loosely timed coding style. The blocking transport interface is appropriate where an initiator wishes to complete a transaction with a target during a single function call, the only timing points of interest being those that mark the start and the end of the transaction. The blocking transport interface only uses the forward path from initiator to target. b_transport method has a single transaction argument passed by non-const reference and a second argument to annotate timing. The b_transport method has a timing annotation argument. This single argument is used on both the call to and the return from b_transport to indicate the time of the start and end of the transaction, respectively, relative to the current simulation time.

Please refer to OSCI TLM2.0 LRM for complete detail on LT modeling style and its class definitions.

11.2.2 Approximately Timed (AT) TLM 2.0 Transaction-Level Modeling

In contrast to blocking transport interface, the approximately timed (AT) interface uses non-blocking transport. The goal of AT modeling is accurate modeling of resource contention and arbitration. It is used to model a system which has strong dependence on timing. The non-blocking transport interface is appropriate where it

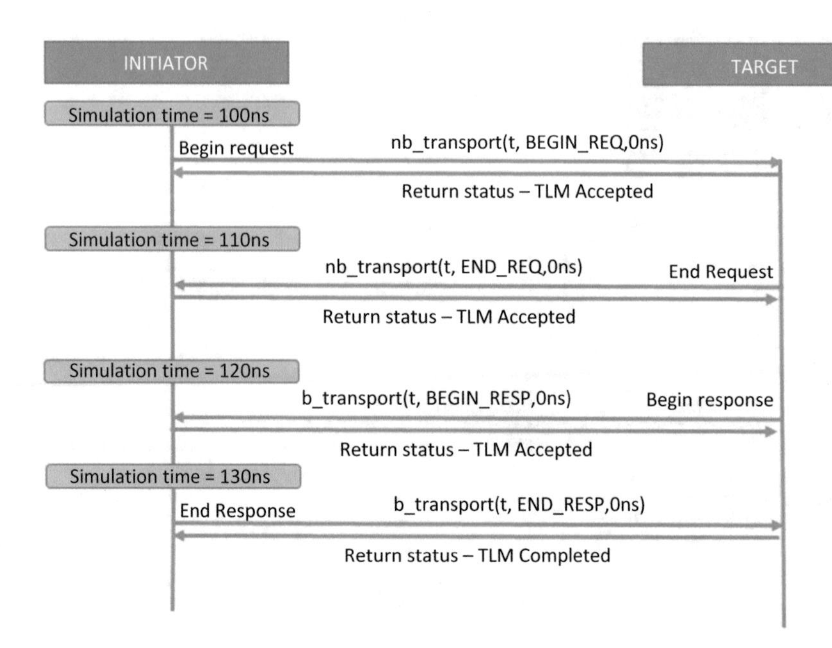

Fig. 11.5 AT modeling style with non-blocking transport calls

is desired to model the detailed sequence of interactions between initiator and target during each transaction, in other words, to break down a transaction into multiple phases, where each phase transition is associated with a timing point (Fig. 11.5). Each call to and return from the non-blocking transport method may correspond to a phase transition. By restricting the number of timing points to two, it is possible to use the non-blocking transport interface with the loosely timed coding style, but this is not generally recommended. For loosely timed modeling, the blocking transport interface is generally preferred for its simplicity. The non-blocking transport interface is particularly suited for modeling pipelined transactions, which would be awkward to model using blocking transport.

The non-blocking transport interface uses a similar argument-passing mechanism to the blocking transport interface in that the non-blocking transport methods pass a non-const reference to the transaction object and a timing annotation, but that's where the similarity ends. The non-blocking transport method also passes a phase to indicate the state of the transaction and returns an enumeration value to indicate whether the return from the function also represents a phase transition. Both blocking and non-blocking transport support timing annotation, but only non-blocking transport supports multiple phases within the lifetime of a transaction. The blocking and non-blocking transport interface and the generic payload were designed to be used together for the fast, abstract modeling of memory-mapped buses. However, the transport interfaces can be used separately from the generic payload to model specific protocols. Both the transaction type and the phase type are template parameters of the non-blocking transport interface.

There are two non-blocking transport methods, nb_transport_fw for use on the forward path and nb_transport_bw for use on the backward path. Aside from their names and calling direction, these two methods have similar semantics. Transactions may be pipelined. The initiator could call nb_transport to send another transaction to the target *before* having seen the final phase transition of the previous transaction.

The AT timing dependencies are modeled using four timing points for a transaction: begin request, end request, begin response, and end response as shown in Fig. 11.5. These act as synchronization points for the models.

Note that TLM2.0 allows both the blocking and the non-blocking modeling styles to simulate together. The recommendation from the OSCI TLM2.0 working group is to use LT modeling for software development and AT modeling for hardware performance verification and hardware functional design verification.

11.3 Virtual Platform Example

Figure 11.6 shows a virtual platform of a system with just enough logic to boot Linux. A few things to note in this figure

ARM A9 Uni Processor TLM2.0 model is directly available from ARM (known as ARM Fast Model) or a "C" version from QEMU (open source).

If you have the resources ($$) to acquire an ARM Fast Model of its CPUs, you are home-free. If you write your own "C/C++" processor model or get it from QEMU, you can still use it in the TLM2.0 virtual platform. The methodology is quite straightforward to take such a "C" or "C++" model and wrap it around with a TLM2.0 wrapper. Every time the CPU ARM A9 issues a load or store, the TLM2.0 wrapper traps it as a transaction to be sent on the on the simple LT BUS. The LT BUS then routes the transaction to the system memory. When read occurs (load), the A9 LT wrapper waits for data to arrive from the system memory, "calls" the A9 "C" model, and provides the data. The A9 models waits until then and before resuming with the next operation. This method works quite well. An engineer does not need to be a "C" or "C++" guru to accomplish the task of building the TLM2.0 wrapper.

Let us discuss the simple Linux boot system depicted in Fig. 11.6.

The system memory is a very simple direct LT model that points to a memory array. The LT model responds to Write/Read requests from the BUS and stores or provides required data.

UART, timer, and interrupt controller were modeled directly in TLM2.0 LT modeling style.

The UART talks to the Xterm emulated Keyboard and Terminal via UNIX Sockets. UNIX sockets communicate directly with the UART LT model.

The BUS is a simple LT Router (address mapped).

Such a platform was created by the author and his team for one of his projects. This platform could boot bare bone Linux in under 5 s (wall clock time). It was faster than the hardware boot of Linux because the Linux used on this virtual platform was a pared down version of full Linux. The virtual platform can boot the operating system and can be used for embedded software development.

11.3.1 Advantages of a Virtual Platform

- Such a platform, as part of a hierarchical verification methodology, will be available way before RTL is ready, a significant advantage to software development teams.
- The platform allows software to be developed "at speed" (i.e., real time execution).
- Such a platform accurately (functionally) models the entire state (registers, local memories, etc.) of an SoC/System. Hence the entire programmer's view of SoC functionality is available to the software developers before RTL is ready.
- Since the BUS is a TLM2.0 transaction level plug & play bus, the virtual platform can be employed by IP vendors to try out their models of IP blocks and run the software stack / device drivers. This guarantees the software compatibility of a vendor's IP.
- Users can simply create the TLM model of their IP employing such a virtual platform, plug in their model to this virtual platform, and access it through the software, running on top of the virtual platform.

11.3.2 Open Virtual Platform (OVP) Initiative

This initiative was taken by the SystemC committees. The idea was to create a base virtual platform (as in Fig. 11.6), TLM2.0 models (such as CPU, Peripheral models, etc.), and make it available for engineering community at large to give them a leg up on creating a virtual platform for embedded software development and design verification. I highly recommend visiting http://www.ovpworld.org website to get an

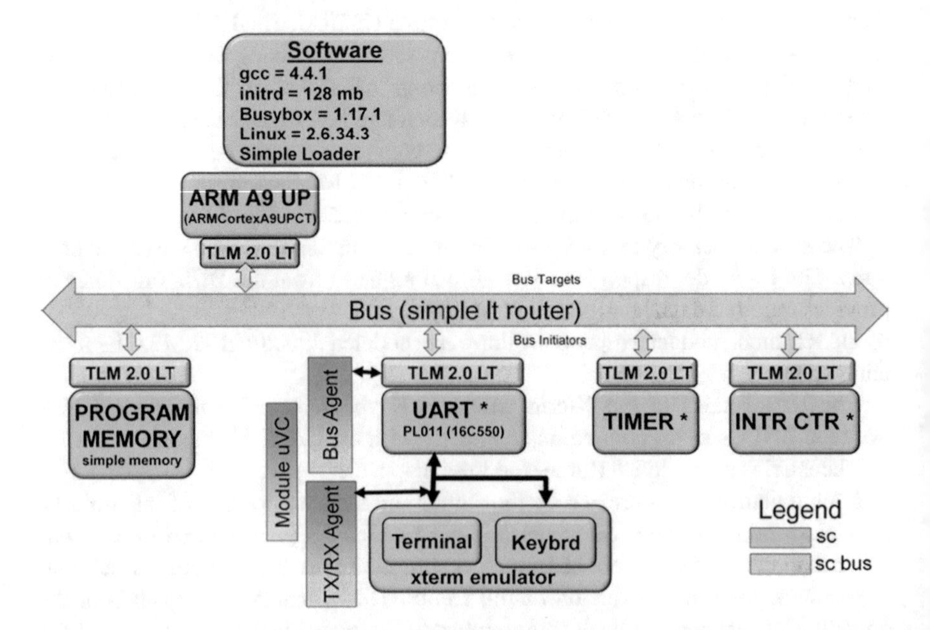

Fig. 11.6 Virtual platform to boot Linux

idea on the full scope of the capabilities of this platform. The entire platform is TLM2.0 plug & play compatible with extensive API support.

Following is a high-level overview of the OVP org as described on the OVP website (OVP n.d.):

> The focus of OVP is to accelerate the adoption of the new way to develop embedded software—especially for SoC and MPSoC/multi-core platforms. If you are developing software to run in an embedded system you will probably already be using an Instruction Set Simulator (ISS) and associated debugger. As you move to having multiple processors or cores in your design then you will need more than just a single ISS. What is needed is a model of your platform that includes models of all the processors or cores and models of the peripherals and behavioral components that the software communicates with. This is a Virtual Platform, or more simply just a simulation model of your design. OVP provides this for you: libraries of processor and behavioral models, and APIs for building you own processors, peripherals and platforms. There are even platform models available as source (we call these Extendable Platform Kits). This is just what is needed to use existing models or build your own, and OVP is easy to use, open, flexible, and importantly, free for non-commercial use.

11.3.3 Rationale for Software Virtual Platforms (OVP n.d.)

The following quote is taken directly from the OVP website, since they are the best source to describe the rationale behind their initiative:

> The most common practice today for developing embedded software is to start to develop initial software in a desktop Windows or Linux development environment and start unit testing—with the software running in the general-purpose Operating System. This development environment is often very different to the final target system—for example using host threads as opposed to separate processors in the real system—requiring much re-writing/ modification/ porting for final deployment.

When a prototype of the embedded system or chip is available, the software is ported to this target environment using cross compilers and related tools targeting the embedded processors, such as ARM, MIPS, Renesas, PowerPC, etc. FPGAs might be used in the prototype to emulate the SoC. A simple debugger is then often connected via a JTAG port.

There are many challenges when using this traditional approach. If you use a hardware prototype of your system, it is often unreliable, not readily available within all your software developments sites (especially those offshore), it can be physically unreliable, and worst of all, it is often available only very near to the end of the targeted product development schedule. All these challenges contribute to real problems in getting software available soon after product hardware availability—the target should be to get the products embedded software up and running very shortly after hardware availability.

Recently there has been much talk of developing software on hardware emulators. Hardware emulators are often very large and very expensive and are very hard to set up. They take the RTL of the design and run it on custom chips or FPGA-based systems, and these execute the RTL which is the source of the design. Yes, they can run significantly faster than the RTL simulators, and yes, they have accuracy of the RTL, but they suffer from two main problems: a) they *require* the RTL, which means they are only useful for software development at the very end stages of a chip project, and b) they are *very* slow when compared to instruction accurate simulators such as OVPsim or Imperas. Hardware emulators are typically 1000 times slower to TLM2.0, for 1000 times more money!

These challenges become acute as more processors interact in the embedded system. Then there are new challenges in multi-core or multi-processor systems where often the hardware prototypes provide limited controllability, observability, and debuggability. When tracking down complex multi-processor issues, the bugs are often very hard to reproduce reliably and isolate in complex real-time hardware.

As a result, development teams are scrambling around looking for a better solution. As more and more chips become multi-core, these teams are looking for a better solution than just awaiting the prototypes... they just cannot afford to be that late to market.

If there was a virtual model of the hardware platform that was available to the software developers at the very earliest stage of the products development and if the initial testing of software is done on a virtual platform, then they could reduce SoC schedules by months and reduce initial development and maintenance costs significantly for SoC embedded software.

This is what OVP is enabling: the availability of freely available virtual platform models early in the product development cycle.

Yes, this methodology of having a model of the system to be used for software development is more critical for an SoC or MPSoC where there is Software on Chip or Multi-Processor Software on Chip, but it is also a benefit for developers of any embedded software. It is far easier to develop software in conjunction with a good simulation of a device than it is on the real embedded device.

OVP is targeting the building of models of embedded components to enable embedded software to be developed efficiently.

Hardware analogy: In the mid-1980s there was a challenge in the chip hardware design business—the chips were getting more and more complex, expensive to build, and taking longer to fabricate—and productivity was a significant challenge. By the end of 1980s, most chips were developed on simulation technology, and you would be hard-pressed to find a chip that was sent for fabrication without significant testing using hardware design simulators like Tegas, HILO, and Verilog-XL.

This move from a “develop prototype” to a “run simulation model-based methodology” dramatically improved hardware development productivity and enabled the hardware teams to harness complexity and manage exploding project schedules. It also allowed the hardware teams to do more and more verification to improve the quality and confidence in their designs.

11.4 ESL/Virtual Platform for Design Verification

11.4.1 Overview

Ok, so we have seen the virtues of virtual platform for software development and high-level view of how ESL. But virtual platform also helps design verification. Let’s look at a few applications and methodologies that help us achieve the goal of design verification with a virtual platform.

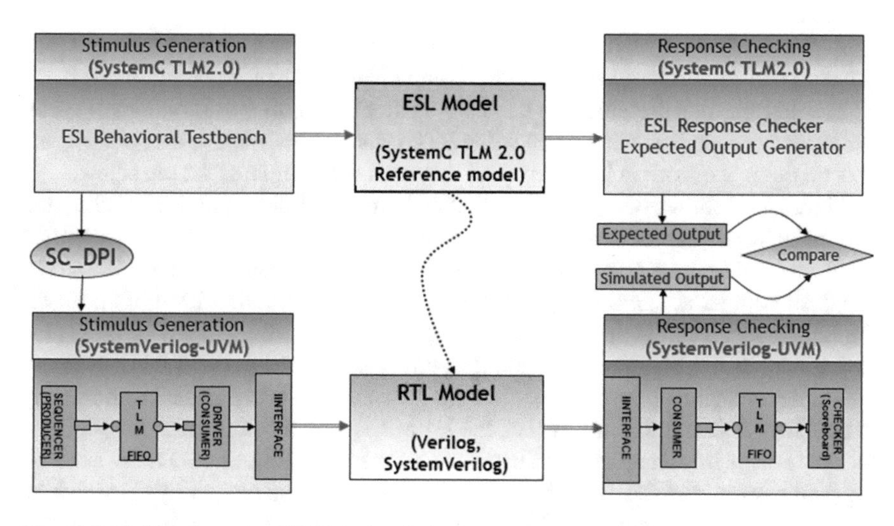

Fig. 11.7 Verification using TLM2.0 virtual platform and RTL co-simulation

Here are some ways in which we can use a virtual platform for design verification:

1. Virtual platform and RTL co-simulation and verification. Compare virtual platform results with RTL results during simulation or during post-processing (Fig. 11.7).
2. Use virtual platform as a reference model as part of UVM scoreboard (Fig. 11.9).
3. ESL/virtual platform refine and reuse methodology. Refine and reuse a virtual platform for test development. In other words, tests/testbenches developed on a virtual platform are reusable on RTL platform.

Let us look at each of these methods in detail. There methods are not theoretical; they have been implemented by the author and his team for use with multibillion transistor SoCs.

11.4.2 *Virtual Platform and RTL Co-simulation and Verification*

As discussed before, one of the main advantage of creating a virtual platform is that it models the architectural (programmer's view) state of the SoC to match 100% with the SoC specs. In other words, the architectural state of virtual platform is directly comparable with the RTL state. While this is not possible at clock granularity (since virtual platform is at transaction level), it is indeed verifiable at transaction level or at CPU instruction boundary.

For example, at the end of a read, the virtual platform will predict the read data, and RTL must match it. Or when interrupt arrives, both the virtual platform and the RTL must service it at the same "transaction" boundary. Another example, when a CPU instruction retires, we can compare the complete register state of the CPU with that of the RTL. They much match. For CPU verification, ISS (Instruction Set Simulator) has been used for many years. ISS is nothing but a virtual platform of the CPU.

In this section, we will see how a virtual platform built in TLM2.0 SystemC standard can co-simulate with RTL and comparisons made at transaction level.

The first question that comes to mind is if a virtual platform runs at billions of cycles per second (yes, this is not a typo) while RTL runs (at most) 500 cycles per second, how do these two work in lock step? The answer is very simple as we see below.

Figure 11.7 shows two paths. The first path on top of the figure is the ESL path. ESL behavioral (transaction level) testbench drives the ESL Model (i.e., the virtual platform) and collects responses from the ESL model. It compares the ESL model response with its predicted response and does what-I-call a first level of check for the ESL model itself. Is the ESL model response correct? If so, it then considers the ESL model response as the Expected Response for comparison with RTL response.

The second path at the bottom of the figure shows the UVM path. The first thing to note here is that the stimulus (i.e., the UVM sequence) comes from the ESL testbench. This is important since this guarantees that both ESL and RTL are simulating their respective DUT with the same stimulus. There are many ways in which an ESL model can interact with UVM testbench. The one shown here is through SC_DPI which is the SystemC Direct Programming Interface (similar to SystemVerilog DPI). The SC_DPI API allows you to send a transaction to the UVM agent model where the transaction will be converted to UVM sequence. After that the UVM simulation proceeds as described under the chapter on UVM (Chap. 4). The RTL model simulates and produces a response which then goes through the UVM monitor and scoreboard. The monitor of the UVM agent sends the RTL response transaction to the scoreboard via analysis port. The scoreboard grabs the expected response produced by the ESL model and compares it with RTL response. Note that this comparison takes place at transaction level.

The speed at which ESL model simulates is hundreds of orders of magnitude faster than RTL. So, with this approach the simulation speed will indeed crawl to the speed of RTL. After every transaction that goes through ESL model, the ESL model will stall until the ESL response is checked against RTL output. Of course, you can deploy deep FIFOs to store ESL results and not stall ESL every so often.

There are many applications (such as checking a single frame output of a video engine) where the ESL model acts as a perfect reference model and you can check the output line by line of the resulting frame.

There is another variation on the comparison of outputs. This is shown in Fig. 11.8. The idea is the same only that the RTL output as captured by the UVM scoreboard uses SC_DPI (SystemC Direct Programming Interface) to send the transaction-level output to the ESL model, and let ESL model compare it with the RTL output transaction.

11.4.3 Virtual Platform as a Reference Model in UVM Scoreboard

This methodology is quite like the one presented in Sect. 11.4.2. It is mainly a different presentation. In Fig. 11.9, a C++ stimulus model is wrapped with a TLM2.0 wrapper which can be accessed by SystemVerilog DPI (or SystemC DPI) interface.

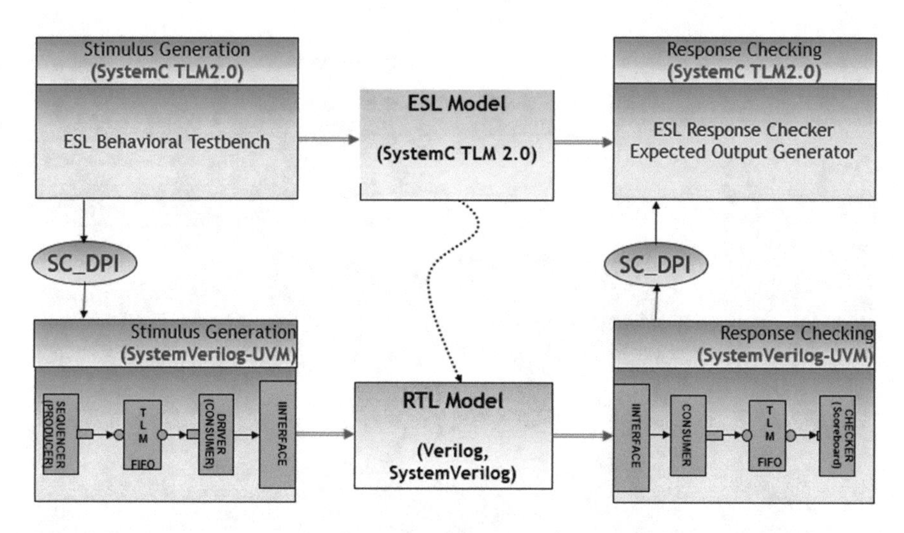

Fig. 11.8 Verification using TLM2.0 virtual platform and RTL co-simulation using SC_DPI

The transaction-level interface TLM2.0 provides transactions to the UVM agent sequencer via the DPI interface. The transaction from the C++/TLM2.0 model is converted to a sequence by the DPI interface. These sequences are fetched by the UVM agent sequencer and driven to the driver. Driver drives signal/cycle accurate protocol to the DUT. Note that the ESL DUT acts both as a reference model for the UVM scoreboard and a DUT.

But the ESL model is at C++ level. How can it communicate at cycle level with the UVM driver? A TLM2.0 wrapper is built around the ESL model to convert ESL model responses to transactions. Another wrapper on top of the TLM wrapper converts each transaction into a cycle accurate transaction that can communicate with the UVM driver.

The same ESLDUT (C/C++/SystemC) with its TLM2.0 wrapper (but not the cycle accurate wrapper) is used by the UVM scoreboard at transaction level. The UVM monitor takes the cycle accurate transactions from the UVM driver and converts them into Transactions. These transactions are then compared with the transactions received by the scoreboard from the ESL/TLM2.0 wrapper interface. Figure 11.9 is quite self-explanatory.

11.4.4 ESL to RTL Reuse Methodology

What is the refine and reuse methodology? We just saw in Sect. 11.4.3 that an ESL DUT model was used both as a reference model and a DUT. This is what we mean by reuse. There are other reuse scenarios that ESL use to help with RTL verification. We need to keep in mind that RTL verification is the eventual goal and *not* the ESL model verification.

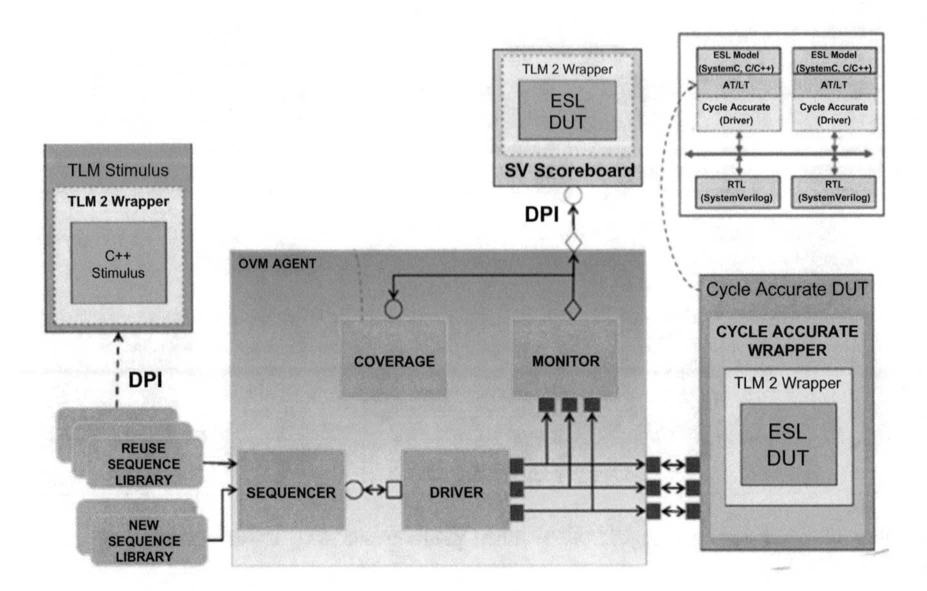

Fig. 11.9 Virtual platform as a reference model in UVM scoreboard

Here are the other use case scenarios of ESL reuse for RTL verification. We will discuss these reuse cases in detail in upcoming sections.

- Reuse:
  1. Reuse ESL stimulus generation testbench to drive RTL design
  2. Reuse ESL response checking logic to check RTL output (at transaction boundary)
  3. Reuse ESL model as a reference model for RTL design verification
- Return on investment (ROI)
  - Reuse allows for a good return on time/resources spent on ESL model development and verification. You don't have to repeat the stimulus generation development all over when RTL is ready.

  Why reuse from ESL to RTL? Here are advantages of the reuse methodology:

- You cannot completely verify at ESL. What this means is that ESL verification provides a good reference model that is functionally accurate with the architecture specs of the design. But at the end of the day:
  - RTL verification is still required before synthesis/gates, tape-out, and silicon.
  - ESL verification with TLM 2.0 LT/AT models does not account for clock level concurrency.
- You create more than twice as much work without reuse:
  - Without reuse of ESL testbench/environment, you will simply reinvent the wheel at RTL level with its longer time to develop, simulate, and debug.

So,

- Reuse TLM model as a reference model for RTL verification.
- Perform stepwise refinement, successively replacing TLM blocks to RTL as described in Sect. 11.4.7.

• Verify functionality as much as possible at ESL level. This means that the entire functional domain verification can be achieved at ESL level. This guarantees the ESL model as the golden reference model against which RTL can be compared. Only the SoC interface (and internal buses) verification remains at the RTL level. A great saving in time (and money).
• To reiterate, reuse is not only for the testbench component but also for the model component.

11.4.5 Design and Verification Reuse: Algorithm ⇔ ESL: TLM 2.0

Let us discuss step by step how do we reuse an ESL model and testbench for RTL verification. First, this requires reuse of algorithmic ESL model to TLM2.0 model. And second, it requires reuse of the TLM2.0 model to cycle accurate interface with RTL.

First, let us tackle ESL (pure C/C++ algorithmic model) to TLM2.0 migration.

Figure 11.10 shows two stages to accomplish going from ESL to TLM2.0.

The first stage shows the ESL algorithmic model coded in C/C++/SystemC. It is not at transaction level yet (think ISS—the Instruction Set Simulator). Both the algorithmic level testbench and the DUT are modeled using procedural purely functional languages like C/C++ or SystemC (even though SystemC has the concept of time). This is the stage where you verify the entire "functional" domain of your SoC. For example, if you want to verify the LRU (Least Recently Used) algorithm of your cache, you can use this model. Or if you want to verify the Ethernet IPV4 (or IPV6) layer processing, you can use this model.

The second stage is to reuse the algorithmic purely functional C/C++ model to develop the transaction-level model, so that the algorithm level functionality can be broken down into transactions. For example, again going back to LRU algorithm, each access to the cache and its access per LRU algorithm can be broken down into transaction-level reads and writes. This is accomplished by getting responses from the C/C++ model and converting them to TLM2.0 compliant transactions as shown in Fig. 11.10. Note that both the testbench and the DUT model are being reused. And that the verification is still at transaction level. The idea is to weed out all the functional bugs before jumping into RTL verification.

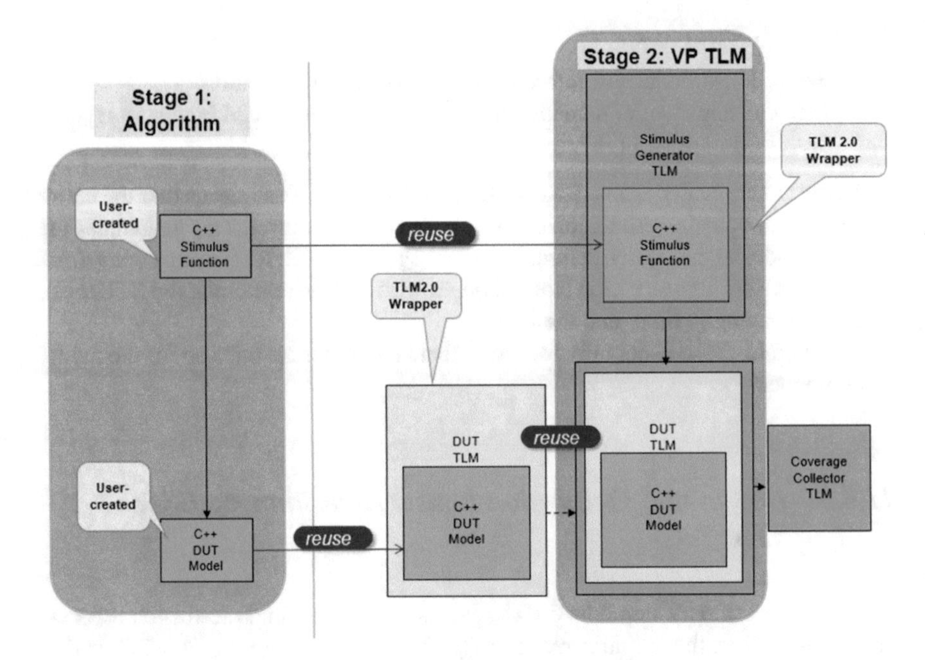

Fig. 11.10 Design and verification reuse: Algorithm ESL—TLM 2.0

11.4.6 Design and Verification Reuse: ESL/TLM 2.0 ⇔ RTL

Now let us see how we go from TLM2.0 to RTL cycle accurate model and the testbench.

As shown in Fig. 11.11, there are three stages involved. We discussed the first two stages in Sect. 11.4.5. Now let us add the third stage.

The third stage is to "convert" (or wrap) the TLM2.0 model to a cycle accurate model. This is shown in Fig. 11.11 as the "RTL Stage." How do you do that? The transactions from the TLM2.0 will mostly end up as read or write from the peripheral devices or the SoC internal registers/embedded memory or the external DRAM. The cycle accurate wrapper will take these transactions and convert them to an appropriate interface protocol. The protocol may be just an internal SoC bus (like AXI or AHB or APB), or it could be an external peripheral device interface. In either case, the transaction from the TLM2.0 will be converted to the protocol of the "bus" to which the cycle accurate model is attached.

The cycle accurate model (wrapper) will be in SystemVerilog (as opposed to the ESL/Algorithmic model in C/C++ and the TLM2.0 model in SystemC). In other words, there are three levels of abstraction in three different languages. But note that the ESL model is still the core of either the TLM2.0 or the cycle accurate model (as shown in Fig. 11.11). In other words, we avoided reinventing the wheel at every

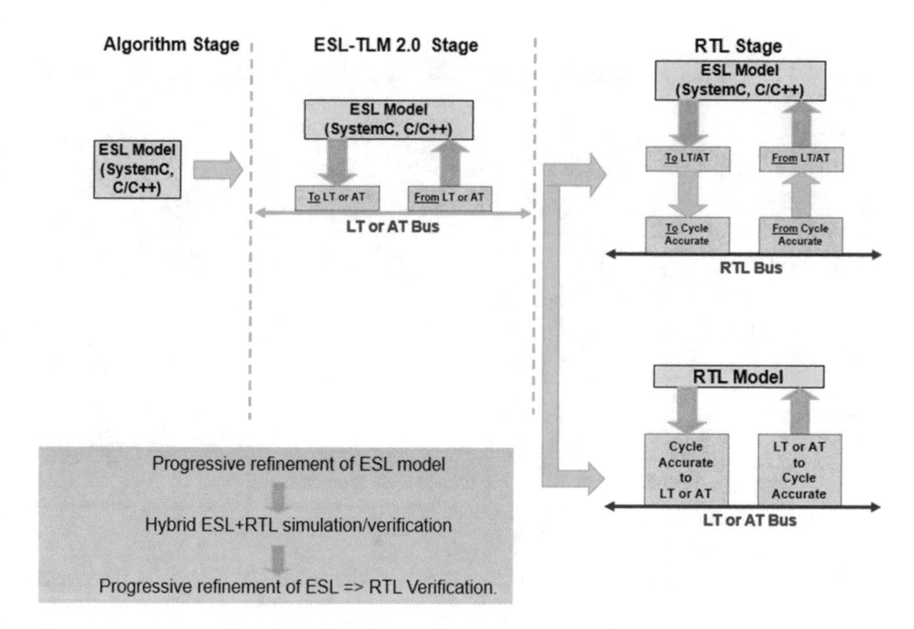

Fig. 11.11 Design and verification reuse: ESL/TLM 2.0—RTL

stage. The high-level ESL model is easy to develop compared to developing the functionality directly in TLM2.0 or the cycle accurate model. The next section describes a very practical methodology that uses these reusable-derived models into your design and verification flow. Such flows are being used by large SoC development projects. The author has successfully deployed such a flow with significant reduction in time/project schedules and accuracy of verification.

11.4.7 Design and Verification Reuse: Algorithm ⇔ ESL-TLM 2.0 ⇔ RTL

Ok, so far, we saw how to "derive" a reusable model from ESL to TLM2.0 to cycle accurate. But having a collection of models has no meaning. What do you do with these models? How do you use them to create a time-saving accurate design and verification methodology?

First, this methodology (Fig. 11.12) has been proven to work as part of TSMC's System Level ESL Reference Flows (RF10 and RF11). Mentor, Cadence, and Synopsys all three vendors have implemented the flow shown in Fig. 11.12. These reference flows are available with detailed application notes for TSMC customers. The author architected and managed development of these working methodology/flows. All three vendors have their own versions of such flows that are commercially available.

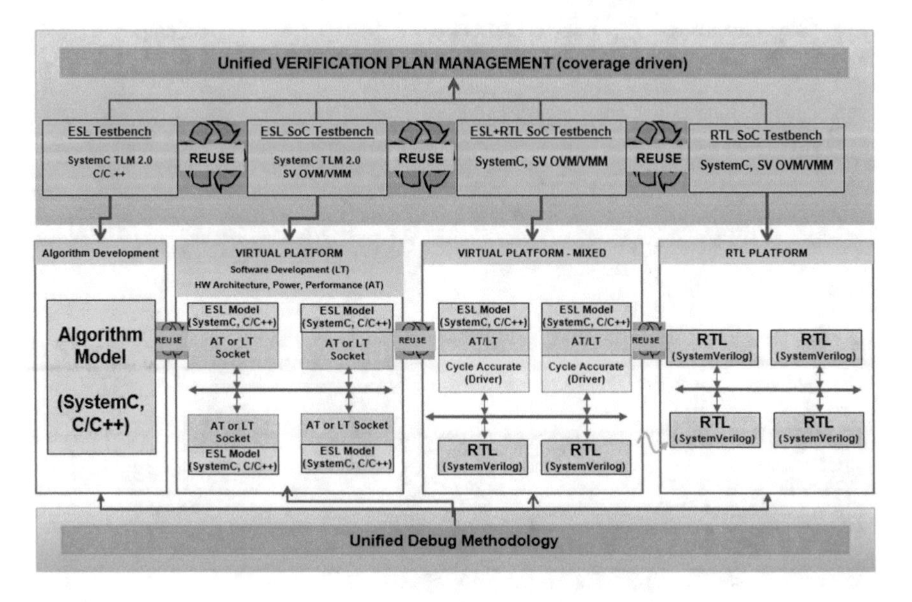

Fig. 11.12 Design and verification reuse: Algorithm—ESL-TLM 2.0—RTL

The figure looks a bit cluttered. So, let's break it down.

- The TOP horizontal flow is for reusable verification methodology.
- The MIDDLE horizontal flow is for reusable design methodology.
- The LAST horizontal box is for a unified debug methodologies.

Let us look at the first two flows, namely, reusable verification and reusable design methodology flows. Note that the verification and design reusable flows go hand in hand. The first stage of the flow is to develop the algorithm model. This is the stage when architects of the design are validating their architecture and not worried about cycle accurate or even transaction-level detail. Corresponding to the ESL/algorithm model development, we also develop a testbench at ESL (C, C++, SystemC) level to verify the algorithm model. The ESL model and the testbench can be considered the "seed" models for the entire flow. Note that the ESL model are designed for each subsystem of the SoC. In other words, there needs to be a subsystem level granularity for them to be reusable. For example, an ISS—Instruction Set Simulator model is an algorithm model for the entire CPU. And this CPU is used as an embedded processor in your single-/multi-core SoC. So, the ISS is considered a subsystem level ESL model which can be reused as you move to more refined lower-level stages.

Once you have modeled and verified the ESL models of your SoC, you need to be able to reuse them at the next stage. The stage is where we move from ESL to transaction-level detail. As shown in Fig. 11.11 and Fig. 11.12, we wrap the ESL models with TLM2.0 transaction-level wrapper to create a TLM2.0 (LT or AT) model of the entire SoC. In other words, the logic of the SoC is reused from the ESL level and converted to write/read/interrupt ack level transactions that will be closer to the final SoC implementation. So, we reuse the ESL model at TLM level, and we also reuse the ESL testbench to TLM level with similar flow. We wrap the ESL testbench with TLM wrapper to make it communicate and verify the design TLM model.

This second stage of evolution is called the virtual platform stage. This is the stage where the entire architectural state of the SoC is now modeled at TLM level. That being the case, this stage is perfect for software development and power and performance evaluation of the system. The TLM level platform simulates thousands of orders of magnitude faster than the final RTL stage. This is also the stage where not only the simulation is fast enough for software development, but the debug is also that much easier since we are not dealing with cycle accurate information of the SoC.

Now the typical dilemma of DV (design verification) teams is that they would like to start verification of the SoC RTL, but the RTL is not ready for the entire SoC. Some subsystems are still at transaction (TLM) level, while others have migrated to RTL. This is where the third stage comes into picture. For those subsystems that are still at TLM level, we apply a wrapper on top of the TLM level to convert the transactions into cycle accurate level (Figs. 11.11 and 11.12). Now you have a system of RTL and ESL+TLM models all simulating together at cycle accurate RTL level. Hence the DV teams do not need to wait until the RTL for the entire SoC is available. The design team can plug & play with the TLM and RTL models to finally move the entire SoC to RTL stage.

The testbench also deploys the same methodology as that of the TLM to RTL progression. The testbench wraps itself around with cycle accurate wrapper that takes transactions coming from the ESL+TLM testbench "core" and converts them to the SoC peripheral cycle accurate activity.

Hence, we have created a completely reusable methodology for both design and verification. Obviously, there will be come caveats to this methodology. For example, if you have very heavily pipelined, superscalar subsystem, the TLM model must be at AT (approximately timed) TLM level; else you won't be able to reuse the TLM level model (with its cycle accurate wrapper) at RTL level. These decisions should be made upfront the design project so that you can achieve maximum reusability throughout the migration from ESL to TLM to RTL stages.

Finally, the debug methodology (supported by all major EDA vendors) crosses the boundary among ESL, TLM, and RTL. In other words, you should be able to see transaction-level activity along with cycle accurate activity or ESL level (e.g., ISS instruction execution) activity along with TLM and RTL level activity. Such unified debug environment is a must for such a methodology to be practical. EDA vendors have taken a notice of this and are offering such unified debug capabilities.

Chapter 12
Hardware/Software Co-verification

Chapter Introduction
An SoC is ready only when both its hardware and software components are ready. You cannot ship silicon until its software is ready because without software, hardware is pretty useless.

This chapter will discuss the methodologies to develop software such that it is ready when hardware is ready to ship. What kind of platform do you need? How does ESL virtual platform play a key role? How do emulators and accelerators fit in the methodology equation?

12.1 Overview

An SoC is ready to ship only when the complete application works, not just when hardware simulations pass regressions. In other words, the ultimate test for a chip is to see that it performs its applications correctly and completely. That means executing the embedded software together with the RTL. Such tests require billions of cycles of execution and usually run at the system level, where design size is the greatest. Simulating such applications even on advanced workstations is simply too slow.

There are few fundamental ways in which hardware–software co-verification takes place. These technologies help (to some extent) the requirement that the co-verification run at meaningful speeds to allow for software execution and RTL running concurrently.

As a sidenote, virtual platform in the following can be virtual platform of the peripheral interfaces of an SoC, or it can be an ISS (Instruction Set Simulator), or it can be a virtual platform with embedded ISS. A common terminology of virtual platform is used throughout this chapter.

A.B. Mehta, *ASIC/SoC Functional Design Verification*,
DOI 10.1007/978-3-319-59418-7_12

1. Virtual platform ⇔ RTL co-simulation (This approach is mentioned for the sake of completeness. It does not yield the required clock speeds to effectively run software. Hence it won't be discussed further. It is feasible but impractical.).
2. Same arguments apply for RTL ⇔ emulation co-simulation.
3. Virtual platform ⇔ emulation.
4. Virtual platform ⇔ hardware accelerator.
5. Prototype FPGA board.

Let us look at these methodologies/technologies in detail.

12.2 Hardware/Software Co-verification Using Virtual Platform with Hardware Emulation

Before we dive into virtual platform and hardware emulation co-simulation detail, let us take a quick look at what hardware emulation entails.

12.2.1 Hardware Emulation and Prototyping

Hardware emulation is the process in which a piece of hardware is made to emulate the behavior of one or more other hardware system under design. It is mostly carried out on very-large-scale integrated circuit designs with the purpose of functionally verifying the system under design. Hardware emulation is a technique that integrates a hardware design into a reconfigurable (e.g., FPGA-based) prototyping platform to allow the functional testing of a design under test including its firmware. This way both hardware and software can be evaluated in a realistic performance setting.

There are main three commercially available emulation systems. Several other vendors also provide FPGA-based prototype systems. Too many to enumerate here.

1. Veloce-2: Mentor Graphics
2. Palladium: Cadence Design Systems
3. ZeBu Server-3: Synopsys

The following technologies are predominantly deployed to build hardware emulation systems.

12.2.1.1 FPGA-Based Hardware Emulator

First used in the 1990s, it lost appeal vis-à-vis to the custom processor-based architecture because of several shortcomings. In the past 10 years or so, the new generations of very large commercial FPGAs have helped to overcome many of the original weaknesses. Its physical dimensions and power consumption are the smallest and lowest for equivalent design capacity. Among the drawbacks, its speed of compilation is low, at least on designs of 10 million gates or more. The full design visibility is achieved by trading off the higher speed of emulation.

12.2.1.2 Custom Emulator-On-Chip Architecture

The custom chip could also contain debug circuitry, visibility mechanisms, and a host of other capabilities.

Each chip can emulate a small piece of a design, and larger designs are handled by interconnecting many of the chips together, again with sophisticated interconnect capabilities.

Pioneered by a French start-up by the name of Meta Systems in the mid-1990s, the emulator-on-chip architecture is based on a highly optimized custom FPGA that includes an interconnect network for fast compilation, which also enables correct-by-construction compilation. Design visibility is implemented in the silicon fabric that assumes 100% access without probe compilation and rapid waveform tracing. It has a few drawbacks; namely, it requires a farm of workstations for fast compilation and has somewhat slower speed and larger physical dimensions than an emulator based on commercial FPGAs of equivalent design capacity.

12.2.1.3 Custom Processor-Based Architecture

Devised by IBM, it has been a proven technology since 1997 and dominated the field in the decade from 2000 to 2010. Advantages include fast compilation, good scalability, fast speed of execution in ICE mode, support from a comprehensive catalog of speed bridges, and excellent debugging. Drawbacks are limited speed of execution in TBA mode, large power consumption, and larger physical dimensions than an emulator based on commercial FPGAs of equivalent design capacity.

12.2.2 Emulation System Compile Time

One of the factors that affect selection of an emulation system is the compile (synthesis) times to put RTL into the box. What good does it do for emulation to provide results in minutes while compile times take hours. Yes, this is true. Hardware emulators can accommodate any design size, but they require a long setup time and are relatively slow to compile, compared to simulation.

While emulators can process billions of cycles in a relatively short time, on smaller designs, the limitations may hinder the benefits of the fast speed. A simulation session of 1 h may lead to higher productivity than an emulation session on the same design running in 10 s. In an eight-hour day, a verification engineer can run more design iterations, including compilation-execution-debug, than with emulation.

Note that in current advances in distributed computation, FPGA compile is now done over CPU/server farms in parallel drastically reducing compile times. The author does not have good, real-life data points to share, unfortunately.

While discussing requirements for fast compile times with an EDA vendor, the following points need to be considered:

- Be 10X to 20X faster than traditional synthesis tools.
- Support for full and block synthesis modes, parallel, and incremental synthesis.
- Support for SystemVerilog, Verilog, VHDL, and mixed language designs.
- Support for user-defined primitives (UDPs) and automatic memory inferencing.
- Have enhanced debugging with support for synthesizable SystemVerilog Assertions (SVA), RTL name preservation, and preload, read, and write support for inferred memories.
- Efficient use of FPGA and emulator resources.
- Comparable area to traditional FPGA synthesis (−10% to +25%).
- Emulation speed equivalent to traditional FPGA synthesis.
- Interoperable with traditional FPGA synthesis tools.

12.2.3 Difference Between Emulator and FPGA-Based Prototype

So, if emulation systems could be based on FPGA, what's the difference between an emulator and a prototype board with FPGAs?

A key distinction between an emulator and an FPGA prototyping system has been that the emulator provides a rich debug environment, while a prototyping system has little or no debug capability and is primarily used *after* the design is debugged to create multiple copies for system analysis and software development. Prototypes has several limitations, primarily due to the difficulty of accessing signals. However, new tools that enable full RTL signal visibility with a small FPGA LUT impact, allow deep capture depth and provide multi-chip and clock domain analysis to allow efficient debug, comparable to the emulator.

FPGA prototypes are designed and built to achieve the highest speed of execution possible. When built in-house, each prototype often is optimized for speed targeting one specific design. They trade off DUT mapping efforts, DUT debugging capabilities (limiting them to a bare minimum that's often useless), and deployment flexibility and versatility. They're used for embedded software validation ahead of silicon availability and for final system validation.

Regardless of the technology implemented in an emulator – custom processor based, custom emulator-on-chip based, and commercial FPGA based – they share several characteristics that set them apart from an FPGA prototype board or system (Rizzati):

- Emulators are targeting hardware debugging and, therefore, support 100% visibility into the design without requiring compilation of probes. Differences exist

between emulators in this critical capability, but they're not significant when compared to those of an FPGA prototyping system.
- Emulators can be used in several modes of operation and support a spectrum of verification objectives, from hardware verification and hardware/software integration to firmware/operating system testing and system validation. They can be used for multipower domain design verification and can generate switching activity for power estimation.

12.2.4 Myths About Emulation-Based Acceleration (Rizzati)

Some of the myths (Rizzati) that prevail in the industry about emulators are listed below. This is to emphasize to the reader that emulation systems are indeed viable and have come a long way from their counterparts a decade ago.

12.2.4.1 Hardware Emulators Are Very Expensive to Acquire and to Maintain

It was true in the old days, not so today. Today, the acquisition cost of a modern emulator pales against the verification power and flexibility of the tool. A hardware emulator is the most versatile verification engine ever developed. It has the performance and capacity necessary to tackle even the most complicated debugging scenarios, which often include embedded software content. Just consider, from five dollars per gate in the early 1990s, the unit cost now hovers around a couple of cents per gate or less.

As strange as it may sound, the tool's versatility makes hardware emulation the cheapest verification solution when measured on a per-cycle basis.

The total cost of ownership also has dropped significantly. Gone are the days when, figuratively, the emulator was delivered with a team of application engineers in the box to operate and maintain it.

The reliability of the product has improved dramatically, reducing the cost of maintenance by orders of magnitude. In addition, the ease of use has simplified its usage.

12.2.4.2 Hardware Emulation Is Used Exclusively in In-Circuit Emulation (ICE) Mode

For the record, in ICE mode, the DUT mapped inside the emulator is driven by the target system, where the taped-out chip would eventually reside. This was the deployment mode that drove emulation's conception and development – namely, test the DUT with real-world traffic generated by the physical target system.

While this mode is still employed by many users, it's not the only way to deploy an emulator. In addition to ICE, emulators can be used in a variety of acceleration modes. They can be driven by software-based testbenches via a PLI or DPI interface, not popular because of the limited acceleration, but still usable to shorten the design bring-up when switching from simulation to emulation or via a transaction-based interface, whose popularity grows since the acceleration factor is in the same ballpark of ICE, at least for some of the current emulators crop. They can be used in stand-alone mode by mapping a synthesizable testbench inside the emulator together with the DUT. And they can accelerate the validation of embedded software stored in onboard or in tightly connected memories. Or, they can be used with combinations of the above.

12.2.4.3 Hardware Emulation Is Useless in Transaction-Based Acceleration Mode

A still widespread misconception is that the transaction-based approach does not work or, in the best case, is limited in performance when compared with ICE. The concept was devised by IKOS Systems in the late 1990s, and it worked. After Mentor Graphics acquired IKOS, it improved upon and pushed the technology as a viable alternative to ICE.

Another and unique benefit of transaction-based acceleration is the ability to create a virtual test environment to exercise the DUT that supports corner-cases analysis, what-if analysis, and more that are not possible in ICE. An example is the VirtuaLab implementation by Mentor Graphics. VirtuaLab models an entire target system, such as USB, Ethernet, or HDMI, in a virtual environment.

12.2.4.4 Dynamic Power Estimation Is a Critical Verification Task, But Hardware Emulation Doesn't Have the Capabilities to Analyze the Power Consumed by an SoC

Another false statement. Dynamic power consumption analysis is based on tracking the switching activity of all elements inside the design. The more granular the design representation, the more accurate the analysis. Unfortunately, higher granularity hinders the designer's flexibility to make significant design changes to improve the energy consumption. This can best be achieved at the architectural level.

Power consumption analysis at the register transfer level (RTL) and gate level of modern SoC designs can be best accomplished with emulation. Only emulation has the raw power to process vast amounts of logic and generate the switching activity of all its elements.

12.3 Speed Bridge

Note that emulators or accelerators are generally not able to run nearly as fast as the real world. Most emulators can only muster a few MHz of clock speed, especially when full visibility is made available. So, it is often necessary to insert a speed bridge that can handle the difference in execution rates each side of the bridge. This may involve data buffering or manipulation of the protocols to artificially slow down the real world to the rate that the emulator can handle.

To tackle the speed issue, the next major way an emulator/accelerator is used is stand-alone. This means that the entire model fits into the emulator or accelerator, along with a set of stimulus to exercise the model. It can run as fast as the emulator is capable of, stopping only when additional stimulus is required, or when captured data must be flushed out of the device. If the design contains a processor, it is also likely that a version of the processor will exist for the emulator.

12.4 Virtual Platform ⇔ Hardware Emulation Interface and Methodology

Figure 12.1 shows virtual platform ⇔ bus interface transactors (SCE-MI: Standard Co-Emulation Modeling Interface) ⇔ emulation methodology for hardware/software co-verification. The communication between the TLM LT (Sect. 11.2.1) adapter and the emulation side AXI master/slave transactor takes place via the so-called SCE-MI. Both the virtual platform and emulator run in MHz speed and hence sustain the performance required for software development, large regressions, etc.

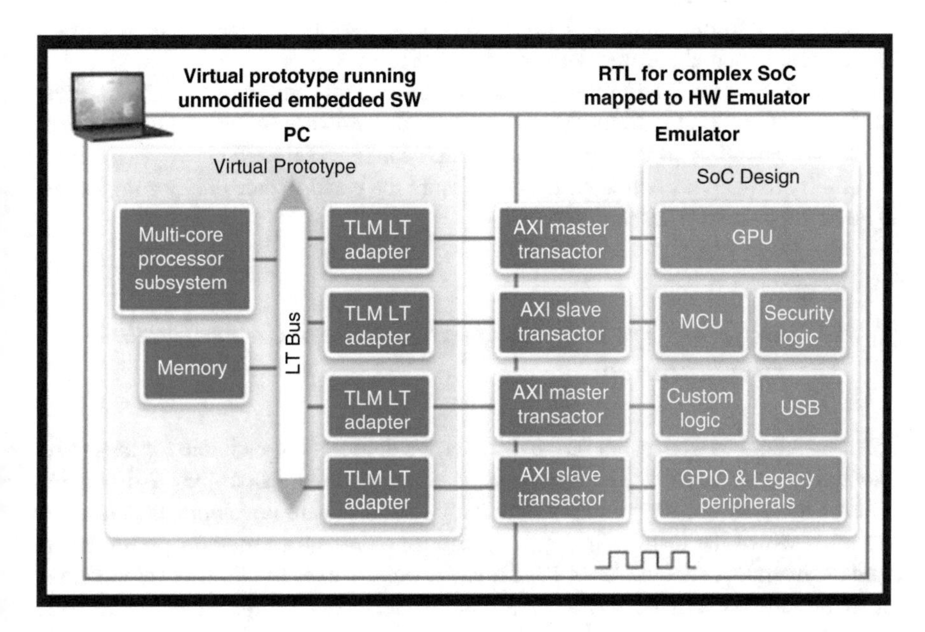

Fig. 12.1 Virtual platform and hardware emulator methodology

Here are some of the requirements for such an environment to work:

- Virtual components that can be assembled and configured easily to build a complete virtual platform with all interfaces, devices, and peripherals are required for application execution.
- Integration with SCE-MI API for TLM ⇔ emulation communication.
- Ability to emulate clocks in the emulation box giving maximum clock speeds during co-simulation.
- Need high bandwidth (data streaming between virtual platform and emulator) and low latency (rate at which messages can be processed by the virtual platform proxies that interface with SCE-MI).
- RTL Master model and Slave BFM (Bus Functional Model) (written in SystemVerilog) that act as transactors to interface with SCE-MI:
  - Emulation transactors should support common protocols and standards' specifications such as PCI Express 3.0, AMBA, USB, MIPI CSI-2 and MIPI DSI, I2C, I2S, Gigabit Ethernet and 10 Gigabit Ethernet, Digital Video, JTAG, etc.
  - The RTL BFMs should be synthesizable into the emulator to make sure that the transactor is always synchronized to the emulated design.
- Memory models: SDRAM, DDR, DDR2, DDR3, DDR4, RLDRAM, Mobile DDR, LPDDR2, LPDDR3, and GDDR5, plus a wide variety of flash and other memory models. These models should leverage onboard memory resources of an emulator and should completely synchronize with the emulated design, eliminating any issues with timing or refresh cycles.
- Software execution and debug while keeping synchronized with the hardware (e.g., waveform dumping), so that you can see software registers or key buses respond to actual software drivers and applications.
- Ability to immediately pinpoint and locate problems and malfunctions in a system simulation of billions of cycles, through hardware monitors and traces, as well as enabling all the standard software debugger features.
- Ability to analyze, benchmark, and measure the performance of key components over long stress test scenarios.

12.4.1 Different Types of Hardware/Software Co-verification Configurations

Figure 12.2 gives a snapshot of the different methodologies available for hardware/software co-verification. The figure shows all methodologies but the co-simulation (signal level) is not a practical methodology for software development. The slow speed at which the testbench simulates will drag down the emulator performance, and the overall speed will be in KHz range at best. This is true for the "simulation" methodology. It will be too slow for hardware/software co-simulation. But the "simulation-only" methodology will allow for 100% visibility into RTL. You would

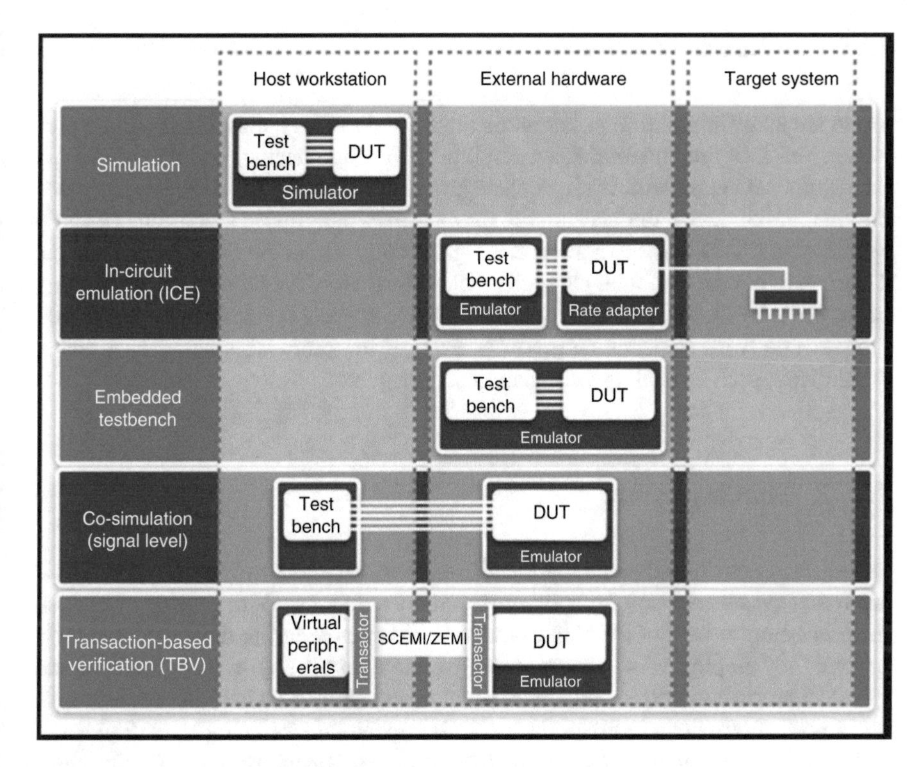

Fig. 12.2 Different types of hardware/software co-verification configurations

run software/hardware co-emulation, and if there is a bug, you can switch to "simulation-only" mode; run a very small snippet of code which found the bug and then debug with full visibility into your design.

12.5 Hardware/Software Co-verification Using Virtual Platform with Hardware Accelerator

Accelerator is the technology product that speeds up software simulation by orders of magnitude. In that sense, its job is to accelerate simulation just as emulation does. Also, line has blurred between acceleration and emulation in recent years. As a matter of fact, all three major vendors namely, Mentor, Synopsys, and Cadence offer hardware products that allow for emulation and acceleration from the same "box." In acceleration mode, the difference is you get excellent visibility into RTL. This allows for quick debug and turnaround time. Emulation is getting there when it comes to debug but cannot provide full visibility into RTL as acceleration does. Acceleration runs in KHz, while Emulation runs in MHz.

Other than debug and compile speeds, emulation and acceleration technologies work hand in hand from the same piece of hardware.

12.5.1 Cadence Palladium

There are some simulation accelerators that contain a large number of simple processors, each of which simulates a small portion of the design, and then they pass the results between them. Each of these processors runs slower than the processor on your desktop, but the accelerator may possess thousands or millions of these smaller processors, and the net result is significantly higher execution performance. They can of course deal with parallelism directly as all of the processors are running in parallel. An example of this type of hardware-assisted solution is the Palladium product line from Cadence (Fig. 12.3). Each of the processors could have many capabilities, such as dealing with visibility, debug, etc.

12.5.2 Mentor Veloce

Within the direct implementation solutions, there are again two main types. These are based on custom solutions or off-the-shelf solutions. With custom solutions, there is going to be an FPGA-like structure somewhere in the device, although in general they employ very different types of interconnect than would be seen in an FPGA. The custom chip could also contain debug circuitry, visibility mechanisms, and a host of other capabilities. Each chip is capable of emulating a small piece of a design, and larger designs are handled by interconnecting many of the chips together, again with sophisticated interconnect capabilities. An example of this type of emulator is Veloce from Mentor Graphics (Fig. 12.4).

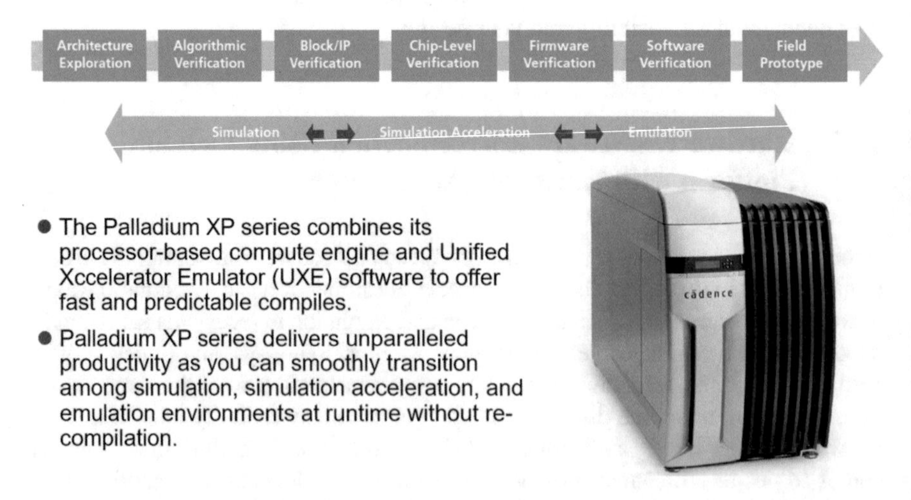

Fig. 12.3 Cadence Palladium XP Unified Xccelerator Emulator (UXE)

Fig. 12.4 Mentor Graphics Veloce emulator/accelerator

Fig. 12.5 Synopsys Zebu emulator/accelerator

12.5.3 Synopsys Zebu

Synopsys Zebu Server (Fig. 12.5) also provides both emulation and acceleration capabilities from the same product. It provides multiple verification use modes, including power-aware emulation, simulation acceleration, in-circuit emulation, synthesizable testbench, transaction-based verification, and hybrid emulation for deployment flexibility based on project requirements.

Chapter 13
Analog/Mixed Signal (AMS) Verification

Chapter Introduction
Current designs invariably have both the digital and analog components within a block and also at SoC level. Without correct verification of analog voltage levels to digital binary and vice versa, the design will be dead on arrival.

This chapter will go into high level discussion of major challenges and solutions, the current state of affair, analog model abstraction levels, real number modeling, SystemVerilog Assertions-based methodology, etc.

13.1 Overview

Earlier SoCs of more than 15 years ago had clear separation of analog and digital blocks. The analog blocks did not incorporate any digital circuitry and vice versa for digital blocks. But as the geometry shrank, more functionality was introduced in the SoC, and the discreet analog logic was incorporated directly on the SoC. The proliferation of multimedia and RF applications was integrated on the same silicon die. This resulted into tightly integrated analog and digital blocks. In earlier days, one could model analog at SPICE level and verify it fully, and once the digital design was ready, they were simulated at SPICE level to verify the IO connectivity with digital blocks as well as the bias generator connectivity.

Analog designers are incorporating more digital techniques into the designs because of the increased variability in smaller geometry manufacturing processes. This means that many precision analog parts are now mixed signal in nature. The complexity of mixed-signal System-on-Chip (SoC) designs is rapidly increasing due to growing analog content, advanced analog and digital interfaces, and tougher requirements for safety and reliability. This is driving a crucial need for advanced verification methodologies and technologies. Figure 13.1 shows the industry trend of AMS designs as part of the overall design starts.

A.B. Mehta, *ASIC/SoC Functional Design Verification*,
DOI 10.1007/978-3-319-59418-7_13

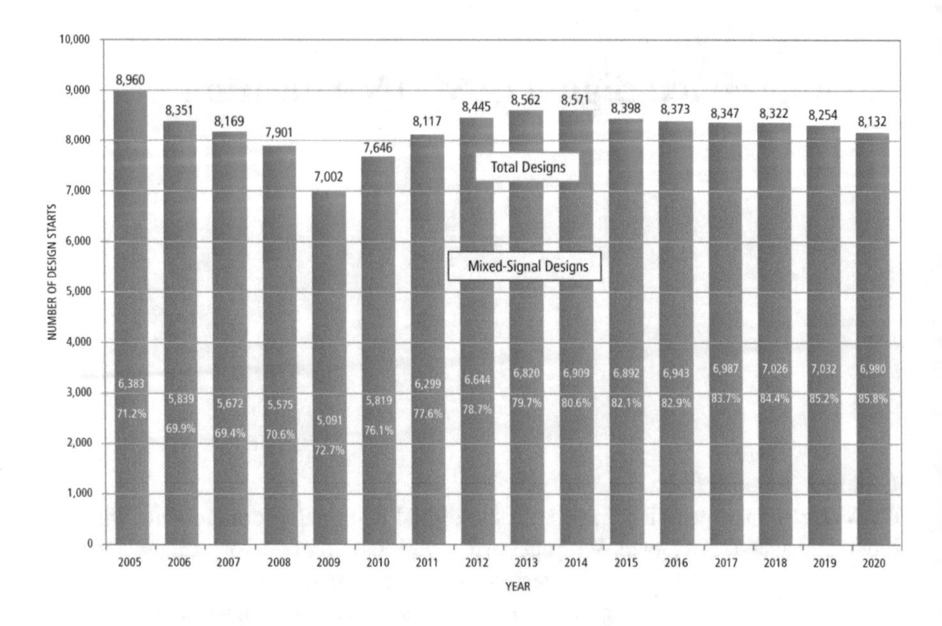

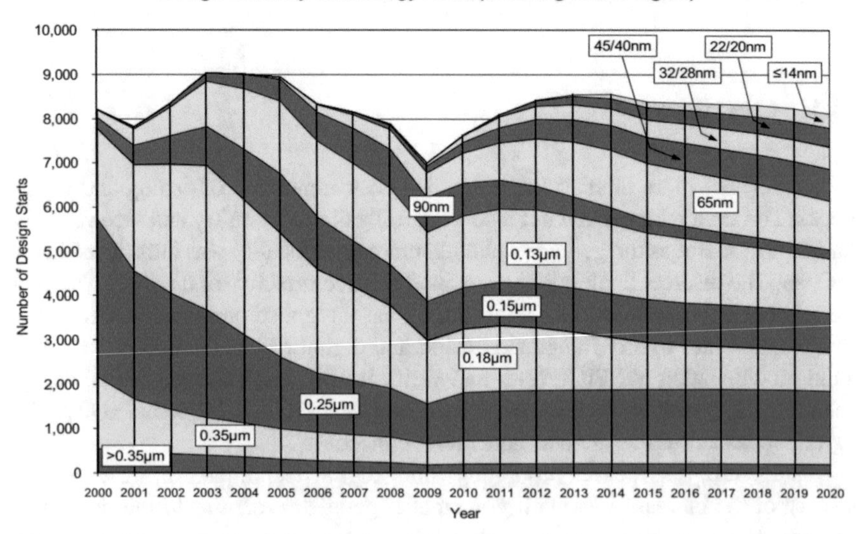

Fig. 13.1 Analog/mixed signal (AMS) design trends

13.2 Major AMS Verification Challenges and Solutions

13.2.1 Disparate Methodologies

As mentioned above, most SoCs are now mixed signal in nature and that brings about a host of verification challenges. A significant part of the challenge is due to different approaches, tools, and methodologies for analog (typically bottom-up) and digital (typically top-down) design and verification. These approaches need to be bridged to handle mixed-signal designs.

The key is to have correct verification plan development based on correct partitioning of the design to allow a top-down verification approach that can execute in parallel to the development of bottom-up analog block. This means that you need a methodology that addresses system-level verification all the way to gate-level verification for AMS to succeed.

This in turn means that you need different abstraction levels of models for both digital and analog blocks to allow verification at different levels. You need to balance simulation speed and simulation coverage in your methodology. Design planning is used to determine which parts of a design specification can be verified using higher levels of abstraction and which parts must be verified using transistor-level abstraction.

In short, you need to bring together the digital top-down and analog bottom-up methodologies to work cohesively in a verification environment. New approaches such as the UVM-MS (Universal Verification Methodology-Mixed Signal) are being introduced to extend the digital approach to the analog and mixed-signal parts of the design.

There are many reasons why top-level verification is mandatory. The primary purpose is functional verification. Though each analog and digital block has been verified and qualified individually with respect to specifications, designers must ensure when assembling those blocks that the full image will work as expected.

Here's what you should expect to find with TOP (system)-level integration of analog and digital blocks:

- Connection errors: wrong signals, wrong power domain
- Incorrect bus wire connections
- Incorrect register bit use
- Misunderstood interface specifications: functional issue mismatch
- Clock phase–frequency mismatch
- Communication/activity during power down
- Current overconsumption
- Stability of IP with a real power supply, especially in startup phases
- Electrical behavior: rise/fall time, loading effects
- Current leakage
- Missing level shifter
- Floating gate
- IP performance, characterization
- Delay and timing issues: signals arriving a cycle or two late
- Bias mismatch

13.2.2 Analog Model Abstractions and Simulation Performance

Figure 13.2 shows the trade-off between simulation accuracy and performance among SPICE, FastSPICE, analog behavioral models (Verilog-A/AMS and VHDL-AMS), RNM (SV-RNM), and pure digital simulation. These numbers are generic

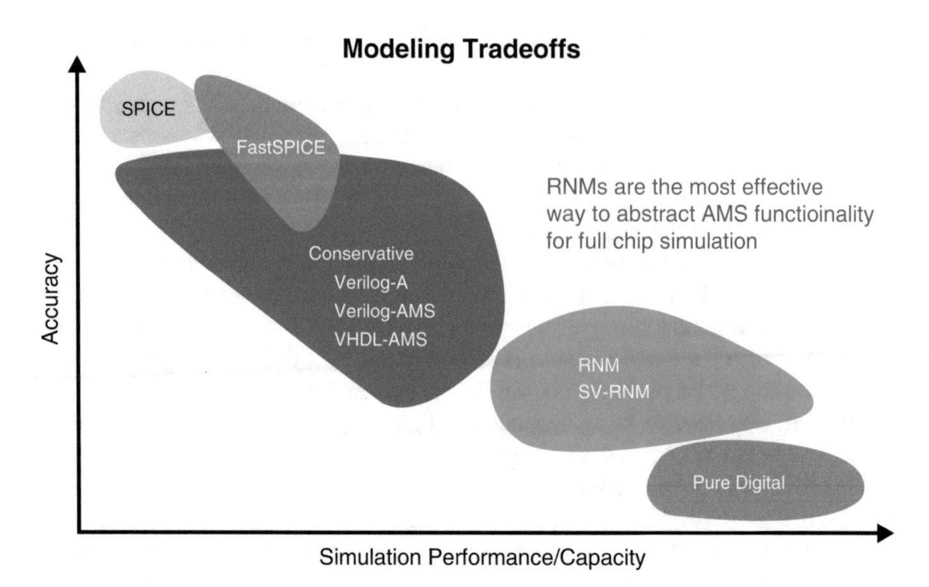

Fig. 13.2 Model abstraction level, accuracy, and simulation performance

and can vary significantly for different applications. Note the wide range of accuracy and performance that is possible for Verilog-AMS and VHDL-AMS behavioral models. Pure digital simulation can only represent an analog signal as a single logic value, but it might be sufficient for connectivity checks in mixed-signal SoCs.

Co-simulating digital discreet event driven with analog SPICE is mostly impractical because of the speed of SPICE (the slowest link in the chain). Analog SPICE (or even FastSPICE) simulators are orders of magnitude slower than digital simulators. Analog designers have traditionally relied on transistor-level circuit simulation to verify their circuits. However, today's designs are so large that it may take several weeks to simulate a single aspect of a complete circuit. With a circuit that has multiple behavioral modes, each must be checked individually to expose all of the functional errors in the design. For a design that implements hundreds of modes and thousands of settings, this may require many months, and perhaps years, of simulation time. Switching to the so-called "fast" simulators can provide some relief, but they are still orders of magnitude too slow to be able to completely verify all modes and settings of a complex analog design. Perhaps the most troublesome aspect of relying on transistor-level simulation is that the regression tests development cannot begin until a first pass of the entire design is complete.

Whereas we used to design op-amps, we now design multi-mode wireless transceivers. And while the scale of the designs has obviously grown fantastically, the character of the designs has changed just as much. Analog designs have hundreds of modes and thousands of settings. They implement sophisticated algorithms and contain blocks that are self-calibrating and self-adapting. In a word, they are complex. And while it is their performance that commands most of the attention of the people that design them, it is now their complexity that is the source of most of the catastrophic failures in their design. While this is a new situation for analog circuits, it has been true for digital circuits for many years. Digital design teams addressed this problem by adopting a strong functional verification methodology driven by verification engineers. Analog designs have now gotten to the point where analog design teams must do the same.

To tackle the simulation throughput issues, designers are turning to behavioral modeling techniques for analog blocks which can increase the simulation speed of AMS simulation. Behavioral modeling includes leveraging Verilog-A, Verilog-AMS, and real number modeling-based event-driven simulation techniques.

Let us look at the model abstraction levels and see how each one can be deployed at different simulation and verification stages.

Following abstraction levels are deployed by design teams to enhance simulation performance. Simulation performance is by far the biggest bottleneck in AMS simulation. Hence, multiple levels of abstractions are invented with language and methodology support from EDA vendors

13.2.2.1 Fully Behavioral, Digital Model of an Analog Block (Modeled in Verilog)

These are the fastest models and have least amount of analog (voltage, current, etc.) detail. These models are suited at architectural levels with digital blocks also at behavioral level. For example, PLL models created at this level still provide full PLL functionality required for digital blocks allowing you to continue behavioral simulation without creating a black box for the analog PLL. Since Verilog doesn't allow real nets or ports, a rather tedious but rewarding workaround uses out-of-module references (OOMRs) (Peruzzi n.d.) to pass signals from one digitally modeled analog block to the next all the way from the input source to the A/D converter. It must be assumed that connectivity and bias integrity are verified in other test cases using more accurate models. This model type is suitable for simulating an extensive digital section of the SoC along with the analog front end.

What Is OOMRs (Out-of-Module References)

Modeling a signal flow of voltages or currents is sufficient for high-level verification, as opposed to modeling physically conservative networks obeying Kirchhoff's laws. Many times, one can use ordinary Verilog for models as mentioned above. An analog "source" model for analog signals uses a fictitious sampling clock with impossibly high sampling rate to create a sequence of real values on a real variable inside that model. Verilog, unlike VHDL, has no concept of real ports or wires, so one can write the module receiving the analog signal to look inside the source module via OOMR.

Using this OOMR approach, the analog signal flows from a source through, say, an amplifier, then a filter, to an analog-to-digital converter. The ports and wires of the netlist cannot conduct the analog signal, so they may be separately used for verifying continuity (Peruzzi n.d.).

13.2.2.2 Fully Behavioral Electrical Model of an Analog Block (Modeled in Verilog-AMS, Verilog-A, or VHDL-AMS)

These have more detail than the fully behavioral digital-only model. And they execute much faster than a SPICE model. Written in Verilog-AMS or VHDL-AMS, these models describe analog behavior in terms of algebraic and differential equations rather than voltage and current. Their electrical I/O exhibit conservation of charge, but internal functional behavior is described by real variables wherever possible. Behavior may be as accurate as required by skewing the modeling style toward the SPICE behavior. This level of model is suitable for detailed SoC verification of the interface, timing, and control of the collection of analog IP blocks in their entirety.

13.2.2.3 Behavioral Model Using Real Number Modeling (RNM) (Using "wreal" and "nettype")

This model style may be written in plain digital VHDL or in Verilog-AMS. The signal path through the model avoids the use of an analog circuit solver. It combines event-driven and self-timed analysis and executes simple mathematical processing of the signal. Plain digital Verilog has no concept of real wires or ports, but Verilog-AMS includes ports and wires of type "wreal" (wire-real). SystemVerilog introduced "nettype" for RNM. There is no feedback path in this style of modeling, and there is no analog solver to choose the sampling points. Op-amps along with their feedback network are replaced by gain blocks. A built-in sampling plan must be written into the model which obeys the Nyquist criteria. As one might expect, this style of model executes blazingly fast and is the top choice for verifying the full SoC signal path including the analog section. Modeling the signal path using "wreal" and using electrical modeling techniques for the bias and reference network results in a nice combination of verification coverage and high execution speed. Section 3 goes into detail of RNM.

13.2.2.4 Transistor: SPICE Level Modeling

As we know this is the most accurate and detailed model of an analog block. SPICE models are the most detailed and exhibit conservation of charge on all electrical I/O and internal nodes. The circuit is described in terms of voltage and current flow. Generally unsuitable for SoC verification, SPICE models are valuable in top-down design flows and in verification of analog subsystems.

Conclusions

Regardless of the kind of abstraction, it involves suppressing detail. It is critical that whoever is formulating this abstraction understands what detail is suppressed. This information must be communicated clearly to anyone who will use the model. Often however, the person who writes the AMS models is not the one using them for verification. This is complicated by the fact that the verification engineer is typically neither an analog expert nor an analog designer. Therefore, the risks of misunderstanding things like the interfaces or the protocols when using AMS descriptions for verification are high.

The communication challenge is compounded by the fact that AMS models written at different levels of abstractions are not equivalent. They represent some ideal effect plus some non-ideal ones. It is not possible to check whether they are equivalent, except by checking whether they describe the same effects with the same level of accuracy. AMS models are designed per the particular effect they will be used to verify.

This is very important to understand because simulations done with AMS models can catch only errors that are described in these models. In other words, these models will not highlight errors they are not designed to find. All of these imply that more than one level of abstraction must be used when verifying a design, according to what needs to be checked. It is important for the model writer to know what to abstract, according to how the model will be used.

Design teams need to make sure to safeguard against such miscommunication. This can be done by using a golden reference model that does not lose any detail, such as SPICE. SPICE descriptions represent the model independently of how it will be used. The SPICE model provides a golden reference against which these various abstractions can be verified. It is also highly recommended to put assertions within models to avoid misusing it during verification.

13.2.3 Low-Power Management

With the advent of UPF (Unified Power Format), the analog design teams are being expected to capture their power intent in UPF so that it can be integrated with the power intent of the digital blocks. The logical to analog and vice versa conversion crossing the digital-to-analog boundary depends on the power state. This is necessary because the simulators must be simultaneously aware of the changing power states of analog and digital blocks (changing power levels, shutoff and isolation conditions, etc.). Industry is addressing these challenges with new tools and technologies, including static and formal methods.

13.2.3.1 Power-Aware Connect Modules

Recent trends in mixed-signal verification tools have centered around the need for each simulation domain (analog or digital) to communicate voltage levels, power shutoff, or state retention concepts across the fence to the *other* simulation domain.

A key technology enabler has been the introduction of power-aware connect modules, which act as interface elements between analog and digital domains. The connect modules dynamically interrogate the power states (which are driven by the UPF description) and present those states to the analog and digital simulation engines in a manner appropriate for both those engines. As a result, the analog simulation engine is aware of the power levels and conditions in the digital part of the design and vice versa.

With the connect module, you can identify problems such as missing level shifters, misconnected power domains, etc.

A connect module is placed on the analog-to-digital and digital-to-analog boundary. When UPF is specified in the design, the connect module should have the capability to not only convert signal values from the logic side to analog side but also convert the correct UPF information from one signal to the other.

The connect modules are of two types: Logical2Electrical (L2E) conversion module and the Electrical2Logical (E2L) conversion module.

Logical (Digital) to Electrical (Analog) Connect Module

The logical to electrical connect module is shown in Fig. 13.3. The conversion process involves, at the least, the following:

1. Conversion of four-state logic to the corresponding electrical voltage values as defined by rules in the connect module. Currently available EDA solutions provide such connect modules and allow users to customize module parameters to meet specs such as operating voltage, L2E reference voltage, etc.
2. Detection of shutoff condition of the PD1 power domain and differentiation between the "X" generated by power shutoff vs. a functional "X" coming in from the digital domain. In the case of the power shutoff condition, the user may wish to specify a certain electrical voltage or a range of voltage values to be able to differentiate between the electrical power shutoff "X" vs. the functional "X."
3. Detection of nominal condition of a power domain and the voltage value associated with that nominal condition. For example, if VDD of PD1 is 1.2v, then a 1'b1 of the digital domain will be represented as 1.2v at the output of L2E connect module.
4. The supply voltage of the connect module L2E is linked with the power domain of the digital instance PD1.

Electrical (Analog) to Logical (Digital) Connect Module (Fig. 13.4)

The need for an electrical to logical conversion arises from the fact that an analog instance, whose behavior is expressed and simulated in the continuous domain, can also reside in a switchable power domain. The following factors need to be noted when performing an electrical to logic value conversion in power-aware fashion.

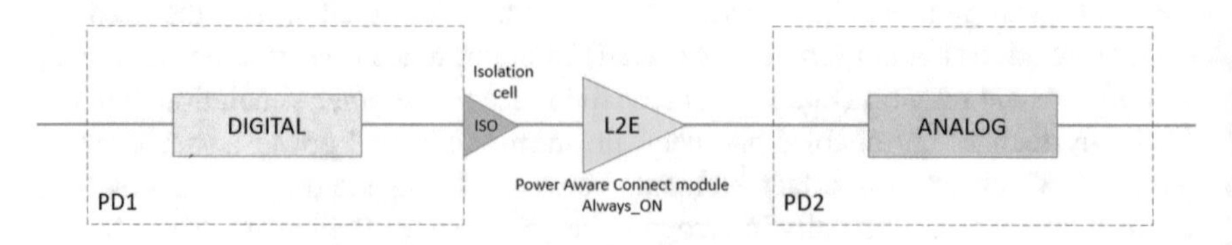

Fig. 13.3 Logical (digital) to electrical (analog) connect module

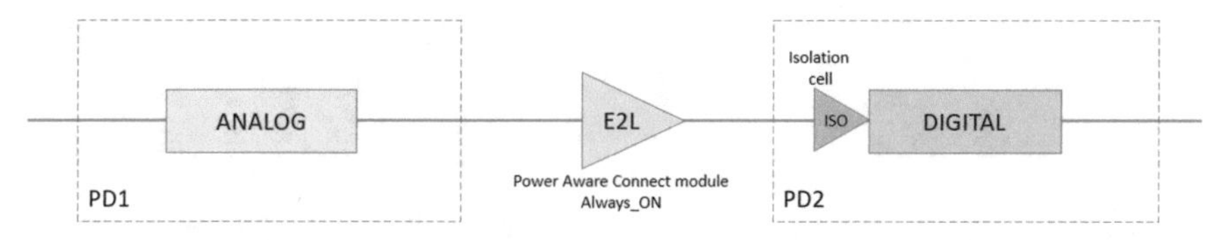

Fig. 13.4 Electrical (analog) to logical (digital) connect module

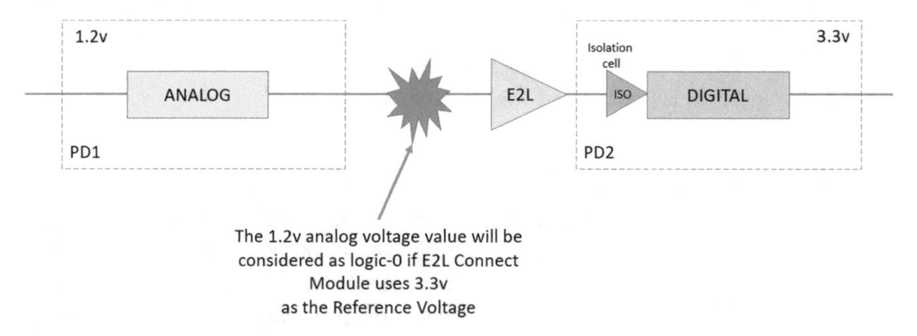

Fig. 13.5 Reference voltage selection for power-aware electrical to logic connect module

1. The supply voltage of the E2L connect module must be linked with the working voltage of the power domain of the digital module.
2. The logic output of E2L will go “X” when PD1 is in shutdown mode.
3. An isolation cell needs to be placed on the input of the digital block to prevent an “X” propagating from the power-off domain into the power-on domain.

Reference Voltage Selection for Power-Aware Electrical to Logic Connect Module (O’Riordan 2012)

A special note on the selection of which voltage reference to select for an E2L connect module. Let us see how the connect module converts a high voltage of analog domain into a 1’b1 for the logic domain OR if it ends up converting this high voltage into a 1’b0 to the digital logic domain. Hence, the reference voltage for the E2L connect module is very important.

The choice of reference voltage for the E2L connect module can lead to unexpected results. In Fig. 13.5, the ANALOG block operates at 1.2v. If a logic “1” that is an output from this analog device is fed into E2L module, it would expect a 3.3v to convert input voltage to a logic “1.” So, if the reference voltage of E2L is connected to the DIGITAL block’s operating voltage, you will get incorrect results. Hence, the AMS Solution tools need to provide a choice of the operating voltage to be connected as the reference voltage of the E2L module. In this case, you should connect ANALOG block’s operating voltage 1.2v as the reference voltage for the E2L module.

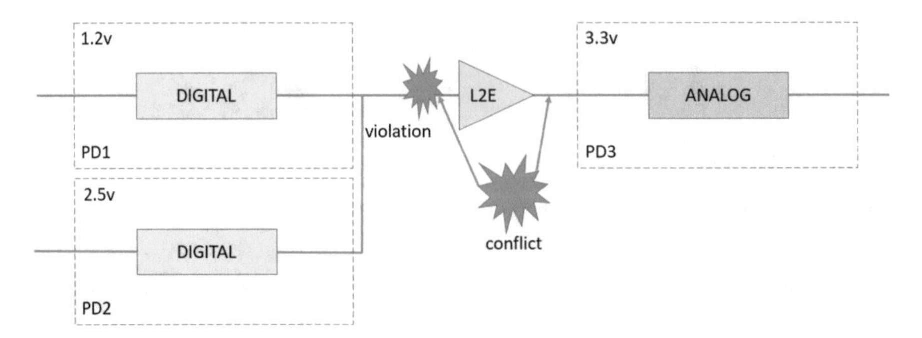

Fig. 13.6 Multiple drivers and nominal voltage conflicts

Multiple Drivers and Nominal Voltage Conflicts (O'Riordan 2012)
When multiple DIGITAL modules are connected to an ANALOG module (as shown in Fig. 13.6), the question of reference voltage consistency must be considered. The need for such consistency applies not only between the digital driving blocks but also between the driving power domains and the receiving power domains. If the driving DIGITAL modules have different nominal voltages, the choice of reference voltage for the L2E connect module becomes ambiguous. That's where the EDA vendor's AMS Solution should come to help and notify the user of such conflicts. Similarly, if the driving domain nominal voltage (digital or analog) is different from that of the receiving domain (digital or analog), we have another source of conflict, and the reference voltage selection for the L2E or the E2L connect modules becomes ambiguous.

13.3 Real Number Modeling (RNM) of Analog Blocks

Real number modeling (*wreal*) represents the second generation of behavioral modeling. It models the analog behavior in the digital domain using discretely simulated REAL values. The result is a considerable speedup in simulation but with less accuracy. RN modeling uses event-driven mechanisms to model analog components using real data values, but it treats time as discrete and manages events instead of equations. In other words, it is a digital model. Simulation speed of several orders of magnitude may be gained (more than 1000x is realistic).

Support for RNM in verification platforms allows the simulation of discrete, floating-point real numbers that can represent voltage levels. RNM enables users to describe an analog block as a signal flow model and then simulates it in a digital solver at near-digital simulation speeds. For analog and mixed-signal block verification, RNM can be used to speedup high-frequency portions of the analog signal path—which take the longest to verify in simulation—while DC bias and low-frequency portions remain in SPICE. But the greatest advantage of RNM is in top-level SoC verification, where engineers can represent all electrical signals as RNM equivalents and stay within the digital simulation environment. Hence, RNM enables SoC-level regressions to cover full-chip functionality while maintaining high-simulation performance.

But first let us look at the differences between analog modeling and real number modeling. Analog modeling has the following features:

- Describes current vs. voltage relationship between nodes in model.
- Newton–Raphson (WIKIPEDIA n.d.) iteration process performs matrix inversion to solve all voltage and currents.
- Time step until next solution is selected based on accuracy criteria.

The real number modeling has the following features:

- No matrix solution–output computed directly from input and internal state. Model defines when to perform each internal computational segment.
- No continuous time operation—only sampled, clocked, and/or event-driven operations. Updates can be performed when inputs change at specific time increments.
- Model analog blocks operation as signal flow model.
- Event driven. No analog solver is used.
- Resistance is not modeled (as of the writing of this book). (Advanced features are being introduced to model impedance.)
- Same format for digital and real modeling—difference is data type.

AMS introduced "wreal" (wire-real) for analog behavioral modeling and SystemVerilog introduced user-defined type (UDT) "nettype" and user-defined resolution (UDR) for real number modeling. Note that "wreal" is not supported in Verilog-A.

13.3.1 "wreal"

Let us first see how "wreal" works in Verilog-AMS.

Verilog-AMS extends the net data types to support a new type called "*wreal*" to model real value nets. The "*wreal*" or wire-real or real net data type represents a real-valued physical connection between structural entities. A "*wreal*" net cannot store its value. A "*wreal*" net can be used for real-valued nets which are driven by a single driver, such as a continuous assignment. If no driver is connected to a "*wreal*" net, its value is zero (`0.0`). Unlike other digital nets which have an initial value of "z," "*wreal*" nets has an initial value of zero.

"*wreal*" nets can only be connected to compatible interconnect and other "*wreal*" or "*real*" expressions. They cannot be connected to any other wires, although connection to explicitly declared 64-bit wires can be done via system tasks $realtobits and $bitstoreal. Compatible interconnect are nets of type wire, tri, and "*wreal*" where the IEEE Std. 1364-2005 Verilog HDL net resolution is extended for "*wreal*." When the two nets connected by a port are of nettype "*wreal*" and wire/tri, the resulting single net will be assigned as "*wreal*." Connection to other nettypes will result in an error.

Here's a very simple example (Accelera, Verilog-AMS LRM Rev. 2.4):

```
module drv (in, out);
        input in;
        output out;
        wreal in;
        electrical out;
        analog begin
                V(out) <+ in;
        end
endmodule

module top ();
        real stim;
        electrical load;
        wreal wrstim;
        assign wrstim = stim;
        drv f1(wrstim, load);
        always begin
                #1 stim = stim + 0.1;
        end
endmodule
```

As mentioned above, Verilog-AMS supports ports which are declared to be real valued and have a discrete-time discipline using "wreal." There can be a maximum of one driver of a real-value net such as "wreal." Here's an example:

```
module top ();
wreal stim;
reg clk;
wire [1:8] out;
        testbench tb1 (stim, clk);
        a2d dut (out, stim, clk);
initial clk = 0;
always #1 clk = ~clk;
endmodule

module testbench (wout, clk);
output wout;
input clk;
real out;
wire clk;
wreal wout;
        assign wout = out;
        always @(posedge clk) begin
                out = out + $abstime;
        end
endmodule
```

```
module a2d(dout, in, clk);
output [1:8] dout;
input in, clk;
wreal in;
wire clk;
reg [1:8] dout;
real residue;
integer i;
        always @(negedge clk) begin
                residue = in;
                        for (i = 8; i >= 1; i = i - 1) begin
                                if (residue > 0.5) begin
                                        dout[i] = 1'b1;
                                        residue = residue - 0.5;
                                end
                                else begin
                                        dout[i] = 1'b0;
                                end
                        residue = residue*2;
                        end
        end
endmodule
```

13.3.2 "nettype"

A SystemVerilog user-defined "*nettype*" without any resolution function can be declared as:

```
nettype myT myNet,
```

where "*nettype*" is the keyword, myT is the user-defined type (UDT), and myNet is the *nettype* identifier.

In general, "*nettype*" provides the following features:

- User-defined types (UDT) that can hold one or more real values
- User-defined resolution (UDR) functions
- Modeling flexibility
- Provides high performance and broad modeling capabilities for faster verification with higher accuracy

Here's a simple example (Ron Vogelsong 2015):

```
typedef struct {
real voltage;
real current;
bit field3;
integer field4;
} myT;

module top;
nettype myT myNet;
myNet w;
assign w = myT'{0.1,0.2,1'b1,10};
initial begin
          $display("Value of w -> %f => %p",$realtime, w);
          #1 $display("Value of w -> %f => %p",$realtime, w);
          #5 $display("Value of w -> %f => %p",$realtime, w);
end
endmodule
```

A type definition of myT is defined and used as the user-defined type (UDT) for the "*nettype*" myNet. "w" is declared a type of myNet. Since the *nettype* myNet is of UDT myT, we can assign the values to the myT struct. The rest of the code is self-explanatory. Here's the output from simulation of this model:

```
Value of w -> 0.000000 => '{voltage:0, current:0, field3:'h0, fileld4:x}
Value of w -> 1.000000 => '{voltage:0.1, current:0.2, field3:'h1, fileld4:10}
Value of w -> 6.000000 => '{voltage:0.1, current:0.2, field3:'h1, fileld4:10}
```

Here's a quick summary of UDT and UDR:

- User-defined types (UDTs):
  - Allows for single-value real nettypes.
  - Keyword used: **nettype.**
  - Allows for multi-value nets (multi-field record style)
  - It can hold one or more values (such as voltage, current, impedance) in a single complex data type (struct) that can be sent over a wire.
- User-defined resolution (UDRs):
  - Functions to resolve user-defined types using keyword: **with**
  - Specifies how to combine user-defined types

Here are a couple of case studies that Cadence presented in CDNLive 2015 (Ron Vogelsong 2015):

Case study 1: A voltage-controlled oscillator

A voltage-controlled oscillator simulated for 5ms	CPU seconds
SPICE	~132 min
VerilogA	~12 sec
'wreal'	~20 msec

Case study 2: 14-bit ADC + 14-bit DAC

14-bit ADC + 14-bit DAC simulated to 2**14 = 16384 steps	CPU seconds
SPICE	Several days
RNM	~3 sec

13.4 AMS Assertion (SVA)-Based Methodology

Assertion-based verification (ABV) is a powerful verification approach that has been proven to help digital IC architects, designers, and verification engineers improve design quality and reduce time to market. But ABV has rarely been applied to analog/mixed-signal verification.

Assertions for Verilog-AMS and SystemVerilog-AMS are still under development with Accelera technical subcommittees (as of writing of this book). Issues such as the following are under discussion.

Several questions need to be answered before AMS assertions are brought into practice: (a) How will the languages for AMS assertions be different from the ones in the digital domain? (b) Should the analog simulator be assertion aware? (c) If so, then how and where on the time line will the AMS assertion checker synchronize with the analog simulator? (d) What will be the performance penalty for monitoring AMS assertions accurately over analog simulation?

While many companies crave a standardized approach (O'Riordan 2012) to such assertion management, some have found even *co-simulation*-based solutions employing the standard PSL/SVA languages to be either excessive in terms of setup cost (for a mixed-signal/digital simulator such as NCSIM or VCS) or simply unpalatable to their heavily SPICE-based design community. Finally, due to these issues, many customers have abandoned (Bhattacharya D. O. 2012) efforts to continuously verify "analog" blocks when integrating them in a mixed-signal/SoC (System-on-Chip) environment, leading to "plug and pray"-based integration attempts (and the subsequent mixed-signal tape-out nightmares).

There is an excellent article by Prabal Bhattacharya that sheds light on possible solutions to deploying SVA for AMS (Bhattacharya P 2012).

Here's what Accelera has to say about the current state of affair with SystemVerilog-AMS:

> The working group is currently working on alignment of Verilog-AMS with the SystemVerilog work of the IEEE 1800, or inclusion of AMS capabilities in a new 'SystemVerilog AMS' standard. In addition, work is underway to focus on new features and enhancements requested by the community to improve mixed-signal design and verification, as well as to extend SystemVerilog Assertions to analog and mixed-signal designs.

Since the dust haven't settled down, many have resorted to homegrown solutions. The author does not feel such homegrown solutions should be publicized in this book. Hence, this important topic will be covered in a second edition of this book as soon as a standard version is published.

Please refer to Chap. 6 for a detailed description of SystemVerilog Assertions for digital systems.

13.5 AMS Simulator: What Features Should It Support?

Having gone through the methodology and planning aspects of AMS simulation, here are the guidelines on what you should look for when deciding on an AMS simulator. Your EDA vendor should be able to explain the following.

13.5.1 Integrated Simulation Solution for Fastest Simulation Throughput

By natively integrating advanced technologies for functional and low-power verification, coupled with analog extensions to the proven UVM methodology, an EDA solution should provide for the rapid development of a coverage-driven, constrained random testbench that can be run in parallel across compute farms to reduce overall regression testing cost. Some EDA vendors offer the so-called multi-core technology in their SPICE or FastSPICE engine. This delivers even higher verification throughput, enabling scalable mixed-signal regression testing with transistor-level accuracy. This allows users to directly use SPICE for small-/medium-sized analog blocks without resorting to behavioral modeling of the analog blocks. These are claims; the author hasn't independently verified these claims!

13.5.2 Support for Wide Spectrum of Design Languages

As different mixed-signal design applications require different configurations of SPICE netlist, RTL, and behavioral models, flexibility in languages and topologies supported is crucial for a mixed-signal verification solution. The simulator should support Verilog-AMS, real number modeling, SystemVerilog, Verilog, VHDL, and SPICE.

13.5.3 Support for Different Levels of Model Abstraction

As we discussed before, you need to deploy different levels of analog models to satisfy the performance needs of verification from system level to transistor level. The simulator should support mix and match of all the abstraction levels at any hierarchical level.

13.5.4 AMS Low-Power Verification Support

EDA solution should provide a comprehensive mixed-signal low-power verification solution, supporting UPF, for mixed-signal designs and provide a system-level solution for low power.

13.5.5 Support for SystemVerilog-Based UVM Methodology Including Coverage-Driven and Assertion-Based Methodologies

The EDA solution should extend the proven SystemVerilog UVM-based methodology for AMS, allow "assertions" on analog nodes, be able to sample analog nodes to monitor incoming traffic, be able to drive constrained random stimuli on analog nodes, support analog coverage ("coverpoints" on analog nodes), and support regression management in a mixed-signal context.

Chapter 14
SoC Interconnect Verification

Chapter Introduction
Today's SoC contains hundreds of pre-verified IPs, memory controller, DMA engines, etc. All these components need to communicate with each other using many different interconnect technologies (cross-bus-based, NoC (Network-on-Chip)-based, etc.).

This chapter will discuss challenges and solutions and interconnect verification methodology and discuss a couple of EDA vendor solutions, among other things.

14.1 Overview

With increasing numbers of CPU cores, multimedia subsystems, and communication IPs in today's System on Chips, the main SoC interconnects, crossbars, or Network-on-Chip (NoC) fabrics become key components of the system. The verification of the SoC bus interconnects faces the challenge of verifying the correct routing of transactions as well as security and protection modes, power-management features, virtual address space, and bus protocol translations while still reaching project milestones.

Remember the days when engineers used to be able to rely on buses to perform the on-chip communication function in chips? Those days are clearly in the past, especially as our increasingly connected world demands so much more functionality from our chips. Today's advanced SoC calls for an interconnect to serve as the communication hub for various IP cores within the SoC. Verifying the functionality and performance of SoC interconnects can be a complex task, given the amount of masters and targets, the different protocols, different types of transactions, and multilayered topology involved. A more holistic approach using tools and technologies can simplify the process of verifying the functionality and performance of SoC interconnects.

A.B. Mehta, *ASIC/SoC Functional Design Verification*,
DOI 10.1007/978-3-319-59418-7_14

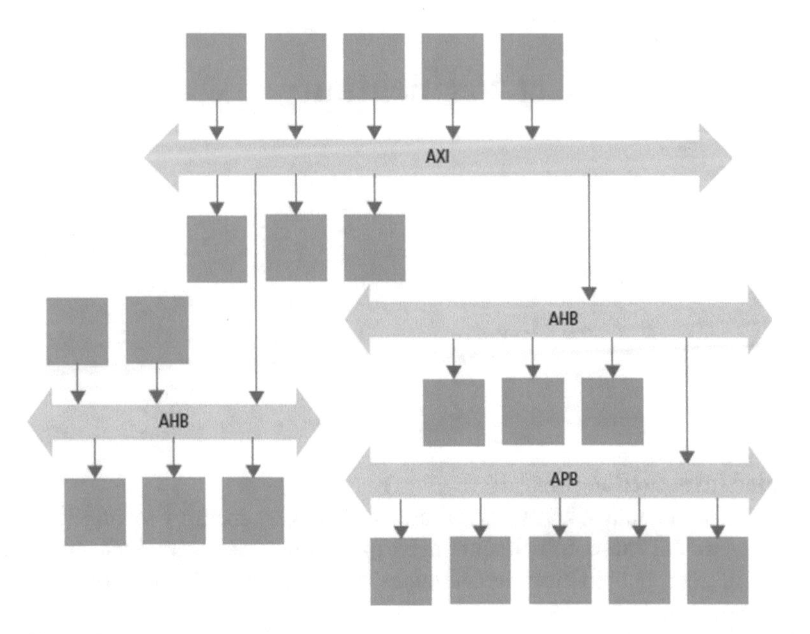

Fig. 14.1 SoC interconnect example

Many SoCs now employ sophisticated interconnect fabric IP to link multiple processor cores, caches, memories, and dozens of other IP blocks. These interconnect fabrics are enabling new generations of low-power servers and high-performance mobile devices. However, sophisticated interconnects are highly configurable and that creates design and verification challenges.

Fig. 14.1 shows a typical SoC interconnect that comprises of three buses, namely, AXI (for internal CPU, memory fast speed), AHB (for internal nonperformance critical blocks), and APB (for peripheral connections). These fabrics are chained together via bridges or a big fabric for overall communication. This adds more challenges for verification, since the scope of verification now increases to multiple interconnects in a system. Also, protocol conversion needs to be considered within a possible hybrid topology.

We need to verify the end-to-end transaction routes from a master interface to a slave interface. Let us discuss the challenges that the SoC interconnect verification poses and solutions thereof.

14.2 SoC Interconnect Verification: Challenges and Solutions

As the complexity of SoC and, thereby, the interconnect increases, we must address the challenges that its verification imposes. Let us look at the challenges from a project point of view.

SoC Interconnect Verification Challenges

1. **Functional Correctness**
   An ideal verification environment should guarantee correct stimulus generation, response/protocol checker, and coverage collection. Our previous BFM and direct test cases are inadequate to achieve this goal.
2. **Verification Completeness**
   This covers the systematic transaction checking around layered interconnects, and it becomes more crucial when the design complexity increases. In a complex SoC, there will be multiple layered interconnects in which different protocols are involved. We need a sophisticated mechanism to check every transaction from point to point, even with different protocols, from different paths, and even in parallel execution. For example, a CPU can issue a transaction across an AXI interconnect to a slave which is tied to an AXI2AHB bridge, which is finally transferred to another AXI interconnect.
3. **Protocol and Protocol Conversion Compliance**
   AXI, APB, AHB, OCP, etc. and AXI to AHB to APB type of bridge connections require protocol conversions, including error response checking.
4. **Stress Verification**
   Random and *concurrent* (all initiators firing at all targets at the same time). There will be more of this later in the chapter.
5. **Security Management**
   Nowadays, interconnects include security management (firewall) features. The basic role of security management is to forbid unsecured transactions targeted to secured/protected memory space. This security feature is to prevent software attacks stemming from illegal instruction execution. The DV environment should check each transaction requested by initiators and abort with error response if that transaction is targeting a protected area. Slaves such as larger memory arrays can have more than one different protection regions, each having their own priority level.
6. **Power Management**
   SoC interconnect can manage power consumption across all the functional blocks on the SoC and all the connections between them. Since the interconnect touches all blocks of a design, it provides the ideal opportunity to enhance the following power-management best practices. This in turn means that we must have verification environment to verify the following:

- Datapath optimization and performance modeling
- Voltage/power/clock domain partitioning
- Power disconnect protocol
- Asynchronous clock domain crossing
- Clock gating—fine grained and unit level
- Partial retention
- Performance probe used as feedback for DVFS

Power-management features add the following constraints to the verification environment:

- The interconnect scoreboard should provide support to the power-management features.
- The power up sequence should therefore be tested while stressing the bus with other transactions. The use of high-level system sequence controlling the different master/target models enables this verification.
- Proper functional metrics should be provided to ensure the conditions have been reached.

7. **Cache Coherency (for Cached NoC)**
   For instance, cache coherency protocols may not complete a transaction to its targeted destination in case the information already exists in interconnect cache (write-back cache). Scoreboard checker implementation needs to be extended to take this into account.
8. **Functional Errors due to Interconnect Latencies**
   For instance, unmatched bandwidth for DMA read and write channel. Due to interconnect routing latencies from request to request and response to response, a DMA engine would not be able to throttle outstanding transactions. The DMA engine FIFO depth could be insufficient for required SoC bandwidth.

14.2.1 Performance Analysis

Performance verification is where designers should make sure that the design will meet its targeted bandwidth and latency levels. Consider an SoC design with multiple interconnects to prevent localized traffic from affecting the rest of the device's subsystems. Interconnect IP plays an important role here, as it can tune each port for unique bus widths, address maps, and clock speed. Usually, there are also mechanisms to adjust bandwidth and latency to tune the interconnect IP in each domain.

However, there are still instances where traffic conflicts will occur. How can traffic in these situations be balanced? Most systems don't have enough main memory bandwidth to accommodate all IP blocks being active simultaneously. What's important is preventing one IP block from dominating and overwhelming the others; otherwise, system performance degrades. Performance analysis can be helpful in this situation, minimizing the impact of system performance degradation.

To make performance analysis as effective and efficient as possible, there are a few aspects you should strive to integrate into the process (Nick Heaton n.d.):

- Cycle-accurate modeling—With cycle accuracy, the logic simulation yields the same ordering of events with the same timing as will be seen in the actual chip. Cycle-accurate simulation models include the RTL-level Verilog or VHDL created during the SoC design process.

- Automatic RTL generation—Automatically generated interconnect RTL is a step toward creating a full SoC cycle-accurate model. To determine the combination that provides the best overall performance, designers need to be able to quickly generate multiple variations of the interconnect IP.
- Verification IP and SystemVerilog Assertions help find protocol violations.
- Testbench generation—Generating testbenches automatically saves several weeks that development can otherwise take to create a test environment for interconnects.
- In-depth analysis—The ability to gather all simulation data—design assessment, the testbench, and traffic—is necessary to debug performance problems and determine how design changes might affect bandwidth and latency.

14.3 Interconnect Functional Correctness and Verification Completeness

Functional correctness and verification completeness (which includes stress verification) go hand in hand. You need to make sure that interconnect is functionally correct and also make sure that you have covered all cases of verification including corner cases, error conditions, etc.

As mentioned above, we need to create a robust Stimulus Generator, a response checker, and a coverage model/methodology to counter the challenges of robust functional correctness of the interconnect.

14.3.1 SoC Interconnect Stimulus Generation

Directed test. First, we need to make sure that the interconnect works well for each subsystem of the design. What's a subsystem? Refer to Fig. 16.1 voice over IP (VoIP) SoC verification and description thereof to get detailed understanding of a subsystem. For the SoC in Fig. 16.1, we identify Ethernet subsystem, TDM subsystem, PCI subsystem, ARM-CPU subsystem, and memory subsystem.

As part of the verification of these subsystems, you are essentially performing an end-to-end verification of the interconnect in a directed manner. For example, the PCI subsystem goes from PCI to internal registers of CAM, TDM, Ethernet, and DDRC modules. This directly covers majority of the interconnect for both *write* and *read* operations. If there were any bugs in SoC interconnect protocol interface among modules or internal modules and peripherals, they would surface during subsystem verification. Please refer to Chap. 16 for a complete description of SoC subsystem verification.

Constrained random concurrent test. Having completed a directed verification of interconnect protocols and protocol conversion bridges, we need to now methodically move toward concurrent but constrained verification to incrementally stress the interconnect fabric for corner cases. For this to happen, you need to simulate multiple subsystems *concurrently*.

This means, each subsystem needs to fire at the *same* time concurrently through a virtual sequencer as shown in Fig. 16.1. This requires careful planning though:

- Address map needs to be divided among each subsystem such that their memory accesses do not clobber each other.
- You cannot write/read from internal registers at the same time.

Example subsystems for constrained concurrency could be:

- PCI *write/read* from external DDR the same time as ARM *write/read* to DDR. This will not only stress the interconnect for simultaneous traffic but also the DDRC as a side benefit.
- Ethernet receive packets interrupt ARM for Rx processing the same time that DMA is transmitting a packet over the Ethernet Tx.
- ARM write/read of the (for example) CAM registers the same time that PCI *writes/reads* from Ethernet subsystem. Swap around modules for register *write/read* in a similar manner.

14.4 Stress Verification: Random Concurrent Tests

Once you have verified a good combination of *collision* among subsystem, it's time to turn on the full force of complete random concurrency. Complete concurrency means all subsystems firing at the same time (which will be the case in real life, anyway). The traffic needs to be concurrent among embedded ARM CPU along with traffic from all the external peripherals. To reiterate, this requires careful planning as noted above.

This will not only stress the interconnect for functional verification but also for its latencies and performance verification. For example, when PCI is accessing DDR the same time as ARM, how will ARM's performance get affected by the traffic from PCI? Will the interconnect prioritize ARM over PCI? Or what happens when DMA is transmitting a packet the same time that the Ethernet/MAC is receiving a packet? Will Ethernet receive subsystem drop packets?

Such full concurrency is a must for interconnect verification as well as for performance verification. Internal dead locks, live locks, stalls, packet drops, incorrect register write/read, etc. will surface.

14.5 SoC Interconnect Response Checker

Ok, now that we have formulated a stimulus generation plan and methodology, how will we make sure that the response provided by SoC under directed and stress verification is correct?

- **Write and read tests**. Write SoC internal programmable registers from each subsystem that can do so. Keep a golden stack of *write* values. Then *read* back the same register and compare with the golden stack. Any discrepancy is either a bug in the *register write/read* logic or a bug in the interconnect. Interconnect bug could be that *write* went through, but *read* did not get translated correctly going from AXI to AHB bridge!
- **Protocol checkers**. Rely on *SystemVerilog Assertions* to do this check. Apply all required protocol checks using SystemVerilog Assertions (Chap. 6). Apply SVA assertions on AXI, AHB, and APB for protocol compliancy. This will catch, among many, issues with interface protocol conversion bridge bugs (e.g., AXI to AHB bridge or AHB to APB bridge).
- **Memory subsystem response checker**. For heavy stress traffic, the *register write/read*, SVA, and an additional memory subsystem checker will cover the entire space. The memory subsystem response checker will *write/read* randomly from the DDR, and a reference model sitting on top of the memory subsystem will do the same. At the end of the simulation, compare the memory dump of DDR with that of the reference model.

14.6 SoC Interconnect Coverage Measurement

Please refer to Chap. 7 for a detailed description on how SystemVerilog functional coverage works and how it can be applied. That chapter discusses a PCI example, which is directly applicable here.

Coverage comprises of code coverage and functional coverage (Chap. 7). For example, PCI subsystem code coverage will tell us if all PCI master and PCI target states were covered (i.e., all different types of PCI cycles were exercised on the interconnect). But code coverage will not tell you if transitions among all the cycles were exercised. In other words, the "transition functional coverage" will tell you if you exercised PCI Write to PCI Read to PCI IO Write transitions or that PCI Config Write followed by PCI Write followed by PCI Read, etc. Such transitions are possible only through functional coverage. To highlight a coverage model for the interconnect, you need to cover the following:

- Interconnect code coverage.
- Interconnect bus cycle *transitions* functional coverage as well as coverage for all bus cycle *types*.
- "Cross" (functional coverage feature) between CPU transactions and external peripheral transactions. Many such "cross" exist in an SoC.
- Interconnect protocol coverage through SystemVerilog Assertions *cover* statement. This includes AXI ⇔ AHB ⇔ APB bridge protocol.
- Interconnect (NoC) register coverage.
- Interconnect (NoC) internal state machine state transitions.
- Cache coherent coverage:
  - Cache protocol functional coverage (e.g., MESI protocol coverage)

- Cache line granularity (byte, word, dword, line) functional coverage
- Snoop conversions, snoop propagation, and snoop filter operation
- Cross-cache line operations
- Cache line false sharing (same cache line at the *same* granularity written by two processors at the same time) and true sharing (same cache line at *different* granularity written by two processors at the same time)
- Write-back operations initiated by the cache coherent NoC

14.7 Cadence® Interconnect Solution (Cadence-VIP n.d.)

The Cadence® Interconnect Solution is designed to meet the needs of verification engineers and system architects by simplifying the verification of interconnect data integrity and identifying performance bottlenecks before they are locked in silicon. The solution includes the Cadence Interconnect Validator and Cadence Interconnect Workbench.

14.7.1 Cadence ® Interconnect Validator (Basic)

The Interconnect Validator (Basic) is used to verify non-coherent interconnect fabrics like ARM's NIC-400 System IP. The solution:

- Supports any number of masters and slaves
- Accommodates independent address forwarding for each master
- Handles data splitting, upsizing, and downsizing
- Supports INCR, WRAP, and FIXED addressing modes
- Supports internal address ranges and unmapped access
- Supports transaction ordering
- Handles slave power-down, interconnect reset, and dynamic address forwarding

Interconnect Validator (Fig. 14.2) is a system-level VIP that serves as a system scoreboard for interconnects. It's a passive component that monitors each transaction behavior within a fabric network and makes sure each transaction behaves correctly during different phases. It verifies the correctness and completeness of data as it passes through the SoC interconnect fabric. Interconnect Validator automatically(!) creates a coverage model of all transactions exchanged between masters and slaves within an SoC. It includes a passive agent to monitor the SoC interconnect as well as an active agent to model interconnect behavior and enable SoC verification in cases where the interconnect design is not yet complete.

Theoretically, Interconnect Validator can work with any number of masters and slaves within any number of layered interconnects, if it is correctly configured regarding ID conversion, address mapping, port definition, etc. It can also handle different kinds of protocol types and transaction types. The tool also has a powerful application programming interface (API) that enables it to support proprietary protocols.

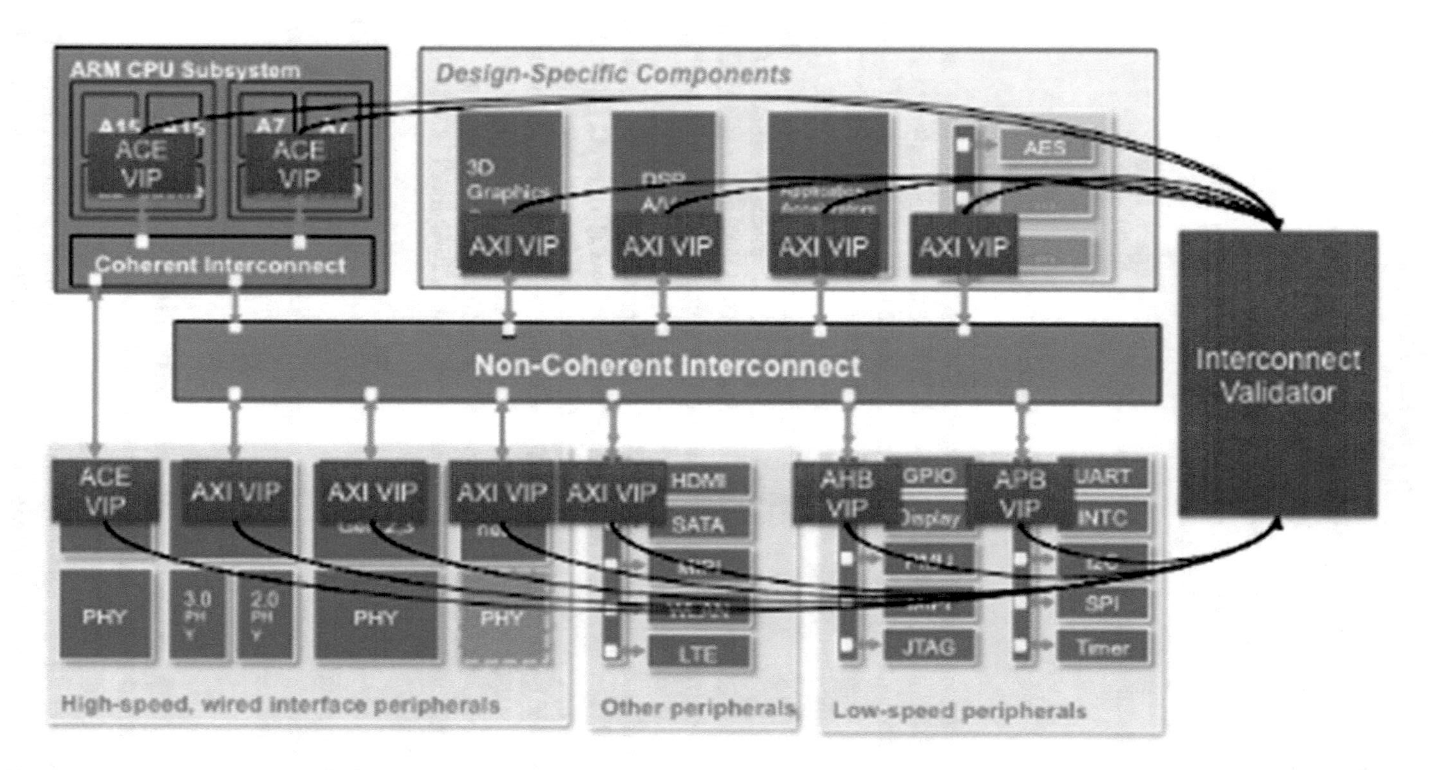

Fig. 14.2 Cadence Interconnect Validator (Basic—non-coherent)

14.7.2 Cadence ® Interconnect Validator (Cache Coherent)

The Interconnect Validator (Coherent) is used to verify coherent interconnect fabrics like the ARM® CCI-400 System IP. The solution:

- Supports any number of outer and inner domains
- Verifies snoop conversions, snoop propagation, and snoop filter operation
- Checks cross-cache line operations
- Verifies barrier transactions
- Supports interconnect-initiated operations

14.7.3 Cadence ® Interconnect Workbench

Your interconnect subsystem might be functionally correct, but are you starving your IP blocks of the bandwidth they need? Is the data from latency-critical blocks getting through on time? With the Cadence Interconnect Workbench (Fig. 14.3), answering these questions becomes much easier. The solution collects cycle-accurate traffic from multiple simulation runs and displays latency and bandwidth measurements in an easy-to-use performance cockpit. Interconnect Workbench saves time by generating testbenches. It imports interconnect fabric RTL and IP-XACT metadata from the ARM CoreLink™ AMBA® Designer product, builds either a performance-oriented testbench or a verification-oriented testbench, and builds a basic test suite.

The tool uses cycle-accurate RTL for interconnect performance validation. However, RTL is not needed for peripheral functions—nor is SystemC. Interconnect Workbench can use highly abstracted traffic profiles, which generate traffic that is not 100% real but is realistic enough for performance analysis. Real software is not required. The use cases specified with traffic profiles on the ports of the interconnect gives users enough information to optimize the performance of the system.

Its features include (product literature from Cadence—author has not validated these claims!):

- Shaping of interface traffic with sequences to assess system performance under various traffic loads
- Performance-sensitivity analysis to compare various implementation options and quality-of-service configurations

14.8 Synopsys Cache Coherent Subsystem Verification Solution for Arteris Ncore Interconnect (NoC)

The Synopsys cache coherent NoC subsystem verification solution generates UVM testbench logic that integrates with Arteris Ncore interconnect testbenches, enabling connectivity of new subsystem level tests, monitors, coverage and performance

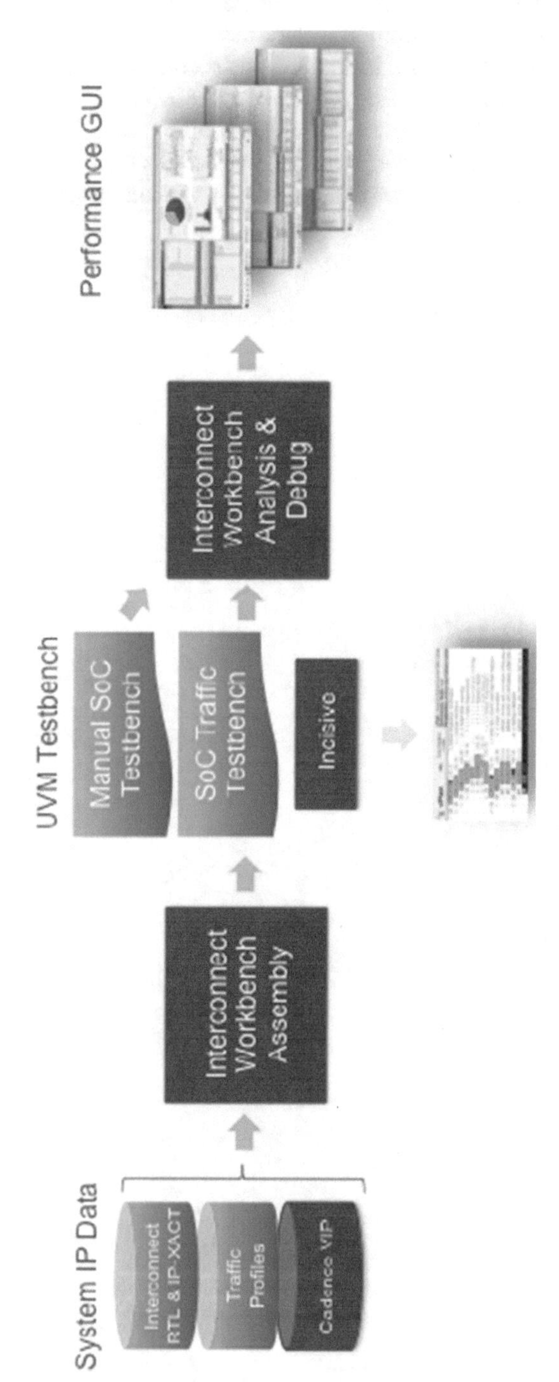

Fig. 14.3 Cadence Interconnect Workbench

tests, and analysis to achieve accelerated verification closure. The cache coherent NoC interconnect subsystem solution includes subsystem level test suites to validate the coherency of the system, in addition to the correctness of data flow across the NoC. Synopsys' Verdi® Performance Analyzer is natively integrated in the cache coherent NoC subsystem verification solution for functional scenarios and provides debug capabilities for performance issues across the SoC.

The author has not verified these claims from the vendor. Please take such claims with a grain of salt.

Chapter 15
The Complete Product Design Life Cycle

Chapter Introduction
In parallel to the development of the verification environment as described in previous chapters, it is equally important to understand other development efforts in the product development cycle. This chapter provides a glimpse into the hardware side of the design and verification and to give the reader an overall view of the complete product design life cycle of a product. Further details will be provided hereinafter.

I'd like to acknowledge Cuong Nguyen of EDA Direct, Inc. for contributing this chapter in its entirety.

15.1 Overview

In a typical product design and development flow, many engineering disciplines are involved spanning from (but not limited to) electrical engineering, mechanical engineering, and software engineering. The discussion here focuses on the electrical engineering and the hardware design side of the product, but similar steps are applicable to other areas.

Within the electrical engineering side, the HW group is responsible for the design, development, and deliverable of the chip (i.e., an ASIC and/or FPGA), the board (i.e., a CPU or graphic card), or the complete system. Each has its own requirements and proceeds in parallel to other efforts. The design specification is formed at the start to enable the implementation phase, and at the end, there is an integration effort between all groups before heading into the product release phase.

Xilinx VCU108 development board is depicted in Fig. 15.1.

Let us look at this flow in more detail. It is important to note that each of the following flow is an involved process with many steps, and to go into the deep details is beyond the scope of this book.

A.B. Mehta, *ASIC/SoC Functional Design Verification*,
DOI 10.1007/978-3-319-59418-7_15

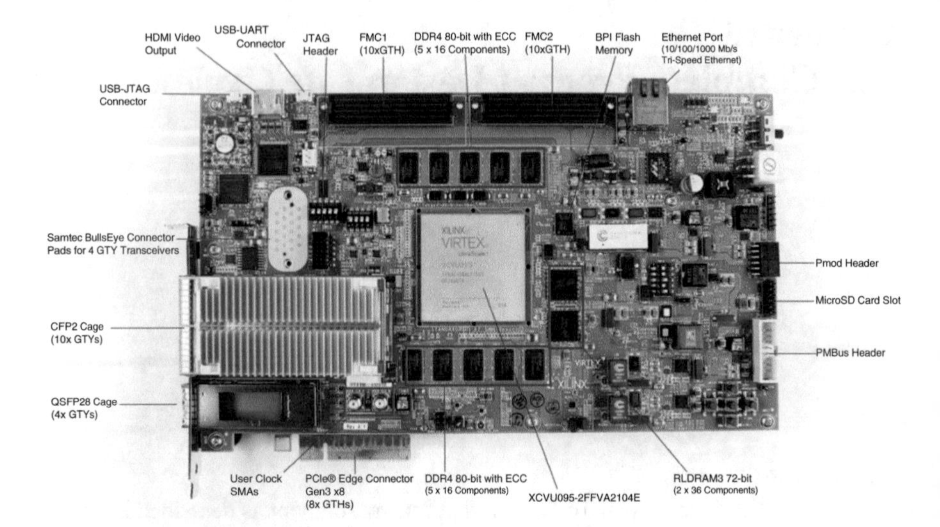

Fig. 15.1 Xilinx VCU108 development board

15.2 Product Design and Development Flow

15.2.1 Design Specification

Referring to Fig. 15.2, the process starts with an MRD (marketing requirement document) or a PRD (product requirement document) that normally comes from sales and marketing. It details the definitions and functional requirements the product needs to have, to target a specific market time window, research data, customer requests, as well as any driving factors for the new product (i.e., lower cost, smaller form factor, lower power consumptions, etc.). Engineering then creates the design specification which further details the deliverables that can be expected from engineering. The deliverables can be at the chip level (i.e., ASIC (application-specific integrated circuit), FPGA (field-programmable gate array), or other electronic components)), at the board level (i.e., a CPU, graphic, or IO boards, etc.), or at the "box" level as a complete product (i.e., a workstation, tablet, Bluetooth headphone, etc.). The design specification describes pertinent information of the system architecture of the product as well as the detail implementations consisting of the microarchitecture of the chip, block diagrams, and bandwidth calculations. If the product is an ASIC, the specification can describe the specific technology that will be used, gate count, packaging info, SPICE simulation/modeling, etc. If the product uses FPGA(s), mechanical, thermal, and verification requirements could also be specified here to drive subsequent implementations.

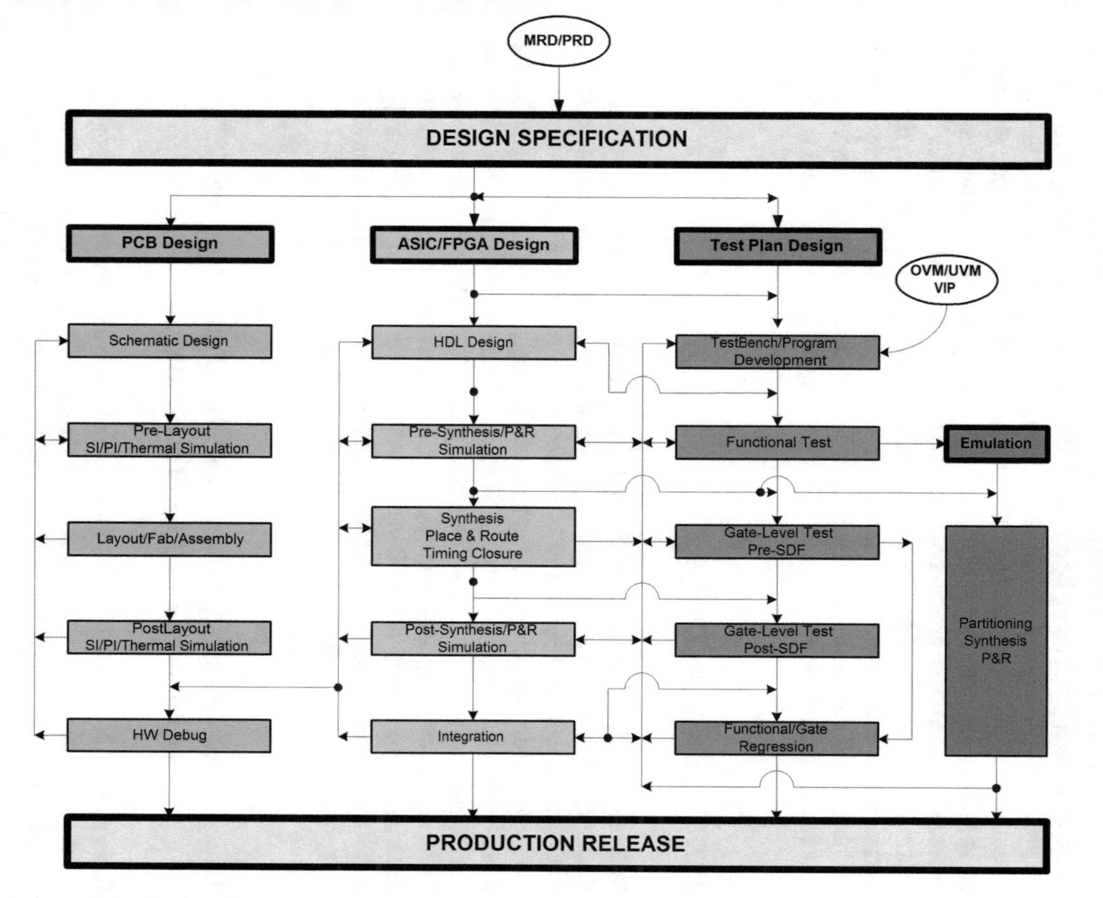

Fig. 15.2 Product design and development flow

15.3 PCB (Printed Circuit Board) Design

The PCB contains the electronics, the mechanical components, as well as other analog and/or RF (radio frequency) devices. Depending on the complexity of the system, the PCB can have a small form factor (i.e., Bluetooth headphone) or a larger board size within a computer, for instance. The PCB consists of layers and layers of materials pressed together along with interconnecting metal traces to connect the components mounted on board such as CPU, memory, switches, connectors, etc. Many designs today are operating at high frequencies, and some design uses components exceeding + pins which create complexities in the routing of these signals. This also causes issues in terms of the electrical signaling and thermal dissipation. EDA vendors such as Mentor Graphics, Cadence, and Altium, etc. all provide tools for the PCB designers to address these design challenges. We will describe the main steps involved in the PCB design flow. A typical PCB is a stack-up design as shown in Fig. 15.3.

15.3.1 Schematic Design

Refer to Fig. 15.4. The key step driving the PCB design is the creation of the schematic capture to represent the components being used and to show the interconnects of these components. Different EDA vendors provide different capabilities to help the designer do the schematic capture of the design. The design specification describes the architecture and block diagrams of the board and specifies the main components to be used in the schematic pages. The schematic capture process involves creating or accessing the electronic symbols from libraries and then connects them according to the function required. The schematic of a typical design can contain a few to hundreds of pages to specify the complete connection of all the components used. An example electronic schematic is shown below.

15.3.2 Pre-layout Signal Integrity (SI), Power Integrity (PI), and Thermal Integrity (TI) Simulation

Refer to Fig. 15.5. In selecting the final technology to implement on the board, often the designers resort to doing simulation for signal quality, power delivery quality, and even thermal simulation to make sure that the product meets all the requirements. Signal integrity simulation is the process to simulate the electrical behaviors between electronic components (i.e., between the CPU chip and memory chip) to ensure that the voltage levels are met, the current driving is sufficient, and there is no extraneous noise when the system is operating. In power integrity simulation, the designer looks at how clean the power delivery subsystem is and if it's able to deliver sufficient power to these devices in times of busy switching activities. If not done properly, the components may not operate reliably leading to system failures in the field. For thermal

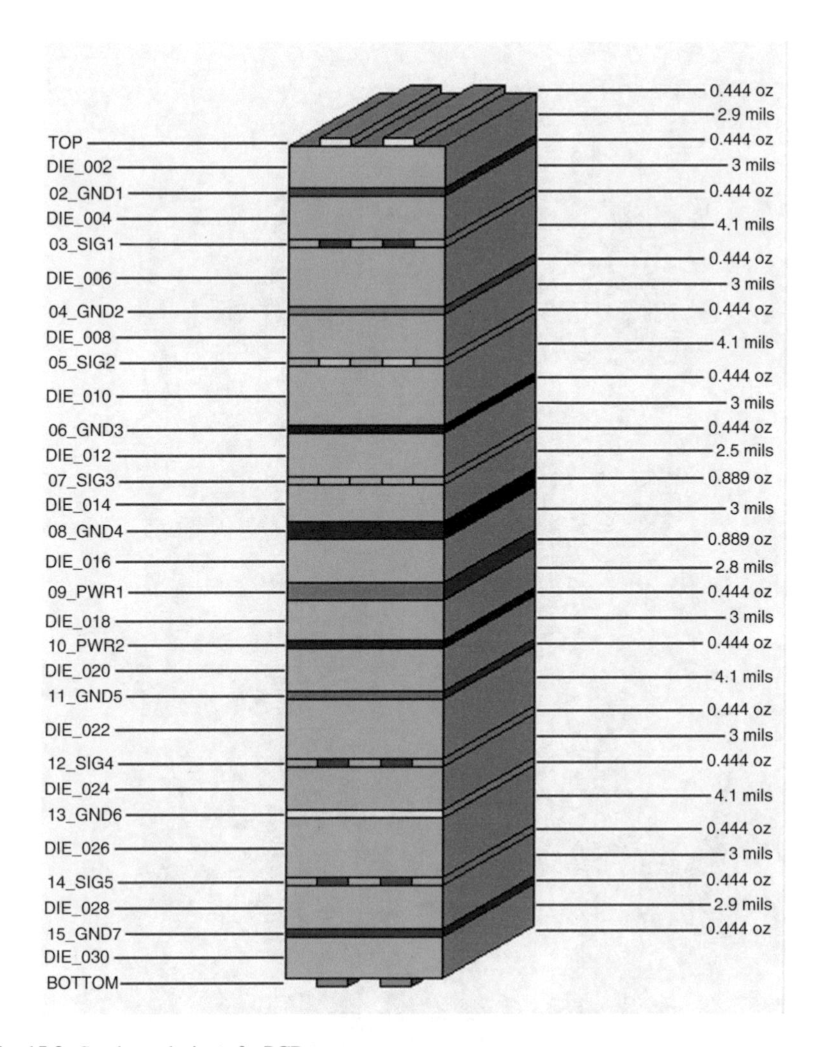

Fig. 15.3 Stack-up design of a PCB

integrity, if there's a requirement that the temperature of the operating product cannot exceed certain level, there is a need to do thermal simulation and to explore different cooling options which may be heat sinks, fans, or even placements of devices. The resulting temperature simulations must meet the overall design criteria not only from a thermal standpoint but also from the system cost as well as physical size.

If the results from the pre-layout SI, PI, and TI indicate potential issues, alternate solutions need to be explored. In some cases, the alternate solutions may require changes in the original design specifications, for instance, if the desire was to use a single large FPGA in the design. However, simulation showed that there can be a potentially high level of cross talk and thermal issues. This might require a change in the partitioning into multiple FPGA devices and the microarchitecture of the implementation.

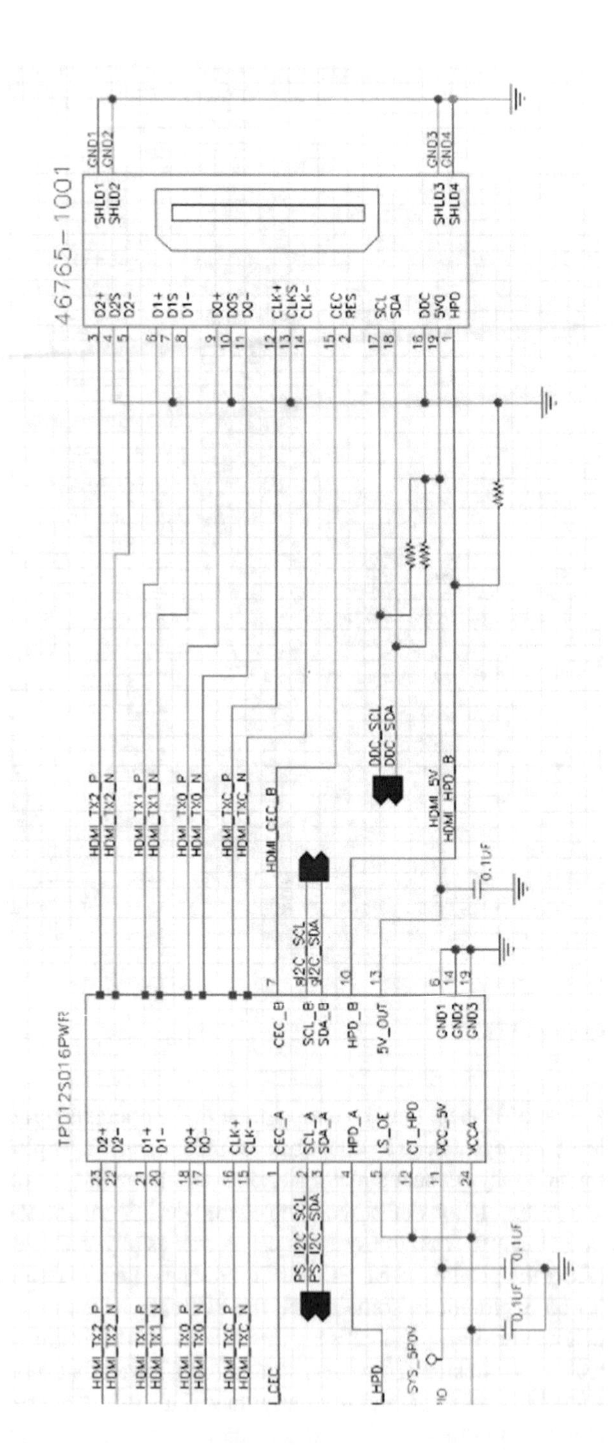

Fig. 15.4 Schematic capture for an HDMI

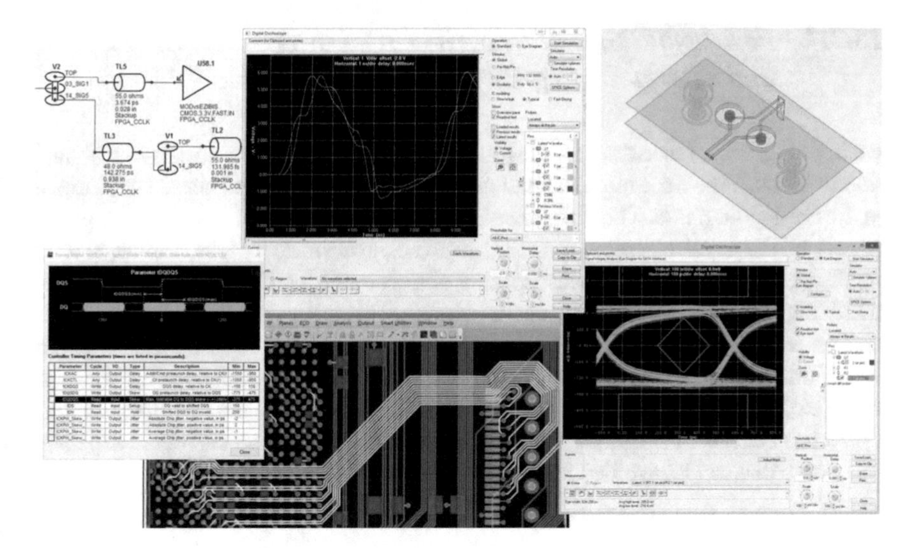

Fig. 15.5 Signal integrity simulation for pre- and post-layout

15.3.3 Layout, Fabrication, and Assembly

Once the schematic is complete, the next step is to start the PCB layout process. In this step, the schematic and symbol library files are used to map into the layout libraries. This is to ensure that the correct footprints of the devices are used and the connections specified from the schematic are valid. In this step, additional physical constraints can be specified so that the electrical characteristics of these connections are met. These constraints can include the lengths, widths, and heights of the traces, the particular layer certain signal has to be in, placements of devices and connectors, spacing of signals, etc. This is an involved process and may require multiple iterations in order to meet the physical size constraints as well as meeting the electrical and thermal requirements. Once a board is "routed," the board can be simulated again (described in the next section) before going to the fabrication and assembly process. In this step, the necessary files are generated (i.e., Gerber files) as well as the BOM (bill of material) file that specifies the manufacture PN, quantity, costs, etc. This step can also trigger the manufacturing process if the design will have a high volume in production (or if it's just a prototype board), and additional DFMA (design for manufacturing and assembly) steps are required. The PCB is now ready to be built along with all the electronic components installed in the assembly process which is usually done by external vendors.

15.3.4 Post-layout Signal Integrity (SI), Power Integrity (PI), and Thermal Integrity (TI) Simulation

Refer to Fig. 15.6. Once the board is routed, it is necessary to do a post-layout simulation to validate the final implementation of the PCB. The process here is the same as in the pre-layout phase. The difference is that now that the board has been completely routed, all the physical structures are in place (i.e., exact trace lengths, widths, heights, number of layers, number of vias, spacing of traces and devices, etc.). Thus, the electrical (and thermal) behaviors are more accurate. This is the chance to predict and find any potential issues before the board is built to start the bench bring-up. Normally, issues relating to high-speed interfaces (i.e., DDR4, PCIe, 10Gbps ENET, etc.) and EMI (electromagnetic interference) violations are found here. Boards that pass functional test but fail EMI checks will have to be redesigned which can be an expensive procedure that can be avoided if simulation is done before built.

15.3.5 Hardware Bring-Up and Debug

When the board has been assembled with all the electronic components, the process of hardware debug and bring-up begins. While not explicitly mentioned, it is important to note that this is the stage where everything comes together from chip design, verification, as well as mechanical design. It is also the first phase of system software integration. To assist the hardware bring-up team, the software team creates a scaled-down "image" of the system software to enable the hardware designer to verify functionality of both internal logic and external system interfaces. Hardware characterization will take place which involves measuring signal quality, capturing data, validating external interface (i.e., disk drives, backplanes, other IOs, etc.). The software team and the hardware team work closely together to validate that the functionality of the board is met. Issues in both HW and SW could surface at this stage, and within the constraints of the design implementation, these issues can be fixed in HW and/or can be mitigated with SW solutions. If major issues are found from bring-up, a re-layout may be necessary (if the issues are related to the board fabrication), a redesign of the chip may be necessary, and the process repeats.

FPGA designers can also take advantage of debugging internal digital logic blocks and examine internal signals on the board by using ChipScope (for Xilinx) or SignalTap (for Altera) via the JTAG (Joint Test Action Group) port built on the PCB which is also used to test connectivity of all electronic components on the board.

System-level debugging (Fig. 15.7) may require additional instruments to generate stimulus and capture responses from the board. Such equipment may include network analyzer, USB 3.0 traffic generator/analyzer, and others.

Once the basic bring-up milestone is reached and if no major issues are found, additional boards may be built to enable the SW team to fully test the system software. On the HW side, additional boards may go through the process of extended temperature testing, some boards may be early delivered to customers for their own testing, and the process of manufacturing begins.

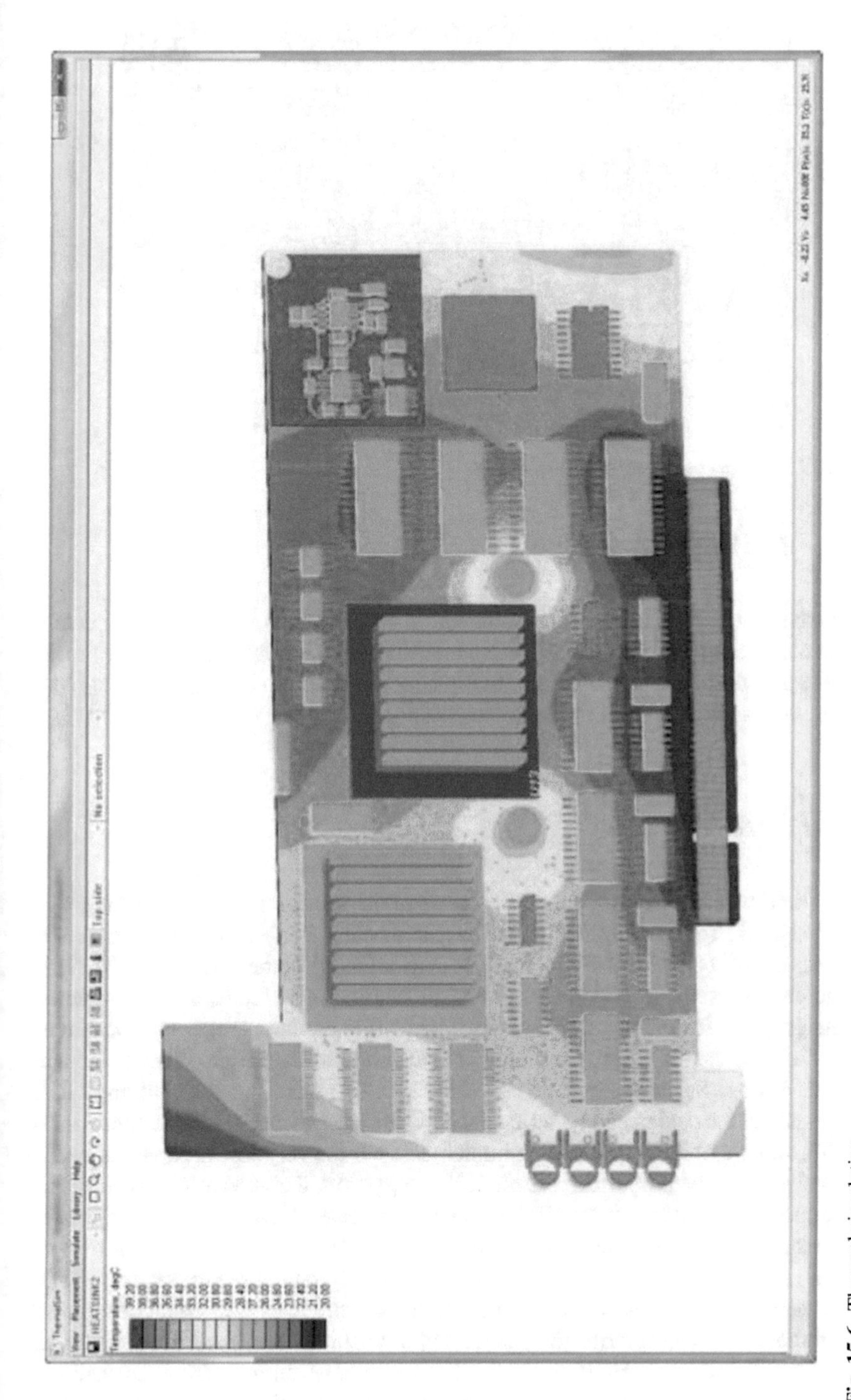

Fig. 15.6 Thermal simulation

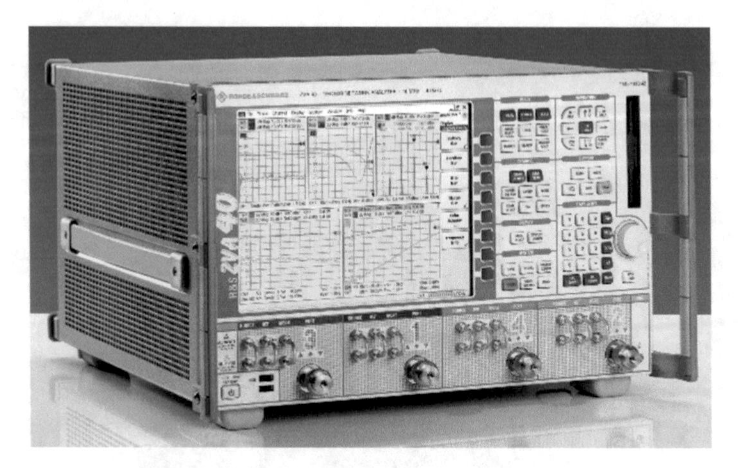

Fig. 15.7 Network analyzer for system debug

15.4 ASIC/FPGA Design

From the design specification and architecture of the system, the microarchitecture of the chip design is formulated to drive the ASIC/FPGA implementation. This can involve partitioning the design into multiple chips, and each chip goes through the design flow as described in the following sections.

15.4.1 HDL Design

HDL (hardware description language) refers to the specialized computer language that designers used to describe the structure and behavior of electronic circuits. The language mostly describes digital logic circuits although some variations of the language can be used for analog circuit design. Verilog and VHDL are the common languages used for many years, but other languages are becoming more commonplace to address the shortcomings of the original language. These include SystemVerilog, SystemC, PSL, SVA, C++, etc. The main benefit of doing HDL design is that the design can be written at a higher level to model certain functionality of a complex block, or it can be implemented at the very low hardware level (if the designer wants to optimize certain technology from the vendor to maximize the performance). In addition, the HDL designs allow easy changes for easy migration and portability across many implementation environments. It is important to note that the HDL language can mimic the behavior of digital logic without the need to implement the final logic in gates. This is often referred to as behavioral design and is commonly used to describe the behavior of a functional block at a high level (i.e., a disk drive's behavior). HDL designs that have synthesizable constructs are referred to as RTL (register transfer level) designs when these structures can be synthesized into digital logic and gates in an FPGA, for instance.

15.4.2 Pre-synthesis Simulation

The HDL code that is written at this stage often is not necessarily optimized for performance and can be a mixture of behavioral as well as structural (i.e., RTL). The code(s) can be used in simulation to validate the functional behaviors between different design blocks. A module-level testbench can be built around the block being designed, and basic stimulus and checking can be performed. As these sub building blocks are being designed and integrated together, there can be many other unfinished blocks (which will have placeholders) as well as assumptions for protocol conditions (which may or may not be supported in the end design). However, this gives the designer a good glimpse into the behavior of how the block being designed will interact with other blocks once the chip design is complete.

Another benefit of doing pre-synthesis simulation is to help drive the verification group in developing the final verification environment.

15.4.3 Post-synthesis/Place and Route Simulation, Timing Closure

Once a large part of the functionality of the design has been verified in the pre-synthesis simulation, the next step is to take the design through physical implementation and validate the synthesized design with timing closure. The RTL design is now represented as digital logic gates in the technology chosen (i.e., for a particular ASIC process node or a particular FPGA device). Depending on the complexity and size of the design, it may or may not route completely in the chip chosen. Even when routed completely, the interconnecting delays may exceed the timing margins budgeted (i.e., the internal trace is too long which exhibited timing delay not meeting setup or hold time). If this is the case, a rewrite of the RTL code may be necessary to redesign the block (i.e., break up a 64-bit counter into eight 8-bit counters), insert pipeline registers, redistribute the combinational logic across registers, use built-in hardware IP (intellectual property) blocks, or even adjust clock timing, etc. These have to be done to ensure that the design can run at the targeted frequencies in order to not impact the system bandwidth requirements. The previous step is referred to as doing timing closure. Once timing closure is achieved, re-simulating the post-synthesized block may be necessary to ensure that no tests fail.

Note that doing simulation for post-synthesized netlist may be a lot longer (i.e., can be >> 10X) due to modeling each delay at each switching point, so running simulation with full back-annotated timing is often skipped. Or the back-annotated netlist can be used to simulate the power-on reset logic. If there are bugs due to logic that does not reset, they will be caught at this step (e.g., incorrectly identified false paths or multicycle paths). No other tests need to be run. One can consider running gate simulation with unit delay (i.e., the delay through all the combination switching points and routing delay using one unit time) to minimize the impact in simulation time.

In general, designers opted to run functional simulation using pre-synthesized netlist and to do a full STA (static timing analysis) in post-routed netlist. This is a widely adopted methodology to minimize debug time and maximize the logic verification.

15.4.4 Integration

In the ASIC/FPGA integration phase, all the design codes are merged into one single design structure. This includes all the IO pin definitions/assignments as well as all the RTL codes for all the modules (whether organized as a flat structure or a hierarchical structure). The RTL code can contain codes written by the designer and/or instantiations of hardware blocks inside the chip (i.e., an Ethernet MAC or PCIe block). In some designs, where prototyping is done in an FPGA but the final design is implemented as an ASIC, it is in the integration phase where the two (potential different) code streams are merged and selected in some way. For instance, assuming that the code stream is the same for the core logic and that between the ASIC and FPGA implementation the only differences are in the IO structures (i.e., pin assignments and signaling), clocks (i.e., PLL vs. DLL), and memory (i.e., RAMs, ROMs, single-/dual-ported memories), a selection mechanism needs to be implemented to ease the full ASIC flow or the FPGA flow if so desired in synthesis, place and route, clock trees, etc.

This stage also ties into the PCB HW debug in terms of electrical connectivity, signaling, power, and debug facilities.

15.5 Verification

Previous chapters have focused on the verification flow in much more details. The discussion on the verification flow in this section centers around how this fits within the context of product design life cycle.

The verification process can start very early, even before the system design specification is formed, and continues even when the product is shipped. The role of verification is very crucial in the design process. It should not be looked at as an engagement after the design is done, but it should be looked at as something that drives how the design is done. It is almost like saying "If you can't verify it, don't design it!"

15.5.1 Test Plan Specification

The key starter in the verification flow is the formation of the test plan from the design specification. The test plan highlights the features of the design and describes in detail how each of the function will be tested, which environment will be tested, and how results are measured against a passing or failing criteria. From the main test plan, subsequent test plans may be created for certain part in the design (i.e., an FPGA), indoor/outdoor environment, or even compliance testing at the system interoperability level.

15.5.2 Testbench and Test Program Development

The term "testbench" (Fig. 15.8) loosely defines an environment where the device under test (DUT) is instrumented along with other blocks to either provide the stimulus and/or to measure the response to/from the DUT. The components included in the testbench can include high-level programs written in C, C++, etc. to behavioral HDL blocks for components and can even include real hardware in the loop to help speed up testing of the communication between the DUT and the real world.

Test programs can then be developed to enable these blocks to generate stimulus into or out of the DUT. Depending on the environment, the test program can be written using one or multiple languages in combination thereof. Once the tests are developed, they can run sequentially in a deterministic behavior or can be dynamic with randomly chosen behavior and stimulus.

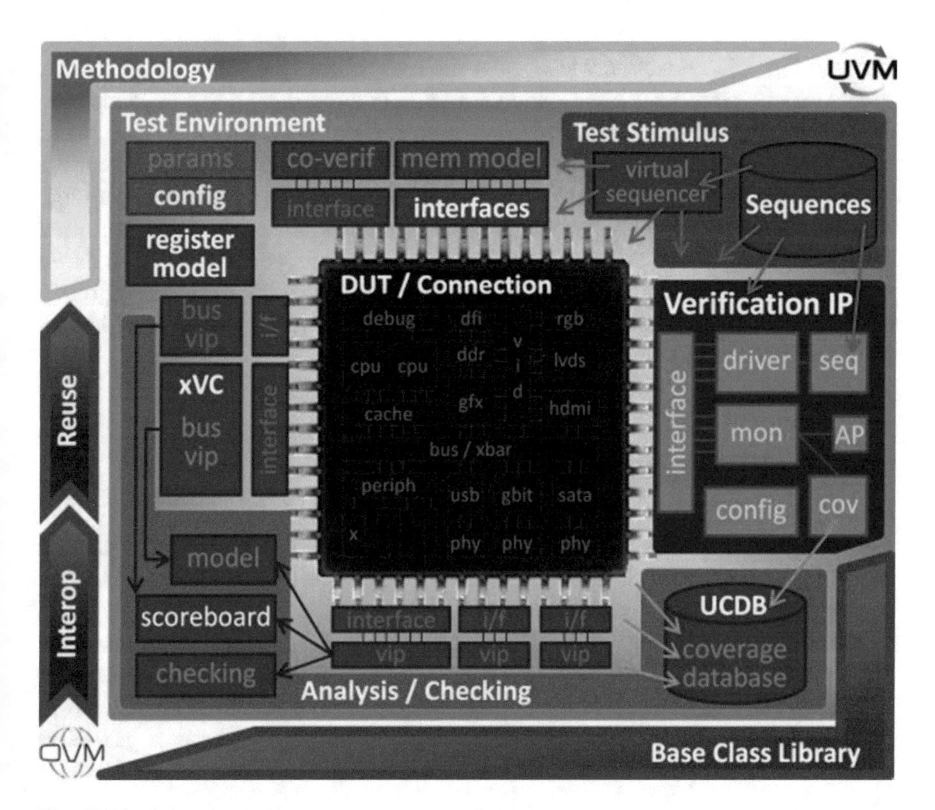

Fig. 15.8 Testbench development

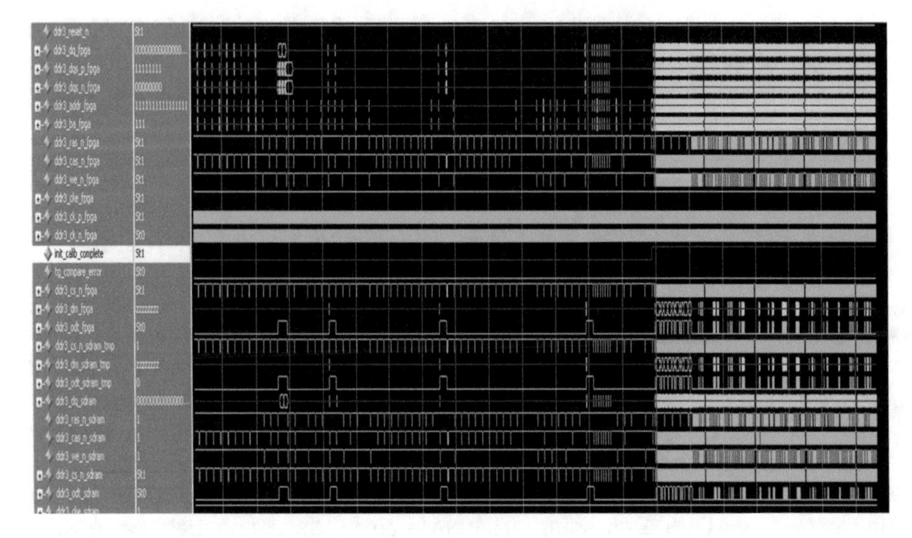

Fig. 15.9 Functional simulation

15.5.3 Functional Test

Refer to Fig. 15.9. Functional tests are written to validate the behavior of the RTL design strictly from pure logic standpoint without any timing delay dependencies coming from synthesis as well as place and route. Functional tests are quicker to run since the design, being described at a higher level of abstraction, does not incur the penalty of logic, timing, and even routing delay computations which, for a large design, can significantly slow down the overall simulation. For instance, for some design that has internal processor (i.e., ARM core), the processor block is replaced with a C++ (or SystemC) model which has the cycle-accurate behavior to greatly simulate the reads/writes to/from the processor without having to deal with the real gates that were used to implement the processor.

15.5.4 Gate-Level Verification

Once the RTL is synthesized into gates and for certain ASIC technology node or a particular FPGA family, the gate netlist can be substituted for the RTL DUT, and the testbench is used to run the gate-level tests.

One question that arises all the time is if we have done functional verification and static timing analysis, and synthesis is guaranteed to work, what's the need for gate-level simulation? For ASIC design and after clock tree and test insertions, the original functional netlist will change. This would require some degree of gate-level testing to validate these changes and insertions. Read on.

15.5.4.1 Unit Delay Synthesized Gate Simulation Without Standard Delay File Timing Annotation

There are two types of gate-level simulation. The first type is synthesized gate simulation where the nets of the design have no delays. This type of gate-level simulation should be done using library cells with "unit" delay and without timing annotation using SDF. This unit delay simulation does not need to run the full set of RTL functional tests. That is simply too slow and impractical and there is no return on investment (ROI). Its utility is in running power-on reset (POR) simulation. This simulation will find if any of the state elements are not resettable and are causing unknowns (X) to proliferate in design. Note that RTL may be resettable but gate may not be because, for example, there could be mismatch in the way RTL code was synthesized (no "default" statement in a "case" statement). Hence, the unit delay gate-level simulation is a good sanity check verification of power-on reset logic.

15.5.4.2 Post-synthesis Gate Simulation with SDF Timing Annotation

The other type of gate-level simulation is post-layout timing-based simulation using the post-layout delays in the standard delay file (SDF). The gate netlist now has timing delays from the logic gates as well as net timing delays from internal routing of the chip. Here also, you need to undertake the power-on reset (POR) simulation. In addition to making sure that all state elements are resettable, there are other advantages. The simulation with SDF annotation will show any timing violations that may exist in the netlist. Why would you have timing violations if you have already done static timing analysis? The main reason could be that the designer incorrectly identified multi-cycle timing paths or false timing paths. In other words, static timing will exclude multi-cycle paths, but SDF-annotated simulation will catch incorrect multi-cycle paths.

In addition to POR simulation, you should also run the critical path (longest path) simulation. If you see timing violations during simulation, that could be again because of incorrect identification (during static analysis) of false or multi-cycle paths.

This is not so much a requirement for FPGA designs as FPGA resources already have tests and scans built in the timing model, and there are no additional penalties from the clock tree implementation.

15.5.5 Functional/Gate Regression

In this regression environment, both the functional tests and gate-level tests are combined and run to test the design. At this stage, both directed tests and constrained random tests are run to exercise the full design. For RTL simulation, for smaller designs, the number of tests can be in the hundreds, but on larger designs (i.e., a CPU), these tests can be in the thousands, even indefinite if random constraints are used. Often, a server farm is used where hundreds and even thousands of workstations are used to run multiple images of the testbench in complete random order, and the results can be staggering when running 24×7. When issues are found, the failing data will be fed back to the designers, and appropriate actions are taken which may involve the fixes and redesign of the particular logic section in the chip and even the PCB designs. Gate simulation with such large set of vectors is impractical. So, use power-on reset and longest path simulation at gate level.

15.6 Emulation

The subject of using emulation as part of the hardware/software co-simulation is discussed in details in Chap. 12. This section highlights the emulation flow in the context of complete product design life cycle.

As the testbench is being developed and tests are written, there might be a scenario where running tests with simulators take too long to progress from one stage to another. This is more so true in designs where there are a lot of mathematical computations required (as in implementing certain DSP algorithm). In these cases, alternate methods need to be implemented to speed up the testing. This can be done in one of two options: either have some hardware in the loop or to port either the whole or partial design in an emulator.

The emulation flow ties in closely with the ASIC/FPGA pre-synthesis simulation and the testbench/test program development in the verification flow. The RTL codes that are written by the ASIC/FPGA designers can be brought over to the emulation environment, and it is here that these codes are subjected to another partitioning, synthesis, and place and route using the logic resources inside the emulator box. The emulation box can be used to mimic the chip being designed or in many cases used in conjunction with the full system software to emulate the real software system in hardware. Note that emulation boxes cannot necessarily run at the full speed of the real system (as is the case for new hardware features in devices that have not made it inside the emulation box). In those situations, a scaled-down speed version of the system can be implemented to support the full software testing.

Chapter 16
Voice Over IP (VoIP) Network SoC Verification

Chapter Introduction
This chapter discusses, in detail, the verification of a complex SoC, namely, a voice over IP Network SoC. We will go through a comprehensive verification plan and describe each verification step with VoIP SoC-based real-life detail.

16.1 Voice Over IP (VoIP) Network SoC

Having discussed many different aspects of design verification, let us now dive into real-world SoC verification. We will use the SoC shown in Fig. 16.1. The figure shows a simplified version of a voice over IP (VoIP) network SoC. It has two embedded ARM processors: one for transmit and one for receive. It has PCI, TDM (for voice), and Ethernet Tx/Rx MAC port. It has internal DMA to route packets from Ethernet to TDM (packet to voice) and TDM to Ethernet (voice to packet). A CAM acts as a lookup table for packet to voice or voice to packet transmission.

The first step in any verification project is to create a verification plan. Let us formulate a verification plan for this SoC. This plan will also serve as the execution plan for the rest of the duration of the project and identify upfront the expertise and tools required for this project. It will also identify the compute resources and physical memory (as well as disk storage) requirements.

16.2 VoIP Network SoC Verification Plan

Here is a comprehensive verification plan to verify the network (VoIP) SoC. This is a high-level view of what needs to go into a verification plan. Each subsequent section will provide detailed verification plan of each step.

A.B. Mehta, *ASIC/SoC Functional Design Verification*,
DOI 10.1007/978-3-319-59418-7_16

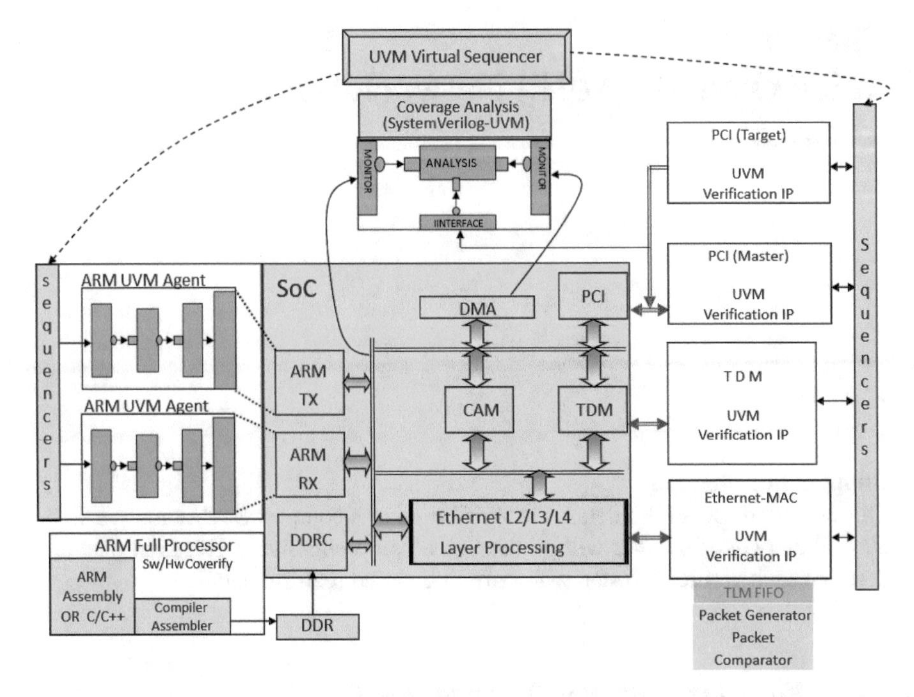

Fig. 16.1 Voice over IP (VoIP) SoC verification

1. Identify subsystems within your SoC.
2. Determine subsystem stimulus and response methodology.
3. SoC interconnect verification.
4. Low-power verification.
5. Static formal or static + simulation hybrid.
6. Assertions methodology (SVA).
7. Coverage methodology (functional—SFC, code, and SVA “cover”).
8. Software/hardware co-verification.
9. Simulation regressions: hardware acceleration or emulation or FPGA prototyping.
10. ESL—virtual platform for software and test development.

16.3 Identify Subsystems Within VoIP Network SoC

A subsystem is one which has clearly defined inputs for stimulus and outputs for response checking. In the SoC of Fig. 16.1, we can identify the following subsystems:

- **Ethernet subsystem**. Ethernet MAC to Layer4 processing subsystem. The MAC (Ethernet) layer receives voice packets over the Ethernet and processes its payload’s (e.g., IPV4) Layer 2, Layer 3, and Layer 4. And once done with Layer 4 processing, it will hand over the packet to the ARM Rx processor. So, this is a subsystem in that there is a clear input and a clear output.

- **TDM subsystem**. Voice comes over the TDM interface and DMA it over to the ARM Rx CPU. ARM searches CAM for voice to packet translation and hands over the packet directly over to Ethernet Tx Logic. Hence, this is a subsystem with a clear input and output points in the logic.
- **PCI subsystem**. PCI subsystem is to program the register space of the SoC (among other functions). So, you can write and read registers from PCI. Hence, this is a closed-loop subsystem.
- **ARM subsystem(s)**. ARM processes the incoming "packet" and the incoming "voice" and converts one from the other. It also boots the system. So, we have Ethernet packet as input and voice as output and voice as input and Ethernet as output. So, in essence, there are two subsystems within the ARM subsystem because for voice to packet, you have a set of voice inputs and packet outputs, and for packet to voice, you have the Ethernet stack as input and voice as output. This can be called "internal" processing subsystems (as opposed to the peripheral processing subsystems identified above).
- **Memory subsystem**. Write (or DMA) and read from the external DDR. Verify the dynamic DDR controller (DDRC), ARM interface, etc.

16.4 Determine Subsystem Stimulus and Response Methodology

- **Ethernet Subsystem**
  - Stimulus

    Stimulus required is Ethernet packet generation software and algorithms. Also, an Ethernet MAC transmit/receive UVM agent. The UVM agent will read in the generated Ethernet packets from a golden packet generator (or a software stack) and transmit those to the MAC controller of the SoC. The SoC will process the received packets for L1/L2/L3/L4 layer processing.

  - Response

    There will be a UVM monitor to track the packets as they go through the L1, L2, L3, and L4 processing hardware. At each stage of layer processing, the UVM monitor will compare a golden L1, L2, L3, and L4 processed packet with its corresponding hardware-processed packet. Any discrepancy will be reported to an analysis port for further processing and error reporting.

    Check to see that error packets do get dropped but the clean packets do get processed.

- **TDM Subsystem**
  - Stimulus

    In a TDM bus, data or information arriving from an input line is put onto specific time slots on a high-speed bus, where a recipient would listen to the bus and pick out only the signals from a certain time slot. The stimulus needs to be provided to comprehensively verify all time slots in all combinations that would be input to the TDM bus. A matrix needs to be created and read in from a file by a UVM–TDM agent.

Send all different types of TDM "frames," and check their response for time–division multiplexing operation.

- Response

Create a reference model for the time slot matrix and predict the signals to be picked up from certain time slots. Use this reference model in the UVM scoreboard connected to UVM monitor via analysis port.

Create a golden reference for expected frames and compare with incoming frames. Use UVM scoreboard.

- **PCI Subsystem**
  - Stimulus

Using UVM master agent; simulate the entire PCI master compliance and target compliance test suite with SoC as the master (i.e., UVM agent as target) followed by SoC as the target and UVM agent as the master. Carefully check for SoC boot after programming PCI Config space for SoC as target from UVM master agent.

Run all error scenarios singularly first and then induce collision on them. For example, SoC issues a master abort on a transaction and target issues a target abort in the very next transaction. Similarly, transaction retry from both master and target in consecutive cycles.

Verify entire register space using UVM register verification methodology. Check for initial values (after reset) of register bits. Write to Read-only bits and see that the write value does not get written.

  - Response

Use external UVM agent as a PCI target (for SoC as master).

Monitor master/target transactions and send those through analysis port to a scoreboard. The scoreboard maintains golden transmit values which are compared with received values (e.g., register write/read).

Scoreboard also maintains the entire PCI compliance suite results for golden comparison.

- **ARM Subsystem**
  - Stimulus

Replace the full ARM CPU core model with an ARM AXI UVM agent. For verification purposes, you don't need the full ARM CPU model since you are not verifying software through the processor. For hardware verification, you need a Bus Functional Model (i.e., an UVM agent) to drive directed and constrained random writes/reads/interrupts to all "targets" of the ARM AXI bus. These BFMs are shown in Fig. 16.1.

To run a simple software boot sequence, you will need the full ARM CPU model. This is shown in the bottom left corner of Fig. 16.1. The compiled software code needs to be loaded in the external DDR. The SoC needs to be reset and program counter should point to the code in the DDR. After this, ARM CPU will fetch instructions from the DDR and execute.

– Response

The UVM agent BFM will monitor the bus activity and send the received read transactions to an external scoreboard through the analysis port. The external scoreboard will have a small memory where the write transactions are stored (i.e., write address, ID, and data). When read returns data, the scoreboard will compare it with the write data that it had stored. The memory where golden write data is stored is shared by both the scoreboard and the sequences so that the data that scoreboard uses as golden is the same data that were sent to PCI, for example, by a sequence.

Check the state of SoC after boot to make sure boot succeeded. The full ARM CPU model accomplishes this. It will compare the state of SoC with the expected state after boot. This is the only time full ARM CPU model is needed.

- **Memory Subsystem**

– Stimulus

First, direct set of tests to drive to DDR covering all different types of protocol (e.g., burst, non-burst, interleaved, etc.). See that DDRC survives the directed tests.

Second, constrain the stimulus to DDR bank crossing boundary and see that operations such as word crossing and word wrap around work correctly.

Third, turn ON the DMA as well as ARM UVM agent BFM and simultaneously blast DDRC with writes and reads.

– Response

Each write from ARM BFM or the DMA will store that write with an ID. When read takes place from the same address and ID, compare write data with read data. These are simple tests but very effective at testing DDRC operations.

Measure read latencies through DDRC and DDR and report to make sure they meet the architectural requirements.

16.5 SoC Interconnect Verification

This is a topic in itself and covered in its entirety in Chap. 13.1.

16.6 Low-Power Verification

Low-power verification with UPF is discussed in detail in Chap. 9.

For the network SoC verification, the methodology with UPF is directly applicable based on the following power domains and the Power State Table defined.

- Power subdomains for SoC in Fig. 16.1
  - ARM
  - Ethernet packet processing block
  - TDM voice processing block
  - PCI subsystem
  - Memory subsystem
- Power State Table for the network SoC (Fig. 16.2)

16.7 Static Formal or Static + Simulation Hybrid Methodology

This topic is covered in its entirety in Chap. 10.

As noted in Chap. 10, static formal (aka static functional) works only on small blocks of logic (a few hundred gates). From that point of view, we identify the following blocks for static formal.

- Based on Fig. 16.1, apply static formal to:
  - SoC synchronous and asynchronous FIFOs
  - A single processing layer (L1 or L2 or L3 or L4) at a time with "assumed" stimulus using hybrid methodology
  - Standalone CAM operation
  - Standalone DMA operation
  - DDRC refresh logic
  - Interconnect bridges (e.g., AXI ⇔ AHB, AHB ⇔ APB, etc.)

State \ Power Domain	ARM_PD	Eth_PD	TDM_PD	PCI_PD	Mem_PD
Normal	ON_10	ON_10	ON_08	ON_08	ON_05
Sleep	ON_10	OFF	OFF	ON_08	ON_05
Hibernate	ON_10	OFF	OFF	OFF	OFF

Fig. 16.2 Power State Table for network SoC

16.8 Assertion Methodology

This topic is covered in its entirety in Chap. 6.

Applying the SVA methodology to the network SoC shown in Fig. 16.1, we need to apply assertions on the following interfaces and modules.

- Apply SystemVerilog assertions to:
  - PCI initiator/target protocol interface
  - Ethernet transmit/receive protocol interface
  - Internal AXI, AHB, APB bus protocol interface
  - Layer processing handoff (from one stage to another) interface
  - DDRC to DDR protocol interface
  - TDM protocol interface
  - All SoC internal FIFOs
  - Misc. FIFOs as part of RTL development (i.e., microarchitecture level assertions)

16.9 Functional Coverage

This topic is covered in its entirety in Chap. 7.

Before we apply functional coverpoints to the network SoC (Fig. 16.1),

- Determine logic that needs to be functionally covered.
- How will you leverage code coverage with SystemVerilog functional coverage?
- What's the strategy to constrain stimulus to achieve desired functional coverage?
- How will you determine that you have specified all required coverpoints and covergroups? This is the hardest (and sometime subjective) question to answer. Continue to strategize as the project progresses.

The functional coverpoints (including *transition* and *cross* coverpoints) need to be applied to the following blocks:

- PCI initiator/target protocol interface
- Ethernet transmit/receive protocol interface
- Internal AXI, AHB, APB bus protocol interface
- Layer processing handoff (from one stage to another) interface
- DDRC to DDR protocol interface
- TDM protocol interface
- PCI, TDM, CAM, DMA, DDRC internal functional coverpoints (e.g., state transitions. Note that states will be covered by code coverage but not the state transitions). ARM is considered a pre-verified IP and hence no coverpoints are needed for it.

16.10 Software/Hardware Co-verification

This topic is covered in its entirety in Chap. 12.

Think about deploying advanced methodologies such as TLM2.0 (ESL) ⇔ emulation or acceleration. This allows you to speed up software running on the SoC. Deploy TLM2.0 models for blocks that are not under verification. TLM2.0 is transaction level and so is UVM. So, integration will not have significant challenges. You will be able to run small pieces of software code with such methodology.

The software that needs to be developed for the VoIP SoC (Fig. 16.1) can be categorized as follows:

- Driver software for Ethernet and PCI subsystems
- Boot software for SoC
- PCI configuration software
- Ethernet transmit logic
- Ethernet receive logic
- SoC programmable register initial value (from reset) programming
- DMA programming

16.11 Simulation Regressions: Hardware Acceleration or Emulation or FPGA Prototyping

This subject is covered in detail in Chap. 12.

- Determine the requirements for software development. How early should the software development start for it to be ready when hardware RTL is ready. This is the key question that needs to be answered before deciding on whether to deploy acceleration or emulation.
- Acceleration, emulation, and prototyping all need RTL to be ready in some shape and form. That is an issue because software development needs to start *before* RTL is ready. See the next section on virtual prototyping.
- If you do decide to go with simulation acceleration tools/methodology, further ponder over following differences among tools and technologies.
- **Acceleration** will have better debug capabilities than emulation.
  - But the speed may be in a few MHz at best.
  - Does acceleration provide enough speed for software development?
  - If the testbench is still in SystemVerilog (i.e., outside the acceleration box), will the SystemVerilog ⇔ acceleration maintain the required speed? Will SCE-MI-II help?
  - Will acceleration work at transaction level with UVM testbench?
  - What about memories? What about multiple clocks? What is the debug strategy?
  - How about assertions? Will they compile into acceleration hardware?

  - How will functional coverage be measured?
  - And finally, as mentioned above, acceleration requires a working RTL so that software development can take place. Without a working RTL, you will spend more time debugging hardware and less time developing software.
- **Emulation** will be orders of magnitude faster than acceleration.
  - But emulation cannot start until RTL is ready. That being the case, will it be too late for software development?
  - How easy/hard will it be to debug since internal node visibility may be poor?
  - What about assertions and functional coverpoints?

16.12 Virtual Platform

This topic is covered in detail in Chap. 11.

In the author's opinion, this is the preferred methodology and wave of the future. In other words, one can start a virtual platform development in parallel (or preceding) RTL development. Here are the high-level advantages of a virtual platform. It is highly recommended that a virtual platform be developed for the VoIP network SoC (Fig. 16.1).

- Virtual platform development requires only the architectural specifications and has no dependency on RTL development. This reason alone allows you to develop the virtual platform and keep ready for software development, way before RTL is ready.
- Virtual platform speeds are at par (or even faster) than an emulated or prototyped SoC. Virtual platforms operate in hundreds of MIPS. As an example, a bare bone virtual platform (Fig. 11.6) will boot a paired down Linux in less than five wall clock seconds.
- The virtual platform methodology is the industry standard ESL/TLM2.0 methodology.
- You will develop software before the RTL is ready.
- You will create and verify tests before the RTL is ready.
- The virtual platform acts as a reference model and will be needed for the following:
  - Match the architectural state of the SoC with that of VP at transaction boundaries.
  - Ethernet layer processing will be verified using the Ethernet subsystem of the virtual platform as a reference model (accessed via UVM scoreboard).
  - TDM crossbar switch and time slot selection logic will need a reference model. This will be part of the virtual platform and available as a reference model toward verification of TDM.

End-to-end packet and voice processing will need a reference model to verify correct translation of voice to packet and packet to voice.

Chapter 17
Cache Memory Subsystem Verification: UVM Agent Based

Chapter Introduction

This chapter discusses, in detail, verification of a cache subsystem of a large SoC. We will go through a comprehensive verification plan and describe, at each verification step, cache subsystem-based real-life detail. This chapter discusses the verification methodology using an UVM agent (as opposed to using an instruction set simulator (ISS) which is discussed in the next chapter).

17.1 Cache Subsystem

This design is a subset of a much larger multimedia SoC. As in the VoIP network SoC, this is a real project that was successfully verified using the environment shown in Fig. 17.1. It is a dual-processor system. CPU_A and CPU_B have their own internal L1 cache. These processors interact with a second-level write back cache L2. The internal bus can be an AXI (for high performance) or a proprietary bus. L2 cache in turn talks to an external memory (not shown) through the AXI bus.

The intent of the verification of this subsystem is to verify the L1 and L2 interaction for write back MESI protocol. The interaction gets complicated because, for example, both the processors could be writing and reading from L2 at the same time. They could be accessing the same L2 line but at different byte granularity. This means we must have transactions going concurrently from both CPUs. In addition, the AXI bus can snoop into L2 for a coherent AXI bus transaction. So, even AXI needs to run concurrently with both the CPUs. All three subsystems need to run concurrently to verify L2 write back MESI protocol along with L1 write through protocol.

As we saw for the network (VoIP) SoC verification, I am reiterating the steps needed to verify the cache memory subsystem (Fig. 17.1):

- Identify subsystems within your SoC.
- Determine subsystem stimulus and response methodology.
- SoC interconnect verification.

A.B. Mehta, *ASIC/SoC Functional Design Verification*,
DOI 10.1007/978-3-319-59418-7_17

- Low-power verification.
- Static formal or static + simulation hybrid.
- Assertion methodology (SVA).
- Coverage methodology (functional—SFC, code, and SVA "cover").
- Software/hardware co-verification.
- Simulation regressions: hardware acceleration or emulation or FPGA prototyping.
- ESL—virtual platform for software and test development

17.2 Identify Subsystems Within the Cache Subsystem

In the cache memory subsystem of Fig. 17.1, we can identify the following subsystems:

- **CPU_A ⇔ L1 Cache ⇔ L2 Cache:**
  - Let CPU_B be idle. CPU_A will write/read from L1 (write through) which in turn will get data from L2 (write back) or external memory. The intent is to verify that for all possible MESI transactions from CPU_A/L1 to L2 and the L2 to AXI that L2 adheres to the MESI protocol.

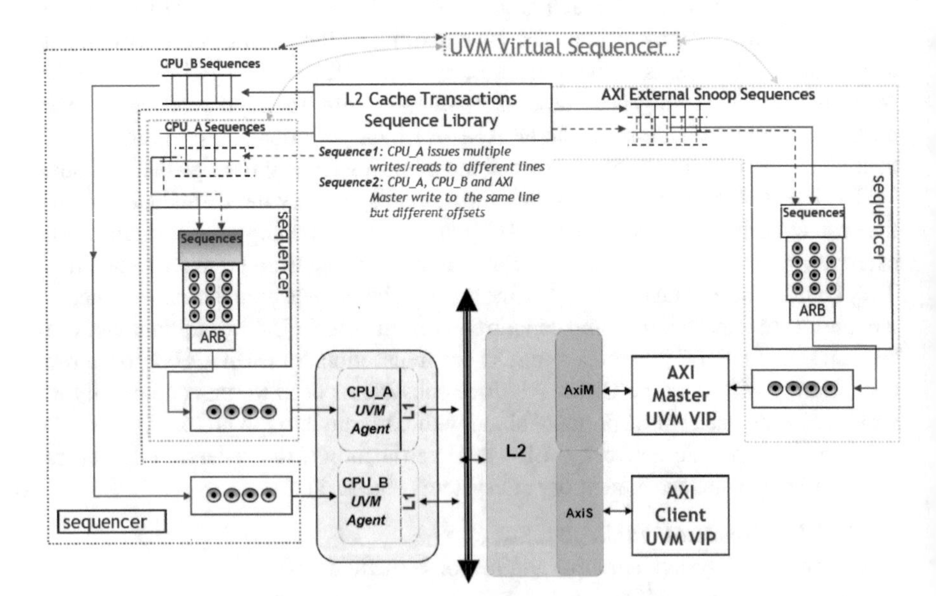

Fig. 17.1 Cache subsystem verification using UVM agents

- **CPU_B ⇔ L1 Cache ⇔ L2 Cache:**
  - Let CPU_A be idle. CPU_B will write/read from L1 (write through) which in turn will get data from L2 (write back) or external memory. The intent is to verify that for all possible MESI transactions from CPU_B/L1 to L2 and L2 to AXI that L2 adheres to the MESI protocol.
- **CPU_A && CPU_B Concurrent ⇔ L1 Cache ⇔ L2 Cache:**
  - Both CPU_A and CPU_B are active. Both write and read from L1 and L2 are constrained randomly; meaning they could be writing/reading one at a time and concurrently. Verify that when both are writing and reading, a single cache line can be accessed concurrently for both false sharing (same cache line and same byte in the cache line) and true sharing (same cache line but different bytes in the cache line). Many such scenarios need to be considered for MESI write back operation.
- **CPU_A, CPU_B ⇔ L1 ⇔ L2 ⇔ AXIM/AXIS:**
  - This is the mother of all verification for this subsystem. Both CPUs concurrently communicate with L1⇔L2, and the external interconnect (memory or other devices) communicates with AXIM and AXIS at the same time through L2 cache snoops.

17.3 Determine Subsystem Stimulus and Response Methodology

Based on above subsystems, let us see what kind of stimulus and response strategy we should deploy.

1. **CPU_A ⇔ L1 Cache ⇔ L2 Cache:**
   - **Stimulus (high level only):**
     - UVM agent for CPU_A. Scoreboard for CPU_A
     - Sequences created for the following:

       Idle CPU_B. No snoops from CPU_B
       Writes/reads from CPU_A to L1 to L2
       Cache write invalidations and other cacheable instructions
       Final cache block flush to flush out entire L1 and L2 cache to the main memory through AXI
   - **Response**
     - A reference model for the memory subsystem that simple writes/reads from the main memory. All writes/reads from CPU_A end up in main memory of the reference model.

- After cache block flush, the RTL memory image should be the same as the memory image produced by the reference model.
- Reference model is connected via DPI interface from the CPU_A and CPU_B scoreboard which communicates with CPU_A and CPU_B agent via analysis ports.
- SystemVerilog assertions to check low-level granularity of cache pipes, FIFOs, register files, L1 ⇔ L2 interface protocol, L2 ⇔ AXI interface protocol, etc.

2. **CPU_B ⇔ L1 Cache ⇔ L2 Cache:**
   - Stimulus and response strategy same as that for CPU_A, but all traffic will now be driven from CPU_B. CPU_A will remain idle.
3. **CPU_A && CPU_B Concurrent ⇔ L1 Cache ⇔ L2 Cache:**
   - **Stimulus:**
     - All singular transactions from CPU_A and CPU_B are reused only that both CPUs will now fire constrained randomly as in the real world.
     - In addition to constrained random, apply direct testing to see that:

       Both processors simultaneously access L2 cache lines at all possible granularities of byte, word, long word, and entire line.
       The CPUs perform false sharing (same line but with different byte addressing) and true sharing (same line and same byte address).
       Tests to see how write before read gets reordered, etc. for corner case testing.

     - Ping pong-style snoops from each processor. This means that each processor continually snoops a line in the other processor.
     - Cache block flush to flush entire L1 and L2 to main memory from both processors.
   - **Response:**
     - Final memory image produced by the memory subsystem reference model. This reference model gets all writes/reads from UVM agent monitor of both processors and determines the final memory image.
     - The final memory image produced by CPU_A and CPU_B cache block flushes should match the one produced by the reference model.
     - SystemVerilog assertions to check low-level granularity of:

       Cache pipes
       FIFOs, register files
       L1 ⇔ L2 interface protocol
       L2 ⇔ AXI interface protocol
       MESI protocol state machine transitions

4. **CPU_A, CPU_B ⇔ L1 ⇔ L2 ⇔ AXIM/AXIS**

   - **Stimulus:**
     - Reuse all the tests created for CPU_A and CPU_B simultaneous operations for this subsystem.
     - Add stimulus from AXIM UVM agent connected to L2 cache. This stimulus must include both cache coherent transactions as well as non-cache coherent transactions.
   - **Response:**
     - Final memory image produced by the memory subsystem reference model. This reference model gets all writes/reads from UVM agent monitor of both processors as well as the AXI UVM agent monitor. The reference model determines the final memory image.
     - The final memory image produced by CPU_A and CPU_B cache block flushes, and UVM AXI agent transactions should match the one produced by the reference model.
     - SystemVerilog assertions to check low-level granularity of cache pipes, FIFOs, REGISTER files, L1 ⇔ L2 interface protocol, L2 ⇔ AXI interface protocol, etc.

17.4 Cache Subsystem Interconnect Verification

This topic is entirely covered in Chap. 14 and directly applicable here.

The interconnects in this system can be coherent or non-coherent. Both cases are covered in Chap. 14.

17.5 Low-Power Verification

This topic is covered in its entirety in Chap. 9. This methodology is directly applicable to the system under discussion here.

For the cache memory subsystem verification, the methodology with UPF is directly applicable based on the following power domains and the State Table defined below.

- Power subdomains for the system in Fig. 17.1:
  - CPU_A and CPU_B including L1 cache
  - L2 cache
  - AXI subsystem

- Power State Table for the system (Fig. 17.2):

State \ Power Domain	CPU_A, CPU_B, L1	L2 Cache	AXI Master/Slave
Normal	ON_10	ON_10	ON_08
Sleep	ON_10	OFF	OFF
Hibernate	ON_10	OFF	OFF

17.6 Static Formal or Static + Simulation Hybrid

This topic and methodology are described in its entirety in Chap. 10.

The following logic blocks (and sub blocks thereof) are identified for static and/or static + simulation hybrid methodology:

- L1 cache MESI protocol state machine.
- L2 cache MESI protocol state machine.
- L1 TAG memory.
- L2 TAG memory.
- AXI master state machine.
- AXI slave state machine.
- AXI ⇔ L2 interface protocol.
- L1 cache ⇔ L2 cache interface protocol.
- Note that the CPU is not a candidate for static formal because it is considered a pre-verified IP.

17.7 Assertions Methodology (SVA)

The methodology and technical detail thereof are described in Chap. 6 in its entirety.

Assertions for our system (Fig. 17.1, cache subsystem verification) need to be deployed for the same candidates that need static formal. Static formal needs assertions, which is why they are candidates for assertions as well. They are listed here for the sake of completeness:

- L1 cache MESI protocol state machine
- L2 cache MESI protocol state machine
- L1 TAG memory
- L2 TAG memory

- AXI master state machine
- AXI slave state machine
- AXI ⇔ L2 interface protocol
- L1 cache ⇔ L2 cache interface protocol

In addition,

- Explicit assertions for synchronous and asynchronous FIFOs of the system
- All inter-block protocols (control logic)

17.8 Coverage Methodology (Functional: SFC, Code, and SVA "cover")

This topic is covered in its entirety in Chap. 7.

Before we apply functional cover points to the cache memory subsystem:

- Determine logic that needs to be functionally covered.
- How will you leverage code coverage with SystemVerilog functional coverage?
- What's the strategy to constrain stimulus to achieve desired functional coverage?
- How will you determine that you have specified all required cover points and cover groups? This is the hardest (and sometime subjective) question to answer. Continue to strategize as the project progresses.

The functional cover points need to be applied to the following blocks:

- L1 write through cache states and state transitions
- L2 write back MESI protocol states and all MESI protocol state transitions
- L1 and L2 Tag comparison logic
- All possible transactions and transitions of these transactions on L1–L2 interconnect
- All possible transactions and transitions of these transactions on L2–AXI interconnect
- L1 and L2 status registers
- Byte, word, and long word access status on L1 and L2 cache

17.9 Software/Hardware Co-verification

The cache memory subsystem is primarily for cache coherency testing. Software testing is not necessary. UVM agents for CPU_A, CPU_B, and AXI will verify L1/L2 cache using directed, constrained random and random transactions. Hence, software/hardware co-verification is not necessary.

17.10 Simulation Regressions: Hardware Acceleration or Emulation or FPGA Prototyping

In this subsystem, cache coherency verification, we don't need full CPU processor models. Only the UVM agents for processors are needed. Hence, no need for a hardware accelerator or emulation.

17.11 ESL: Virtual Platform for Software and Test Development

A virtual platform for the cache subsystem is not necessary. What is necessary is to develop a TLM2.0 memory subsystem reference model. This model will interact with UVM agent monitor analysis ports. It will predict the final memory image with which the RTL-produced memory image will be compared.

Chapter 18
Cache Memory Subsystem Verification: ISS Based

Chapter Introduction

This chapter discusses, in detail, the verification of a cache subsystem of a larger SoC. We will go through a comprehensive verification plan and describe at each verification step, cache subsystem-based real-life detail. This chapter discusses the verification methodology using an instruction set simulator (ISS) (as opposed to an UVM agent which is described in Chap. 17).

The CPU and memory subsystem platform shown in Fig. 18.1 use the cache memory subsystem of Fig. 17.1. The entire cache memory subsystem is plugged into the CPU and memory subsystem. The idea is that we can verify the cache memory subsystem of Fig. 17.1 using either the CPU_A and CPU_B UVM agents as we saw in the previous chapter or the complete CPU ISS (instruction set simulator) models for these two processors.

Figure 18.1 subsystem shows the use of ISS as the full processor model for CPU_A and CPU_B. These ISS models can execute instructions from a real C/C++ program to allow for real-world applications to run on the cache subsystem. You don't need to "manually" generate transaction traffic as you would have to, in the case of UVM agents for CPUs. In addition, this ISS also includes the write back L2 reference model. The L2 reference model does not have to be fed with write/read transactions from the UVM agents, rather they interact directly with the ISS model and maintain the L2 reference model's memory image.

As shown in Fig. 18.1, a random instruction test generator first creates a test memory image which is loaded in the external DDR of the cache subsystem and into the ISS internal memory. Instruction execution by ISS starts after reset which creates a transaction stream of writes and reads on the L1 and L2 interconnect. The program execution continues until the end and won't stop (except when an SVA assertion fires and stops simulation) until the end of the program. At the end of the program execution, the ISS will create an expected memory image (not of the L2 cache but of the external DDR). The RTL will also have the final state of memory in the DDR (after executing the final full cache block flush instructions). The expected DDR image is compared with the simulated DDR image and failures noted thereof.

A.B. Mehta, *ASIC/SoC Functional Design Verification*,
DOI 10.1007/978-3-319-59418-7_18

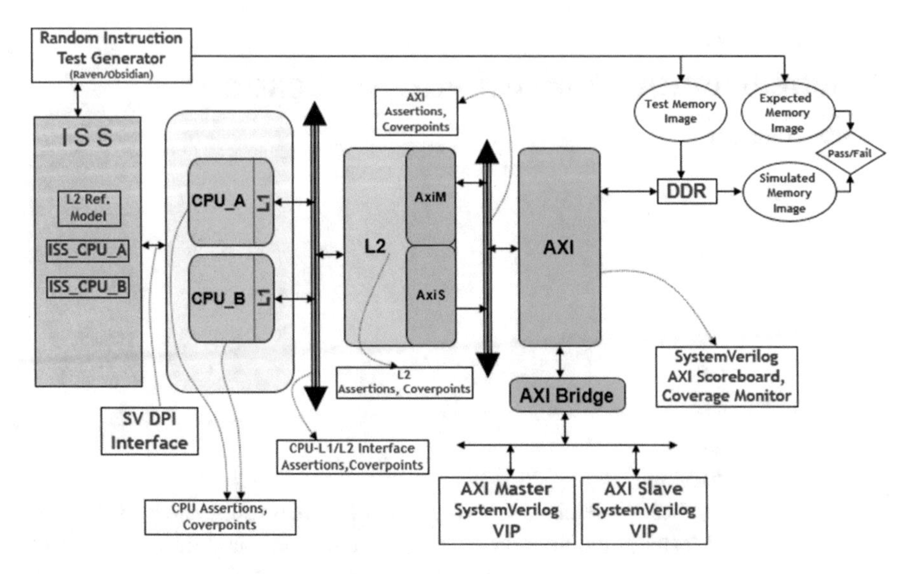

Fig. 18.1 Cache memory subsystem verification using ISS CPU models

Note that if you do find a discrepancy between the simulated and expected memory images, it will be very hard to recreate the failing scenario. This is because the final memory image does not translate directly into the activity that took place between L1 and L2. There are two ways to encounter this. One is to create an L2 memory image at the end of the program execution and compare it with simulated L2 image. The other is to heavily use SystemVerilog assertions for all corner cases (as described in Sect. 17.7) and see that the simulation stops as soon as an assertion is violated. That's the easiest way to know where exactly the bug arises from.

There are distinct advantages and disadvantages of using either the UVM agent approach or the ISS full processor model approach.

Advantages of verification using UVM agents for CPU_A and CPU_B:

- No need to generate CPU instructions. No need for an instruction generator.
- CPU transactions can be easily targeted to stress verify L1/L2 cache coherency.
- Easy to create directed, constrained random and full random transactions on L1/L2 interconnect.
- Easy to target corner cases.
- Easy to target cases that miss functional coverage.

Disadvantages of verification using UVM agents:

- Need to manually create transactions to mimic transactions created by real-life applications. In other words, create transactions to mimic instruction source, destination dependencies, cache block flushes, cache block invalidations, etc.
- Need to manually collide transactions from the two processors to create corner cases and stress verification.

Advantages of using ISS for CPU_A and CPU_B processors:

- Real-life scenarios for cache collision and cache coherency verification.
- Random instruction generator can be tuned to create exhaustive corner cases.
- Easy to create the test memory image for program execution. Easy to create final memory image of the subsystem (as expected memory image) for comparison with the simulated memory image.
- Run real-life applications for a final sanity check/verification of the subsystem.

Disadvantages of using ISS-based verification:

- Hard to direct traffic on L1 and L2 interconnect from an instruction stream.
- Hard to target cache coherency corner cases since no direct way to target a memory location from both processors at the same time. So, stress verification is difficult.

So, which of the two approaches would you want to use? The short answer is both. Here's how.

1. Start with the UVM agent approach. This approach should be used to weed out 90%+ bugs of the memory subsystem because the UVM agents provide the best way to direct your tests for corner cases as well as standard testing. It is very hard to write an assembly program to execute on the full ISS processor model and direct the traffic the way you want. UVM agent allows you to target a specific scenario deterministically. The ISS will have a tough time targeting such a scenario.
2. When you are done with the UVM agent-based verification, switch over to the ISS-based CPU models. This approach is simply to see that we haven't missed any cases that would appear in real-life applications. If you did your job well with the UVM agents, the chances of major surprises with ISS models should be less. But the confidence you get by running small snippets of real code is not achievable by the UVM agent approach.

Bibliography

Accelera. (2015). *UVM 1.2 User's guide.* Accelera.
Accelera. (n.d.-a). *Universal Verification Methodology (UVM) 1.2 User's guide.*
Accelera. (n.d.-b). *Verilog-AMS LRM Rev. 2.4.* Accelera.
Agarwal, R., et al. (n.d.). An insight into layout versus schematic. *EDN Network.*
ASIC-WORLD. (n.d.). Retrieved from asic-world.com: http://www.asic-world.com/systemverilog/random_constraint1.html
Bailey, B., & Martin, G. (n.d.). *ESL models and their application.* 2010, Springer
Bembaron, F., et al. (n.d.). Low power verification methodology using UPF , TI, DVC on 2009.
Bhattacharya, D. O. (2012a). PSL/SVA assertions in SPICE. *DVCON.*
Bhattacharya, P. (2012b). Retrieved from EETimes: http://www.eetimes.com/document.asp?doc_id=1279150
Cadence. (n.d.). *Encounter conformal low power.* Retrieved from www.cadence.com
Cadence-VIP. (n.d.). Cadence VIP: Interconnect Validator.
Cummings, C. E. (2000). Simulation and synthesis techniques for asynchronous FIFO design. *SNUG 2000.*
Cummings, C. E. (n.d.). Clock domain crossing (CDC) design & verification techniques using SystemVerilog. Paper presented at *SNUG 2000.*
Donohue, R. (n.d.). Synchronization in digital logic circuits. Paper presented at *SNUG 2000.*
Erich Marschner, P. Y. (n.d.). Static verification for complex design. Paper presented at *SNUG 2000.*
IEEE Standard for Design and Verification of Low Power Integrated Circuits,. I.-2. (n.d.). IEEE standard for design and verification of low power integrated circuits, IEEE Std 1801–2009, 2009.
John Brennan, T. Z. (n.d.). The how to's of AMS verification. *DVCON.* Accelera.
Jones, A., & Sonander, J., S. T. (n.d.). An introduction to property checkers for functional verification. Paper presented at *SNUG 2000.*
Kaiser, S. (n.d.). ESL solutions for low power design. *IEEE.*
Litterick, M. (2006). *Pragmatic simulation-based verification of clock domain crossing signals and jitter using SystemVerilog assertions.* Retrieved from www.verilab.com/files/sva_cdc_paper_dvcon2006.pdf
Mathur, A., et al. (n.d.). Functional equivalence verification tools in high-level synthesis flow. Paper presented at *SNUG 2000.*
Mehta, A. (2016). *SystemVerilog assertions and functional coverage. A comprehensive guide to methodologies and applications.* Los Gatos: Springer.
MentorGraphics. (n.d.). Verification Academy. https://verificationacademy.com/
Modh, H. (n.d.). http://hardikmodh.blogspot.com/

A.B. Mehta, *ASIC/SoC Functional Design Verification*,
DOI 10.1007/978-3-319-59418-7

Nick Heaton, A. B. (n.d.). Functional and performance verification of SoC interconnects. *Embedded Computing Design*.

O'Riordan, P. B. (2012). Mixed signal verification methodology. In J. Chen et al. (Eds.), *Mixed-signal methodology guide*. San Jose: Cadence Design Systems.

OVP. (n.d.). *Open virtual platform*. Retrieved from http://www.ovpworld.org

Peruzzi, R. (n.d.). *https://www.design-reuse.com/articles/23018/verification-virtual-prototyping-ams-behavioral-model.html*. Retrieved from https://www.design-reuse.com

Report, S. T. (2008, September). Low power design, special technology report. *SCDsource*.

Rizzati, L. (n.d.). 11 myths about hardware emulation.

Ron Vogelsong, A. H. (2015). Practical RNM with SystemVerilog. *CDNLive*.

Synopsys. (n.d.). *Logic equivalence using formality*. Retrieved from YouTube: https://www.youtube.com/watch?v=LfqNlRfpVWo

SystemVerilog_LRM_1800-2012. (n.d.). SystemVerilog LRM 1800-2012.

TLM2.0, O. (n.d.). OSCI TLM-2.0 language reference manual.

Turpin, M. (n.d.). The dangers of living with an X. ARM (2003).

Wen, H., & Chen, J., S. a. (n.d.). Tackling verification challenges with interconnect. *Cadence White Paper*.

Wikipedia. (n.d.-a). *Wikipedia*.

Wikipedia. (n.d.-b). *Wikipedia*. Retrieved from https://en.wikipedia.org/wiki/Newton's_method

Yeung, P. (n.d.). Five steps to quality CDC verification. *Mentor Graphics White Paper*.

Index

A.B. Mehta, *ASIC/SoC Functional Design Verification*,
DOI 10.1007/978-3-319-59418-7

Printed in the United States
By Bookmasters